AF685278

About an Important Phenomenon—Water Hammer

About an Important Phenomenon—Water Hammer

Editors

Kamil Urbanowicz
Helena M. Ramos

MDPI

Basel • Beijing • Wuhan • Barcelona • Belgrade • Novi Sad • Cluj • Manchester

Editors
Kamil Urbanowicz
West Pomeranian University
of Technology
Szczecin
Poland

Helena M. Ramos
IST, CERIS, University of Lisbon
Lisbon
Portugal

Editorial Office
MDPI
St. Alban-Anlage 66
4052 Basel, Switzerland

This is a reprint of articles from the Special Issue published online in the open access journal *Water* (ISSN 2073-4441) (available at: https://www.mdpi.com/journal/water/special_issues/water_hammer).

For citation purposes, cite each article independently as indicated on the article page online and as indicated below:

Lastname, A.A.; Lastname, B.B. Article Title. *Journal Name* **Year**, *Volume Number*, Page Range.

ISBN 978-3-0365-9827-7 (Hbk)
ISBN 978-3-0365-9828-4 (PDF)
doi.org/10.3390/books978-3-0365-9828-4

© 2024 by the authors. Articles in this book are Open Access and distributed under the Creative Commons Attribution (CC BY) license. The book as a whole is distributed by MDPI under the terms and conditions of the Creative Commons Attribution-NonCommercial-NoDerivs (CC BY-NC-ND) license.

Contents

water

MDPI

Editorial

New Advances in Water Hammer Problems

Kamil Urbanowicz [1,*] and Helena Margarida Ramos [2]

[1] Faculty of Mechanical Engineering and Mechatronics, West Pomeranian University of Technology in Szczecin, 70-310 Szczecin, Poland
[2] Department of Civil Engineering, Architecture and Georesources, CERIS, Instituto Superior Técnico, University of Lisbon, 1049-001 Lisbon, Portugal; helena.ramos@tecnico.ulisboa.pt
* Correspondence: Kamil.Urbanowicz@zut.edu.pl

Citation: Urbanowicz, K.; Ramos, H.M. New Advances in Water Hammer Problems. *Water* **2023**, *15*, 4004. https://doi.org/10.3390/w15224004

Received: 5 November 2023
Accepted: 13 November 2023
Published: 17 November 2023

Copyright: © 2023 by the authors. Licensee MDPI, Basel, Switzerland. This article is an open access article distributed under the terms and conditions of the Creative Commons Attribution (CC BY) license (https://creativecommons.org/licenses/by/4.0/).

1. Introduction

When the flow within pressurized pipes experiences abrupt stoppages, initiation, or directional alteration, it gives rise to the phenomenon of water hammer, characterized by the propagation of waves. This phenomenon, which was relatively obscure in the past, has gained significant prominence in contemporary times. Given its profound relevance in practical engineering, the body of literature concerning this intricate subject matter has exhibited a consistent and substantial growth over the years.

From a historical vantage point, engineers have grappled with the challenges posed by water hammer since the inception of pipe systems for the conveyance of liquids. More than two millennia ago, Marcus Vitruvius Pollio [1] had already delineated the repercussions of water hammer and cavitation in the context of clay and lead conduits supplying water to the water distribution systems of ancient Rome. On a mathematical front, the formal description of this phenomenon commenced with the early works of von Kries [2,3], Joukowsky [4], and Allevi [5]. Joukowsky [4] delved into the intricacies of wave reflections within pipes and the deployment of air chambers, such as equalizing tanks and spring-loaded safety valves, ultimately furnishing a fundamental formula for the calculation of pressure surges resulting from rapid valve closures, which remains applicable even in modern times.

Contemporarily, the discourse surrounding this subject has expanded considerably. The central issues associated with water hammer problems encompass various dichotomies, including single versus multiphase flow, laminar versus turbulent flow, the elastic versus viscoelastic behavior of pipe materials under strain, the presence of gaseous or vaporous cavitation, Newtonian versus non-Newtonian flow, the distinction between rigid and flexible pipe walls, and the differentiation between fast (impulsive) and slow–transient flow. The deliberation of these aforementioned issues often necessitates the consideration of related phenomena, such as mechanical energy dissipation due to fluid friction, the manifestation of viscoelastic delayed deformations in pipe walls resulting from cavitation-induced liquid column separation, and fluid–structure interaction.

This Special Issue is dedicated to all aspects of modeling water hammer phenomena and the experimental verification of this particular form of unsteady flow. We have examined works that encompass (a) the exploration of accompanying phenomena (such as unsteady friction, delayed strain, cavitation in both vapor and gas phases, and fluid–structure interaction); (b) methods and devices aiming to safeguard pipe systems (such as cushion surge chambers and air valves) against the adverse consequences of water hammer (e.g., noise, vibrations, leakages, etc.); (c) advancements in numerical methodologies; and (d) the overall expansion of our understanding of water hammer.

In this Special Issue, entitled "About an Important Phenomenon—Water Hammer", we have been privileged to receive contributions from esteemed authors hailing from diverse corners of the globe, each of whom has devoted many years of professional expertise to the exploration of transient pipe flow challenges.

2. Review of New Advances

Tasca et al. (contribution 1) analyzed the identification of the typical water hammer stages associated with the pump trip scenario. The two key power-loss behaviors were found: (a) attenuated and (b) water-hammer-dominated. Deep research of the sensitivity of the transient response to the air valve outflow capacity and location was carried out. The impact of different sources was taken into account in the analyses of (a) variations in pipeline length; (b) the initial steady water velocity; (c) the elevation difference between downstream and upstream reservoirs; (d) changes in both wave speed and friction on the transient response; and (e) negative pressures. The realized study explored (a) maximum hydraulic grade (HG) values; (b) maximum head values; (c) air pocket collapse times; (d) the timing of peak transient HGs; and (e) critical orifice sizes. Finally, the paper investigated how non-slam air valve design parameters impact the transient response.

This work is of great practical importance because the selection of proper air valves is very problematic to many system designers and pipeline operators. This is mainly "because their behavior is highly non-linear, sometimes showing high sensitivity to assumed conditions and sometimes displaying robust and stable results". The authors defined four stages occurring just after a pump trip-off: depressurization, air admission, air expulsion, and the creation of a secondary wave. The minimum hydraulic grade line (HGL) is formed during the initial two stages while the maximum HGL is often only completed after the first instance of air pocket collapse. Three main transient behaviors have been identified: attenuated (type 1 behavior) for small d/D values, intermediate (type 2 behavior) for moderate d/D values, and water-hammer-dominated (type 3 behavior), which typically occurs for large d/D values (d—diameter of the outflow orifice of the air valve; D—pipe diameter). An analysis of variation of certain key initial parameters (initial velocity, elevation difference, pressure wave speed, and friction coefficients, which were initially locked in for the main set of numerical simulations) revealed that the influence of these parameters on the transient response is more evident in relation to type 3 behavior than in type 1 behavior. It was found that a well-sized non-slam air valve can mimic the behavior of a large inflow orifice and small outflow orifice air valve during the pump trip scenario. For a non-slam air valve to be effective in mitigating the transient event, the selected values should be sufficiently small, both those of the transition head (to enable the utilization of the small outflow orifice for expulsion) and those of the large outflow orifice.

Kim (contribution 2) wrote governing transient flow equations in dimensionless form and used them for two widely adapted water supply systems: a reservoir pipeline surge tank valve reservoir system and a reservoir pump check valve pipeline surge tank valve reservoir system. Inertia (lumped) and expression (integrated solution) for surge tank (located in specific section of the pipeline) were taken into account in a simplified, dimensionless way.

Both frequency and time domain solutions of the pressure response for the analyzed systems are presented and compared (partially with experimental results). The proposed solution is interesting as eliminates the main restriction: the discretization problem of numerical methods based on the method of characteristics (MOC). In MOC, the requirement of equality of the Courant number to one is hard to fulfill for existing pipeline systems.

Kim's (contribution 2) analysis of the surge tank in pipeline systems with and without pumping stations and check valves indicated that "the resonance of the system can be explained by the pipeline length and the locations of the pipeline structures, such as the surge tanks, pumps, and check valves". The difference in the time domain response between the two analyzed systems indicates that the amplification and mitigation originated from the boundary conditions of the interaction with the surge tank. The proposed normalization of the main equations (written in a dimensionless form) provides an intuitive understanding of the system response.

The air cushion surge chamber is a useful device that protects large-diameter pipeline systems (e.g., long-distance water transfer and hydropower systems). It is a closed chamber that is partially filled with water and compressed air [6,7]. "As compared to an open-type pressure regulating chamber, it is rarely restricted by geological or topographical features, and offers numerous advantages, such as shorter construction supply, lower excavation volume, cost-effectiveness, and minimal ecological impact" (contribution 3). Liu et al. (contribution 3) improved a second-order finite volume method (FVM) taking into account unsteady friction models. With the use of FVM, a novel methodology to simulate the dynamic behavior of the air cushion surge chamber in a water pipeline system was developed, and its usefulness was validated through comparisons with experimental pipe system results. The pipeline was a 582-meter-long system based on a coiled brass pipeline with a small diameter (21 mm). Four tests were conducted (two in laminar regime, Re = 1284, and two in a low-Reynolds-number turbulent one, Re = 4334).

The performed comparisons reveal that the steady friction model only accurately predicts the first pressure peak and it seriously underestimates pressure attenuation in later stages. The incorporation of unsteady friction models resulted in much better predictions of the complete pressure attenuation process. Among the two tested friction models, in particular IAB (instantaneous acceleration-based) and CBM (convolution-based models), the CBM unsteady friction model with acceptable accuracy reproduced pressure peaks and whole-pressure oscillation periods. In these systems, air cushion surge chamber energy attenuation is primarily due to pipe friction (occurring in elastic pipes) and air cushion. Some differences in peak values of subsequent amplitudes were noticed that the authors (contribution 3) related to the absence of wall heat exchange in the mathematical model.

Pezzinga (contribution 4) analyzed the results obtained using numerical 1D and 2D models with experimental results (long-duration experimental tests—about 70 periods) in which no vapor areas appeared. A numerical solution named MOC-Z by the author was used, which operates without interpolation for flows with liquid and gas. Pezzinga's experimental test stand contained several air-release valves. According to the author, it was very difficult to completely eliminate the air from the circuit. The gas release effect is considered a possible reason for the further oscillation damping noticed in the experimental results. To overcome the above problem, a proper mass balance equation involving gas release was taken into account. The calibration of the proposed model parameters was carried out with a micro-genetic algorithm. A different turbulence model [8] was also considered for comparison.

Both constant and variable gaseous mass for water hammer flow were taken into account. "Taking into account the mass of gas, considered as constant, reduces the MAE because it allows to phase the computed oscillations, approaching it to the observed one. If the mass of free gas is considered as a variable, taking into account a gas release and solution process, the oscillation damping is caught altogether, provided that a proper calibration of the parameters of the model is made" (contribution 4). Pezzinga noticed a significant improvement in the modeling of the head oscillation damping results from the 2D flow schematization with respect to the 1D one with quasi-steady friction and concluded that the oscillation damping observed in water hammer flow is mainly due to unsteady friction, but other mechanisms of dissipation exist; for example, the thermic exchange between bubbles and the surrounding liquid. These mechanisms can influence the values of the calibrated parameters, but they do not seem capable of fully reproducing the observed pressure traces. Finally, Pezzinga mentioned that more complex models of gas release could be considered in future studies.

Neyestanaki et al. (contribution 5) analyzed the literature in the direction of modeling different types of valve closure with a transient water hammer with the help of three-dimensional computational fluid dynamics (CFD). In their opinion, 3D models are more accurate than numerical simulation with use of 1D solutions. The performed literature survey showed that there is no study comparing the different methods for modeling valve closure in terms of modeling accuracy and computational cost. Moreover, a sliding mesh has not been used before for modeling axial valve closure. In their paper, water hammer in a straight 3D pipe (36 m long with relatively large inner diameter D = 0.3 m; Reynolds number Re = 7×10^5) during an axial gate valve closure was modeled using CFD. Three methods were used to model the valve closure: dynamic mesh, sliding (motion) mesh (Ansys Fluent), and immersed solid (Ansys CFX) methods were used for modeling the valve closure. The calculation results were compared with experimental results provided by Sundstrom and Cervantes [9] that include the variation in the differential pressures between two cross-sections and the wall shear stress. The performed result analysis showed that the immersed solid method has a delay in flow rate reduction and therefore underestimates differential pressure rise and overestimates wall shear stress close to the end of the valve closure. The dynamic mesh method models are more time-consuming and three times more expensive in terms of computational cost than other methods. Additionally, the dynamic mesh was unstable, with the possibility of divergence. The best results were received with help of the sliding mesh method, which is an inexpensive and stable technique that provides results closest to those of the experimental method. The authors proved that, for a thin gate valve, the axial movement of the valve can be modeled using mesh movement without any mesh deformation (predicting more physical results). The three-dimensionality of the flow after the valve closure was addressed as a non-symmetrical recirculation region that appears near the gate valve with its movement. The computational cost of the mentioned methods was discussed. "Immersed solid method and sliding mesh have quite the same computational cost. For dynamic mesh, the process of re-meshing is added to the calculation. Moreover, the lower quality of regenerated mesh makes it more possible to diverge. Therefore, a lower under-relaxation factor and, consequently, a higher number of iterations per time step is applied in the simulation. The mentioned drawbacks made the computational cost of the dynamic mesh method around three times more than the sliding mesh and immersed solid method".

Ferreira and Covas (contribution 6) rightly pointed out that, in any numerical model (1D, 2D, or 3D), the mesh size and configuration strongly affect the computational effort, as well as the accuracy of the results. The 2D mesh adaptation is very important for the geometrical boundaries. Flow physics needs ensure that the velocity variation is of the same order along the numerical mesh—a non-uniform grid should be generated (higher resolution near the pipe wall) due to the high gradients observed. The calibration of the grid (according to the velocity gradient history) can reduce the number of mesh points, maintaining the same accuracy level. An in-depth assessment of the 2D radial mesh's influence on the computation of unsteady energy dissipation in pressurized pipes was conducted. An extensive numerical analysis of the effect of the numerical schemes and of the radial mesh (several radial meshes were defined) on the computation of unsteady energy was carried out. The simulation results with use of 1D and 2D models were compared to the experimental method for laminar flows (Zielke solution) and for two valve closure maneuvers (i.e., an instantaneous and an S-shaped closure).

A new optimized equal area cylinder (OEAC) radial mesh (with 40 cylinders) was proposed using 2D simulations. It is defined by a high-resolution grid near the pipe wall and a lower-resolution grid in the pipe core. Its main advantage is achieving good model accuracy without increasing the computation effort. For an instantaneous valve closure scenario, this mesh reduced the calculation time by four times with a similar simulation error to that of the standard equal area cylinder (EAC) mesh. The comparisons of the results of the new mesh compared to those of the traditional mesh geometries (GS, ETC, and EAC) for a calibrated S-shaped valve closure revealed that (a) compared to the geometric sequence (GS) mesh, the new geometry achieved a significant improvement in accuracy for meshes with fewer cylinders ($NC \leq 60$); (b) the other two radial meshes (ETC and EAC) do not provide the same accuracy; and (c) new mesh correctly describes the experimental data with only 20 cylinders. One of the notable advantages of the Ferreira and Covas approach is its adaptability to turbulent flows, achieved by incorporating a suitable turbulent model into the fundamental equations.

Urbanowicz et al. (contribution 7) presented an extensive review of two modeling techniques of unsteady friction related to IAB and CBM. The filtering method of weighting function used in CBM proposed in Urbanowicz's earlier paper [10] were further analyzed.

Its usefulness was checked for cases without vapor (cavitation) zones forming in the pipe. The second objective of this paper was the verification of the effectiveness of Johnston's lumped friction model [11], according to which the unsteady friction can be concentrated only at the boundary nodes of the numerical grid (in inner nodes the simplified quasi-steady assumption was used). The weighting function in this proposed computationally effective and accurate model was simply composed of two exponential terms, and its values are chosen to be dimensionless and time-step-dependent. The simplification of the weighting function in conjunction with the corrected effective method for solving the convolution integral enabled the proper determination of resistances. The proposed method's mathematical complexity is similar to that of the IAB model. Contrary to the IAB model, in the proposed method, there is no need to calibrate any parameters describing wall shear stress. The simulations carried out with the use of Johnston's model (lumping unsteady friction in boundary nodes of the MOC grid) showed that the analyzed transient histories of pressure can be simulated with sufficient compliance, maintaining acceptable simulation accuracy even more quickly.

Kubrak et al. (contribution 8) numerically (using a fixed-grid method of characteristics and IAB unsteady friction model) and experimentally studied the water hammer phenomenon in a serially connected steel–plastic pipeline system with significantly different inner diameters and for several different lengths of each section (maintaining a constant total length of the pipeline system). The experimental data revealed that the maximum pressure increase linearly depended on the share of the steel section in the total pipeline length. The combination of the IAB model applied to the steel section and the one-element Kelvin–Voigt model accounting for the viscoelastic properties of the HDPE pipe made it possible to obtain a satisfactory agreement between the calculated and the measured pressure signals.

Parameters defining the IAB model and the creep compliance of the HDPE plastic pipe were calibrated for single-pipeline systems. They were then introduced into a proposed numerical model to simulate the water hammer in different configurations of lengths of steel and HDPE sections. It was demonstrated that this approach failed to reproduce pressure head histories for a pipeline system with significantly different lengths of each section, as any change in the configuration of the pipeline system caused the creep parameters to be recalibrated.

Vardy (contribution 9), in his review paper, pointed out that during unsteady pipe flows, many different phenomena can occur simultaneously and interact with one another. Because of this, it is beneficial to have a clear understanding of the potential sources of damping (of pressure waves propagating in pipes) and their likely importance in any particular situation. It is common that two or more causes of damping exist simultaneously, but usually one of these is dominant and the others are of secondary importance.

In events forced by valve closure or pump trip-off, low-frequency components of waves are dominant. In studies of acoustics or leak detection in pipelines (where acoustic sensors are used) where waves are of a sufficiently high frequency (their wavelengths are shorter than a few pipe diameters), cross-sectional variations in pressure become significant (radial disturbances dominate axial ones). The most well-known cause of damping is skin friction (SF). Vardy explains popular rough assumptions in detail, according to which SF is approximated in a quasi-steady way. Notably, shear stresses are almost never measured directly except in highly specialized laboratory experiments designed expressly for this purpose. In practical applications, the SF influence is less strong than seen in the Holmboe experiment. Vardy suggests that there are three reasons for this: (a) the measurements were made in a pipe with a very small L/D ratio (time intervals between successive steps are short; L—pipe length); (b) the valve closure was as rapid as the experimenters could achieve; and (c) the Reynolds numbers of the initial steady flows were smaller than usual in large-scale engineering.

After analyzing the skin friction effect, the pipe wall properties are discussed. Here, typical steel pipes (and other metal pipes) are conventionally treated as linearly elastic, while in water supply systems, plastic pipes play the main role. The wall material of such pipes behaves in a viscoelastic manner, which physically means that the pipe diameter responds relatively slowly to changes in the fluid pressure (a delayed strain occurs). The influence of the assumed viscoelastic (VE) properties of the pipe wall are sometimes so dominant that little useful purpose would be served by taking account of unsteady components of skin friction. Vardy's comments about wave speed (WS) state that it would not be safe to infer the WS from the time intervals between the pressure histories maxima.

In the following sections, Vardy discusses the role of:

(a) Axial and lateral structural movement; he discusses its widely fluid structure interaction effect (FSI). In FSI, structural movement can cause a behavior that complicates attempts to make reliable inferences from the measurements of pressure alone. FSI's importance depends significantly on the interactions between internal pressure forces and forces due to structural movements. This effect can decrease the pressure amplitudes in some locations, but increases amplitudes in others.
(b) Variable wave speed. A small amount of gas—free (bubbles of undissolved gas) or vapor (as an effect of cavitation)—can cause changes in the WS, which also causes dispersive behavior and, hence, influences damping (either increasing or countering it).
(c) Porous surfaces. "Strong damping can exist when fluid can discharge laterally through pipe walls. It will be rare for this to be desirable in the case of liquid flows, but it can be beneficial in some gas flows" (sonic boom-like disturbances occurring in tunnels exit portals).
(d) Delayed reflections. "Disturbances propagating from the pipe into the reservoir radiate in a spherical-like manner whereas the reflections along the pipe approximate closely to planar. The time required for the changes in pressure at the outlet to die away is short—typically in the order of the time required for a wave to travel one pipe diameter—so the phenomenon is justifiably neglected in many practical applications".
(e) Experimental measurements. Wave superpositions can have a strong influence on pressure histories at any particular location and can complicate the interpretation of the physical measurements exhibiting damping. The author also discussed problems with possible unidentified oscillations.
(f) Numerical damping, which can be caused by interpolation algorithms or by incorrect estimations of flux across interfaces between adjacent cells. It can also be introduced intentionally for special reasons; for instance, through the use of "artificial viscosity" to suppress unrealistic oscillations close to locations of especially rapid change.

Lu et al. (contribution 10)'s paper concerned problems related to modeling the load-rejection process of a hydropower plant with an air cushion surge chamber. The main aim of this paper was to develop an accurate and efficient water hammer numerical model, which is significant for the proper design and safe operation of hydropower plants. The authors worked on a second-order finite volume method (FVM) based on a Godunov-type scheme (GTS) to solve the main set of equations, motivated by the fact that with its help, results would be more accurate and more stable with less numerical dissipation than in MOC. This FVM-GTS model was validated earlier in simple scenarios. In their paper, hydraulic transients of the load-rejection process occurring in the hydropower plant with an air cushion surge chamber (rarely analyzed previously) were studied. "The results calculated by the proposed second-order FVM GTS models were compared with the exact solution and the measured values as well as predictions by the MOC scheme. The accuracy and efficiency of the proposed approach were discussed. Another important purpose is that the proposed accurate model was used to explore the possible computation error caused by the MOC scheme in a hydropower plant with a complex pipe system". As sometimes complicated boundaries need to be taken into account, virtual boundaries were introduced (upstream, downstream, and hydraulic component connection sections) to provide a connection between the air chamber and the unit to achieve uniformity in the calculation of the control cells inside the pipeline and at the boundaries.

The performed numerical comparisons revealed that for a Courant number Cr = 1, both models, FVM and MOC, were consistent with the exact solution. When Cr < 1, both computational results had numerical dissipation. For a Cr number with a gradual decrease, the second-order FVM simulation results were more stable as the numerical dissipation of MOC was more serious. For complex pipe systems, the proposed FVM model better reproduced the experimental data, which were more accurate than those of MOC. For the load-rejection process of hydropower units containing an air chamber, the results calculated using the proposed FVM model were basically consistent with the measured rotational speed variation. The second-order FVM does not need to adjust the wave speed for the pipes; it only needs to reduce the Cr condition appropriately. The error in the MOC calculation was associated with the air chamber parameters. The second-order FVM is robust in simulating the water hammer problems in a simple or complex pipe system. "Considering the higher accuracy, stability, and efficiency, the high-order FVM is feasible and suggested for water hammer simulation in real hydraulic systems with more complicated pipe components and devices".

Paternina-Verona et al. (contribution 11) developed a study in which they explored various issues related to transient flows during different filling events through the use of 3D computational fluid dynamics (CFD) models. The main findings and issues addressed in the study are as follows:

Gauge Pressure Oscillations: The 3D CFD models accurately predicted gauge pressure oscillations in the air pocket during filling events. The entrapped air pocket experienced cyclic changes in volume, and this pressure variation had a significant impact on the hydraulic behavior.

Water flow velocities: The study examined water flow velocities during transient events. The authors observed that the pressure damping of the entrapped air pocket led to water flow velocity transitions, with higher inlet gauge pressures resulting in more intense mixing of air and water.

Thermodynamic phenomena: Temperature changes occurred during the filling processes due to the compression of the trapped air pocket. The study found non-uniform temperature distributions, with the highest temperatures away from the air–water interface. This adiabatic behavior in the air pocket can affect system efficiency and safety.

Backflows and hydraulic efficiency: Transient flows can lead to backflows towards the pumping source, potentially causing a loss of hydraulic efficiency during filling events. The study showed that these backflows were more critical with higher inlet gauge pressures and gradually dissipated over time due to the damping pressure of the entrapped air pocket.

Detailed visualization: The use of 3D CFD models with an unstructured mesh allowed the detailed visualization of various variables, including streamlines, velocity contours, and temperature distributions, providing a comprehensive understanding of hydraulic–thermodynamic phenomena.

Selection of pipe class: The study highlighted the importance of evaluating the selection of pipe class in large-scale water installations, as the compression of entrapped air pockets can result in absolute pressure values higher than those observed with monophasic fluids. Current regulations often do not consider such two-phase flow models when designing systems, and the use of 3D CFD models can improve system reliability during filling operations.

The study suggests directions for future research, including the investigation of the three-dimensional behavior of hydraulic events with air expulsion orifices, which has not been extensively studied using 3D CFD models. Additionally, the simulation of large-scale hydraulic scenarios, with and without air expulsion orifices, is recommended for further exploration. In summary, this study delves into the complex interplay of hydraulic and thermodynamic phenomena during filling events in pipes with entrapped air pockets. It emphasizes the utility of 3D CFD models for understanding and predicting these phenomena and suggests potential improvements and areas for future research.

The investigation presented by Wu et al. (contribution 12) employed the integer total energy method to formulate the energy equation within a quasi-2D model representing viscoelastic pipes. Subsequently, a comparative analysis of work variations pertaining to the friction term and viscoelastic term in both 1D and 2D models was conducted across different initial Reynolds numbers (Re). The key findings are as follows: (i) When the initial Reynolds number was less than 3.0×10^5, the 1D model exhibited a tendency to underestimate the work associated with the friction term, a trend not observed in the 2D model. This discrepancy lessened as the initial Re exceeded 3.0×10^5. (ii) The work performed with the viscoelastic term of the pipe wall demonstrated a relative constancy across varying initial Reynolds numbers in both the 1D and 2D models. (iii) As the Reynolds numbers increased, both the viscoelastic and the frictional work within the 1D and 2D models showed a progressive increase over time. However, for cases with a high initial Re, the work due to friction surpassed that of the viscoelastic term. (iv) The dissipation of energy related to the friction term displayed an initial significant increase, followed by a gradual deceleration, eventually reaching a constant value. (v) The energy transformation associated with the viscoelastic term exhibited sinusoidal fluctuations during the initial phase of transient flow. In cases with lower initial Reynolds numbers, these oscillations persisted for an extended duration, portraying an overall upward trend before converging to a steady state. (vi) With an escalation in initial Reynolds numbers, the proportion of energy dissipation attributed to the friction term within the total energy transformation exhibited continuous growth, while the proportion originating from the viscoelastic term decreased proportionally.

Zeng et al. (contribution 13) presented hydraulic systems encompassing both pipelines and open channels, posing a challenge for hydraulic transient analysis. This study introduced a novel coupling method that combines the method of characteristics (MOC) for pipeline modeling and the finite volume method (FVM) for open-channel modeling. The interface between these two simulation domains was established using Riemann invariants. Parameters were exchanged between the MOC and FVM regions through the coupling boundaries in both directions. To validate the method, the authors developed tests on a simple tank-pipe system, and the results were compared with those obtained through 3D computational fluid dynamics (CFD) analysis. Subsequently, the method was applied to a practical hydropower station featuring a sand basin situated between the upstream reservoir and the turbines. The sand basin was treated as an open channel, integrated with the pipes within the system, and transient processes were simulated by modeling the sand basin as a surge tank. Comparing the results obtained through the new coupling method with those from the MOC-FVM coupling method, it became evident that the new approach offers increased reliability and accuracy. This improved performance stems from considering horizontal flow velocity in the sand basin, a factor that was previously neglected when modeling the sand basin as a surge tank within the MOC framework.

Cao et al. (contribution 14) enhanced hydraulic transients with significant operational risk due to the potentially destructive surge waves they generate. This study delved into the analysis of hydraulic surge phenomena and the control of surge damping in the context of pipe flow modeling and valve optimization. The study involved the development of a one-dimensional transient model using the modified instantaneous-accelerations-based (IAB) model, which incorporates considerations of energy dissipation, specifically the compression–expansion effect. This model was subsequently solved through the method of characteristics (MOC).

In a manner analogous to addressing valve operations via the traveling salesman problem (TSP), an innovative surge damping strategy was proposed, leveraging an improved artificial fish swarm algorithm (AFSA). To validate the unsteady model and the optimization algorithm, the effectiveness of surge wave damping was evaluated through case studies encompassing various pump operation scenarios.

The results revealed that the nonlinear optimized control method, as presented, is capable of reducing surge amplitudes by 9.3% and 11.4% in pipe systems, with and without a centrifugal pump in operation, respectively. Moreover, the method can yield a substantial 34% increase in time margin or a maximum surge reduction of 75.2% when employing a positive displacement pump. The optimized nonlinear valve closure exhibits distinct profiles in scenarios involving both rapid and gradual closure.

The surge damping strategy presented in this study not only holds practical implications for guiding real-time valve control but also provides valuable insights for valve design geared toward safeguarding against wave surge occurrences.

Liu et al. (contribution 15) presented a study to assess the sensitivity of input parameters concerning output results in hydraulic transient simulations employing the method of characteristics (MOC). In the context of a gravity flow water delivery project, six primary parameters influencing hydraulic transient simulations were identified. The study focused on the maximum pressure as the output parameter to conduct a sensitivity analysis. Two distinct approaches, namely Morris sensitivity analysis (Morris) and the partial rank correlation coefficient method based on Latin Hypercube Sampling (LHS-PRCC), were employed for this analysis. The findings indicated that the sensitivity of each parameter was generally consistent, with the exception of the friction factor. Specifically, the flow rate and Young's modulus exhibited a positive correlation with the maximum pressure, whereas the pipe diameter, valve closing time, and wall thickness demonstrated a negative correlation. It was noted that the variability in the friction factor was primarily attributed to the functioning of the flow and pressure regulating valve. In situations where other conditions within the gravity flow project remained constant, an increase in the friction factor corresponded to an elevation in maximum pressure. Significantly, the flow rate, pipe diameter, and valve closing time emerged as the critical parameters influencing the model. Moreover, the study underscored that both the Morris and LHS-PRCC methods proved effective in assessing parameter sensitivity in hydraulic transient simulations.

In conclusion, this study focused on analyzing parameter sensitivity in hydraulic transient simulations employing the method of characteristics (MOC) within the context of gravity flow. The examination was conducted using a single engineering case, demonstrating the effectiveness of the two sensitivity analysis methods. To broaden the scope and achieve a more comprehensive understanding, further analyses involving additional examples are warranted.

3. Conclusions

The papers included in the *Water* Journal's Special Issue titled "About an Important Phenomenon—Water Hammer" provide a contemporary snapshot of research trends in this field. All fourteen research papers in this collection involve a comparative analysis between simulation results and experimental findings. Among these papers, the method of characteristics (MOC) was employed in eight instances to model unsteady flow, while two papers utilized the second-order finite volume method (FVM). Additionally, one paper introduced a coupled approach combining MOC and FVM, which was also found applicable for open-channel flows. Notably, one paper presented a frequency-domain solution, while two other works relied on computational fluid dynamics software, specifically Ansys and OpenFoam. This reaffirms the ongoing significance of the MOC in the numerical analysis of water hammer problems.

In all the research papers, the transient water hammer flow was induced through valve closures (both hypothetical–instantaneous and realistic valve closing performance) and/or by deactivating the pump power, mirroring the conditions often encountered in real-world operational systems. In the majority of these papers (eight in total), the instantaneous-accelerations-based (IAB) unsteady friction model was employed by the authors to simulate friction losses using a 1D modeling approach. This preference suggests a desire for friction models that are relatively straightforward in their application. An interesting model meeting these expectations (contribution 7) focuses on unsteady friction mainly at the boundary nodes of the numerical mesh, based on the assumption proposed by Johnstone [11], and utilizes a convolution-based model (CBM) with a weighting function relying on only two exponential terms. The future, however, is likely to belong to meshless methods [12].

The majority of the comparative analyses (twelve in total) centered on elastic (metal) pipelines, while only three papers extended their simulations to scenarios found in contemporary water supply systems employing plastic pipes with viscoelastic wall behavior characterized by delayed strain.

Returning to notable observations made by Tasca et al. and Vardy, Tasca et al. (contribution 1) aptly emphasized the inevitability of real-world systems deviating from numerical idealizations due to unforeseeable future conditions. As such, they stressed the importance of design and operational engineers assessing the potential impact of real-world variability in system parameters on system performance. Vardy (contribution 9) highlighted the positive influence of damping in the context of water hammer in pipelines. Damping can facilitate the dissipation of strong disturbances before subsequent disturbances arrive, ultimately mitigating the potential consequences of superpositions.

The ongoing relevance of water hammer, particularly with the consideration of accompanying phenomena such as cavitation, unsteady friction, delayed strain, and fluid–structure interaction, underscores the continuous need for advancements in modeling methods. However, it is essential to exercise caution during analyses, as dissipative phenomena typically induce damping, and dispersive phenomena often do the same, though exceptions can arise where energy redistributions lead to superpositions that would not occur otherwise (contribution 9).

Author Contributions: Conceptualization, K.U. and H.M.R.; investigation, K.U. and H.M.R.; writing—original draft preparation, K.U. and H.M.R.; writing—review and editing, K.U. and H.M.R.; All authors have read and agreed to the published version of the manuscript.

Funding: This research received no external funding.

Conflicts of Interest: The authors declare no conflict of interest.

List of Contributions:

1. Tasca, E.; Besharat, M.; Ramos, H.M.; Luvizotto, E., Jr.; Karney, B. Exploring the Sensitivity of the Transient Response following Power Failure to Air Valve and Pipeline Characteristics. *Water* **2023**, *15*, 3476. https://doi.org/10.3390/w15193476
2. Kim, S. Dimensionless Pressure Response Analysis for Water Supply Pipeline Systems with or without Pumping Station. *Water* **2023**, *15*, 2934. https://doi.org/10.3390/w15162934
3. Liu, Y.; Lu, J.; Chen, J.; Xia, Y.; Liu, D.; Hu, Y.; Feng, R.; Liu, D.; Zhou, L. Finite Volume Method for Transient Pipe Flow with an Air Cushion Surge Chamber Considering Unsteady Friction and Experimental Validation. *Water* **2023**, *15*, 2742. https://doi.org/10.3390/w15152742
4. Pezzinga, G. Gas Release and Solution as Possible Mechanism of Oscillation Damping in Water Hammer Flow. *Water* **2023**, *15*, 1942. https://doi.org/10.3390/w15101942
5. Neyestanaki, M.K.; Dunca, G.; Jonsson, P.; Cervantes, M.J. A Comparison of Different Methods for Modelling Water Hammer Valve Closure with CFD. *Water* **2023**, *15*, 1510. https://doi.org/10.3390/w15081510
6. Ferreira, P.L.; Covas, D.I.C. New Optimized Equal-Area Mesh Used in Axisymmetric Models for Laminar Transient Flows. *Water* **2023**, *15*, 1402. https://doi.org/10.3390/w15071402
7. Urbanowicz, K.; Bergant, A.; Stosiak, M.; Deptuła, A.; Karpenko, M.; Kubrak, M.; Kodura, A. Water Hammer Simulation Using Simplified Convolution-Based Unsteady Friction Model. *Water* **2022**, *14*, 3151. https://doi.org/10.3390/w14193151
8. Kubrak, M.; Kodura, A.; Malesińska, A.; Urbanowicz, K. Water Hammer in Steel–Plastic Pipes Connected in Series. *Water* **2022**, *14*, 3107. https://doi.org/10.3390/w14193107
9. Vardy, A.E. On Sources of Damping in Water-Hammer. *Water* **2023**, *15*, 385. https://doi.org/10.3390/w15030385
10. Lu, J.; Wu, G.; Zhou, L.; Wu, J. Finite Volume Method for Modeling the Load-Rejection Process of a Hydropower Plant with an Air Cushion Surge Chamber. *Water* **2023**, *15*, 682. https://doi.org/10.3390/w15040682

11. Paternina-Verona, D.A.; Coronado-Hernández, O.E.; Espinoza-Román, H.G.; Fuertes-Miquel, V.S.; Ramos, H.M. Rapid Filling Analysis with an Entrapped Air Pocket in Water Pipelines Using a 3D CFD Model. *Water* **2023**, *15*, 834. https://doi.org/10.3390/w15050834
12. Wu, K.; Feng, Y.; Xu, Y.; Liang, H.; Liu, G. Energy Analysis of a Quasi-Two-Dimensional Friction Model for Simulation of Transient Flows in Viscoelastic Pipes. *Water* **2022**, *14*, 3258. https://doi.org/10.3390/w14203258
13. Zeng, W.; Wang, C.; Yang, J. Hydraulic Transient Simulation of Pipeline-Open Channel Coupling Systems and Its Applications in Hydropower Stations. *Water* **2022**, *14*, 2897. https://doi.org/10.3390/w14182897
14. Cao, Z.; Xia, Q.; Guo, X.; Lu, L.; Deng, J. A Novel Surge Damping Method for Hydraulic Transients with Operating Pump Using an Optimized Valve Control Strategy. *Water* **2022**, *14*, 1576. https://doi.org/10.3390/w14101576
15. Liu, J.; Wu, J.; Zhang, Y.; Wu, X. Sensitivity Analysis of Hydraulic Transient Simulations Based on the MOC in the Gravity Flow. *Water* **2021**, *13*, 3464. https://doi.org/10.3390/w13233464

References

1. Pollio, V. *The Ten Books on Architecture*; Morgan, M.H., Translator; Harvard University Press: Cambridge, MA, USA, 1914; Book 8, Chapter 6, Sections 5–8, pp. 245–246.
2. Tijsseling, A.S.; Anderson, A. A precursor in water hammer analysis—Rediscovering Johannes von Kries. In Proceedings of the 9th International Conference on Pressure Surges, Chester, UK, 24–26 March 2004; pp. 739–751.
3. Tijsseling, A.S.; Anderson, A. Johannes von Kries and the history of water hammer. *J. Hydraul. Eng.* **2007**, *133*, 1–8. [CrossRef]
4. Tijsseling, A.S.; Anderson, A. *The Joukowsky Equation for Fluids and Solids*; CASA Reports; Department of Mathematics and Computer Science, Eindhoven University of Technology: Eindhoven, The Netherlands, 2006; pp. 1–11.
5. Allievi, L. General theory of the perturbed motion of water in pipes under pressure (water hammer). *Ann. Della Soc. Degli Ing. Ed Archit. Ital. (Ann. Soc. Ital. Eng. Archit.)* **1902**, *17*, 285–325.
6. Pandey, M.; Winkler, D.; Vereide, K.; Sharma, R.; Lie, B. Mechanistic Model of an Air Cushion Surge Tank for Hydro Power Plants. *Energies* **2022**, *15*, 2824. [CrossRef]
7. Xu, T.; Chen, S.; Zhang, J.; Yu, X.; Lyu, J.; Yan, H. Comparison on Hydraulic Characteristics of Vertical and Horizontal Air-Cushion Surge Chambers in the Hydropower Station under Load Disturbances. *Energies* **2023**, *16*, 1501. [CrossRef]
8. Lam, C.K.G.; Bremhorst, K.A. Modified Form of the k-ε Model for Predicting Wall Turbulence. *J. Fluids Eng.* **1981**, *103*, 456–460. [CrossRef]
9. Sundstrom, L.R.J.; Cervantes, M.J. Transient Wall Shear Stress Measurements and Estimates at High Reynolds Numbers. *Flow Meas. Instrum.* **2017**, *58*, 112–119. [CrossRef]
10. Urbanowicz, K. Fast and accurate modelling of frictional transient pipe flow. *Z. Angew. Math. Mech.* **2018**, *98*, 802–823. [CrossRef]
11. Johnston, D.N. Efficient methods for numerical modelling of laminar friction in fluid lines. *J. Dyn. Syst. Meas. Control* **2006**, *128*, 829–834. [CrossRef]
12. Xu, Y.; Deng, Y.; Jiao, Z. Fast Meshless Solution with Lumped Friction for Laminar Fluid Transients. *J. Press. Vessel Technol.* **2023**, *145*, 061401. [CrossRef]

Disclaimer/Publisher's Note: The statements, opinions and data contained in all publications are solely those of the individual author(s) and contributor(s) and not of MDPI and/or the editor(s). MDPI and/or the editor(s) disclaim responsibility for any injury to people or property resulting from any ideas, methods, instructions or products referred to in the content.

water

MDPI

Article

Exploring the Sensitivity of the Transient Response following Power Failure to Air Valve and Pipeline Characteristics

Elias Tasca [1,*], Mohsen Besharat [2], Helena M. Ramos [3,*], Edevar Luvizotto, Jr. [1] and Bryan Karney [4]

1 School of Civil Engineering, Architecture and Urban Design, State University of Campinas, Campinas 13083-889, Brazil; edevar@unicamp.br
2 School of Civil Engineering, University of Leeds, Leeds LS2 9JT, UK; m.besharat@leeds.ac.uk
3 Department of Civil Engineering, Architecture and Georesources, CERIS, Instituto Superior Técnico, University of Lisbon, 1049-001 Lisbon, Portugal
4 Department of Civil and Mineral Engineering, University of Toronto, Toronto, ON M5S 1A4, Canada; bryan.karney@utoronto.ca
* Correspondence: elias.tasca@gmail.com (E.T.); hramos.ist@gmail.com (H.M.R.)

Abstract: Air valves are protective devices often used in pressurised water pipelines, ideally admitting air to limit sub-atmospheric pressures and controlling the release of entrapped air. This work summarises a comprehensive sensitivity analysis of the transient behaviour in a rising water pipeline with an air valve following a pump trip. The paper examines the water hammer stages associated with a pump trip, namely, the initial depressurisation, followed by air admission, then air expulsion, and finally the creation of a secondary pressure wave. For each air valve location and specific set of design conditions, the relationship between the transient magnitude and air valve outflow capacity is found to be non-linear, but to roughly follow the shape of a logistic curve having a lower left plateau for attenuated (type 1) behaviour and transitioning through type 2 behaviour to a higher right plateau for water-hammer-dominated (type 3) behaviour. Through an extensive set of simulations covering a wide range of conditions, the study identifies the size of the critical outflow orifices associated with both type 1 and type 3 responses and assesses the influence of the location of the air valve on the transient magnitude and on the timing of air pocket collapse. Furthermore, the paper highlights that a non-slam air valve is capable of effectively mitigating transient magnitudes provided that its design parameters are judiciously chosen and account for both the system's attributes and the characteristics of the transient event.

Keywords: air valve; air pocket; water hammer; hydraulic transient; pump trip; pumping system; water supply

Citation: Tasca, E.; Besharat, M.; Ramos, H.M.; Luvizotto, E., Jr.; Karney, B. Exploring the Sensitivity of the Transient Response following Power Failure to Air Valve and Pipeline Characteristics. *Water* **2023**, *15*, 3476. https://doi.org/10.3390/w15193476

Academic Editor: Stefano Alvisi

Received: 30 June 2023
Revised: 25 September 2023
Accepted: 26 September 2023
Published: 2 October 2023

Copyright: © 2023 by the authors. Licensee MDPI, Basel, Switzerland. This article is an open access article distributed under the terms and conditions of the Creative Commons Attribution (CC BY) license (https://creativecommons.org/licenses/by/4.0/).

1. Introduction

The presence of air in pressurised water pipelines is often problematic. The inability to effectively manage entrapped air can lead to increased power consumption and can compromise conveyance capacity [1]. When a line is initially pressurised, the presence of entrapped air can permit rapid acceleration of the water column and can thus lead to higher transient pressures compared to lines containing water only [2]. Yet, paradoxically, the presence of sufficient air can also cushion transient events, as in the case of either an unvented system or when venting is controlled by directing air flow through a relatively small aperture [3]. Moreover, air can play a broad and beneficial role in mitigating hydraulic transients if strategically contained, as it is in hydropneumatic tanks [4].

Air valves are a frequently employed strategy for addressing air management in pressurised pipelines. Air-release valves exhaust accumulated air during regular operation, while air/vacuum valves are often selected to allow for air exchange during controlled filling and draining operations, as well as for air exchange during water hammer events [5]. Yet, in reality, air valves often underperform, with their inadequacies sometimes leading to severe consequences, including system disruption or even failure although, at other times and in other systems, their underperformance may have entirely benign outcomes [6]. Thus, a practical puzzle is created: in some systems, the presence of air valves is crucial for reliable operation, while in other systems, air valve removal may have little or no discernible effect on performance. It is to this puzzle that this paper is primarily addressed. The task of evaluating consequences raises many complexities: these include the challenge of predicting the air exchange through an air valve [7], the interaction between air and water within the pipe system [8], the often contradictory sizing criteria for different hydraulic events [9], and the non-linear impact of the air valve on transient behaviour [10].

Complications related to air exchange through the air valve include unreliable performance data from product catalogues and the unpredictability of the dynamic closure of the air valve during expulsion [11]. Accurately modelling transient air–water interactions is challenging without a CFD approach, especially for complex scenarios, such as the filling procedure of horizontal or irregular profile pipelines. Regarding air valve sizing, often the air valve must be sufficiently large to enable nearly unrestricted entry of air during drainage [12] or down-surge events [10], as well as for the efficient release of air during controlled pipeline filling [13]. However, oversizing an air valve's outflow orifice can lead to excessive over-pressures upon air valve closure during filling operations or following a pump trip event [14]. Notably, the pipeline transient response exhibits non-linear behaviour in relation to air valve capacity [15], valve location, and system characteristics [10].

Research relating to air in pressurised pipelines can be categorised into three classes: experimental studies [16], combined experimental–numerical investigations [17], and purely numerical explorations [18]. Due to space limitations in controlled labs, experimental and experimental–numerical studies often focus on small-scale systems, which hinders their applicability to large-scale applications [19]. For instance, although the Bergant et al. (2012) study [20] conducted an experimental–numerical investigation using typical air valve and pipe sizes, the pipeline lengths employed were quite short. Regarding pipeline filling, studies often concentrate on either horizontal [21] or vertical pipes [22], which are prevalent in few large-scale systems [6]. Furthermore, the common use of orifices for air exchanges in these studies, rather than the more commonly employed air valves, introduces an added challenge when translating the findings into practical applications [3]. Certainly, numerical studies exploring the transient behaviour of large-scale pipelines influenced by the presence of air can be found in the literature. These studies include, for example, the works by Pozos et al. (2010) [23], Ramezani and Karney (2017) [10], and Li et al. (2022) [18]. The current paper is in this tradition and provides a comprehensive numerical analysis of the transient behaviour in a pressurised rising water pipeline with an air valve during a pump trip scenario.

This paper is structured as follows. A concise overview of recent related research is first provided, followed by the identification of the typical water hammer stages associated with the pump trip scenario. The two key power-loss behaviours, namely, attenuated and water-hammer-dominated events, are identified, as well as the related transitional states. A comprehensive investigation into the sensitivity of the transient response to the air valve outflow capacity and location follows. This involves analysing variations in pipeline lengths experiencing negative pressures, maximum hydraulic grade (HG) values, maximum head values, and air pocket collapse times. The timing of peak transient HGs is explored, as are the critical orifice sizes for achieving attenuated and water-hammer-dominated events. The work specifically assesses the influence of the initial steady water velocity, the elevation difference between downstream and upstream reservoirs, and the impact of changes in both wave speed and friction on the transient response. Finally, the paper investigates how non-slam air valve design parameters impact the transient response and concludes by summarising all the insights gained.

2. Overview of Related Research

Ramezani (2015) [24] studied the numerical sensitivity of transient severity to air valve capacity and pipeline configuration in a pump trip scenario. The simulations utilised an elastic model through the method of characteristics, and Monte Carlo filtering was employed to statistically evaluate the simulation results. This study found that both the air valve inflow capacity and initial steady water discharge affect the minimum transient pressures, while the air valve outflow capacity and initial steady water discharge affect the maximum transient pressures. The analysis prioritised statistical insights over the consideration of the individual physical processes that constitute the hydraulic transient phenomenon.

To complement the earlier work, Ramezani and Karney (2017) [10] developed semi-analytical formulas to elucidate the fundamental wave processes that occur following sudden depressurisation. This work emphasised the significance of air valve location to the transient response, suggesting that the downstream branch of the pipeline, in relation to the air valve, has a greater influence on severity than the upstream branch. Longer downstream branches lead to longer air pocket collapse times and more attenuated secondary pressures. The semi-analytical formulas incorporate several simplifications, including the omission of macro-cavitation, the assumption of free air inflow and outflow, and the assumption that the pressure in the air pocket is always atmospheric.

Coronado et al. (2018) [25] conducted a numerical sensitivity analysis to investigate how the air valve capacity and pipeline configuration (specifically, a single straight pipeline) can influence pipeline transient behaviour during both filling and draining procedures. The simulations used a rigid column model for the water phase and the polytropic transformation equation for the air phase. For filling, transient behaviour displayed sensitivity to pipe slope, air valve size, pipe diameter, and the friction factor. For draining, transient behaviour showed sensitivity to air valve size, air pocket size, pipe slope, and pipe diameter.

Tasca et al. (2021) [14] conducted a provisional numerical study into the impact of the air valve capacity and pipeline configuration on pipeline transient response following pump trip. While the air valve location was emphasised as significant, the importance of the air valve capacity was acknowledged, but its relevance was less apparent due to the limited range of tested air valve sizes. Tasca et al. (2021) [14] noted that the transient magnitude can be affected not only by the size of the air valve orifice but also by its discharge coefficient. Increasing the elevation of the air valve was found to attenuate the secondary waves created upon air valve closure and led to larger admitted air volumes. Tasca et al. (2022) [11] clarified the influence of the air valve outflow capacity on the pipeline transient response during the pump trip scenario; a significant finding was that the critical orifice sizes previously identified as resulting in attenuated events during pipeline filling also have considerable applicability to the pump trip scenario.

Li et al. (2022) [18] and Li et al. (2022) [15] numerically investigated the impact of air valve capacity on pipeline transient behaviour during a water hammer event resulting from the closure of the downstream valve in a gravity pipeline. It was found that restriction of the air valve outflow capacity tended to attenuate secondary pressures and that increasing an air valve's inflow capacity tended to limit negative pressures. Li et al. (2022) [18] noted that the significance of discharge coefficients becomes less pronounced when dealing with small outflow orifices. Additionally, a ratio of 0.10 [15] or 0.05 [18] between the outflow and inflow orifice sizes was often found to be effective in simultaneously mitigating negative pressures and secondary over-pressures. The numerical simulations involved some simplifying assumptions, including the absence of macro-cavitation and isothermal air pocket evolution. In reality, small orifices often lead to attenuated transient behaviour, but may also lead to a substantial increase in air pocket temperature [26].

3. Methodology

The current paper examines a pump trip scenario in a typical rising water pipeline equipped with an air valve at a distinct high point. Figure 1 illustrates the key components of the analysed pumping system, with certain attributes allowed to vary. Similar to many operational systems, the layout includes an upstream reservoir, a pumping station (consisting of two pumps in series, each equipped with a check valve), a pipe segment featuring an air valve at a local high point, and a terminal pipe section connected to the downstream reservoir. The lengths of the two pipe segments downstream of the air valve are kept equal in this setup.

The main point of the selection of this system is not due to its universal applicability, though the configuration does typify many operational systems. Rather, this specific system is chosen as it nicely characterises the general nature of the non-linear response that even relatively simple systems can display to air valve, system, and transient attributes. Thus, the goal is to give system designers, owners, and operators an essential "sense" of where and when high sensitivity to assumed system and valve characteristics can constitute a threat to system reliability, performance, and behaviour.

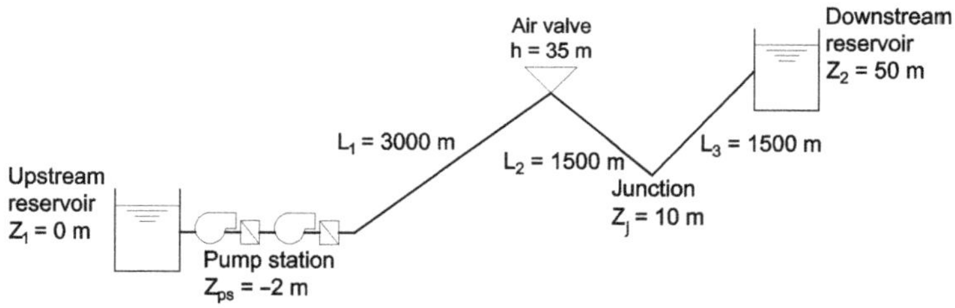

Figure 1. Possible test pipeline configuration including an air valve at the line's distinct high point, as considered in the numerical simulations.

3.1. Simulation Scenarios

The current study considers three sets of numerical simulations: (i) the main or foundational set of "basic" simulations, (ii) a set of simulations incorporating variations in key initial parameters, and (iii) a set of simulations involving non-slam air valves. A great many variations of the system parameters are then considered within these three cases.

The primary simulations set the initial steady water velocity to $v_0 = 1.5$ m/s, the elevation difference between reservoirs as $\Delta Z = 50$ m, the wave speed $a = 1000$ m/s, the Hazen–Williams friction coefficient $C = 130$ (with corresponding absolute roughness of 0.259 mm), ratio between the air valve inflow orifice and the pipeline diameter $d_{in}/D = 0.20$, pipeline length $L = 6000$ m, and pipe diameter $D = 0.50$ m. Moreover, the simulations consider five longitudinal L_1 air valve locations (1000, 2000, 3000, 4000, and 5000 m), five vertical air valve locations (i.e., h value is 20, 25, 30, 35, or 40 m), and 17 d/D values unequally spaced from zero to 0.40—between 0.02 and 0.10, the d/D values are spaced in increments of 0.01; between 0.10 and 0.20, the d/D values are spaced in increments of 0.02. The smallest d/D value considered with $d/D > 0$ is $d/D = 0.02$. The values $d/D = 0$ and $d/D = 0.40$ are considered. The system response without air valve protection is also considered, to bound the results.

Significantly, the current paper explores variations in the following parameters: v_0, ΔZ, a, and C. More specifically, three v_0 values are used, namely, 0.5, 1.0, and 2.0 m/s; three ΔZ values are considered, namely, 25, 75, and 100 m; and three a values are simulated, namely, 400, 600, and 800 m/s; moreover, three C values are tested, namely, 90, 110, and 150. For all these cases, $L_1 = 3000$ m and $h = 40$ m. In scenarios where a variation in ΔZ is incorporated, the air valve is positioned at an elevation equivalent to 80% of that of the downstream reservoir. This leads to h values of 20, 60, and 80 m for ΔZ values of 25, 75, and 100 m, respectively.

This paper also investigates the effect of having a non-slam air valve at the high point instead of a regular air valve. Herein, a non-slam air valve refers to a device with a single inflow orifice but with both a small outflow orifice and a large outflow orifice, while a regular air valve refers to a device with a single inflow orifice and a single outflow orifice.

Variations in the following parameters have been considered: transition head ($\Delta H_{\text{transition}}$), large outflow orifice (d_{large}), and h. The transition head refers to the threshold head that triggers the change from the large outflow orifice (d_{large}) to the small outflow orifice (d_{small}). Six $\Delta H_{\text{transition}}$ values are considered: 0.05, 0.10, 0.15, 0.20, 0.25, and 0.30 m. Eight d_{large}/D values are considered: 0.06, 0.08, 0.10, 0.12, 0.14, 0.16, 0.18, and 0.20. Two h values are considered: 20 and 40 m. For all these cases, $L_1 = 3000$ m and $d_{\text{small}}/D = 0.04$.

3.2. Modelling Approach and Limitations

Each numerical simulation consists of two stages: the initial steady flow and the hydraulic transient. The initial steady flow is calculated using the Darcy–Weisbach equation, while the hydraulic transient is modelled using the 1D elastic model employing the conventional method of characteristics. The pipeline elements involved in the transient simulations include pipeline segments, upstream and downstream reservoirs, pumping station, air valve, and air pocket admitted by the air valve. Appropriately selecting the simulation time step and the simulated time is important to ensure that the numerical results capture the primary aspects of the transient.

Various numerical choices are also relevant to the transient simulation, including the calculation mesh, how short pipes are integrated into the mesh, the methodology for accounting for friction losses, and the approach used for modelling cavitation. The simulations employ the conventional method of characteristics (MOC) on a rectangular computational mesh, with the Courant number fixed at unity. Pipe reach adjustments are allowed for this purpose. Further information about the elastic model and the MOC can be found in Chaudhry (2014) [27]. Unsteady friction is incorporated into the simulations following Vitkovsky's methodology. Additional details regarding unsteady friction methods can be found in Abdeldayem et al. (2021) [28]. The discrete vapour cavity model (DVCM) is utilised for macro-cavitation modelling. In the DVCM, vapour cavities form at specific computational nodes and remain localised. The absolute pressure in the system is constrained by the vapour pressure of water. The release of air resulting from a decrease in pressure and changes in wave speed due to the presence of vapour cavities are not taken into account. Further information about cavitation models can be found in Bergant et al. (2006) [29].

Both reservoirs are treated as constant-level reservoirs. The main set of simulations includes two pumps, each equipped with a check valve. The pump curve from the manufacturer is employed, and the simulations incorporate pump inertia with instantaneous check valve closure upon reverse flow. Further details about the simulation of turbo-machines can be found in Wylie and Streeter (1983) [30]. The air flow through the air valve is modelled according to the isentropic air flow model, considering discharge coefficients of 0.6 for both admission and expulsion. Circular air exchange orifices are assumed. More information about the isentropic air mass flow model can be found in Tasca and Karney (2023) [9]. The air pocket admitted by the air valve is assumed localised, and its evolution is modelled adiabatically, i.e., the evolution of the air pocket follows the polytropic transformation equation with a polytropic exponent of 1.4. The assumption of a stationary entrapped air pocket is reasonable since the pipe segments forming the high point with the air valve are ascending towards the air valve. When an air pocket is present in the system, the upstream and downstream branches in relation to the air valve become hydraulically separated; thus, water column separation is assumed. The current study does not consider the occurrence of air bubbles in the water phase or two-phase flows. Atmospheric pressure is assumed to be sea-level atmospheric pressure. Further details about the polytropic transformation equation in the context of hydraulic transients can be found in Fuertes et al. (2016) [31].

The time step utilised is 0.04 s. Only a slight sensitivity to the time step was observed. For the main set of simulations, a simulated time of 635 s (1.5 times the collapse time for $d/D = 0.02$, $L_1 = 1000$ m, and $h = 40$ m) is employed for $0.02 \leq d/D \leq 0.04$; for $d/D \geq 0.05$, a simulated time of 190 s (1.5 times the collapse time for $d/D = 0.05$, $L_1 = 1000$ m, and $h = 40$ m) is employed. An analogous approach is adopted for the set of simulations incorporating variations in key initial parameters, with each variation having its own simulated time, categorised into two brackets: $0.02 \leq d/D \leq 0.04$ and $d/D \geq 0.05$, with $h = 40$ m, but $L_1 = 3000$ m. For the set of simulations involving a non-slam air valve, a simulated duration of 440 s is utilised (1.5 times the collapse time for $d/D = 0.02$, $L_1 = 3000$ m, and $h = 40$ m).

The numerical simulations were conducted using the application HAMMER from Bentley Systems (version 10.1.1.4).

3.2.1. Pumping System Configuration

For the main set of simulations, which considers $v_0 = 1.5$ m/s, $\Delta Z = 50$ m, and $C = 130$, the pumping system consists of two pumps in series of the model KSB ETA 250-33 330/330. For simulations with $v_0 = 0.5$ m/s, the pumping system consists of one pump of the model KSB ETA 125-40 395. For simulations with $v_0 = 1.0$ m/s, the pumping system consists of two pumps in series of the model KSB ETA 200-33 300. For simulations with $v_0 = 2.0$ m/s, the pumping system consists of two pumps in series that are in parallel with two other pumps in series of the model KSB ETA 200-40 340. For simulations with $\Delta Z = 25$ m, the pumping system consists of one pump of the model KSB ETA 250-40 380. For simulations with $\Delta Z = 75$ m, the pumping system consists of two pumps in series of the model KSB ETA 250-40 380. For simulations with $\Delta Z = 100$ m, the pumping system consists of two pumps in series of the model KSB ETA 250-40 400. For simulations with $C = 150$, the pumping system consists of two pumps in series of the model KSB ETA 250-33 330/300. For simulations with $C = 110$, the pumping system consists of two pumps in series of the model KSB ETA 250-33 330/330. For simulations with $C = 90$, the pumping system consists of two pumps in series of the model KSB ETA 250-40 380. For all these cases, the nominal rotational speed is 1760 rpm. Since the check valve is assumed to close immediately in flow reversal, and then to remain closed, Suter curves for 4-quadrant pump characterisation and for energy dissipation zones were not required.

3.2.2. Limitations of the Study

Regarding the current examination of the general transient response of pipelines in the context of a pump trip scenario, it is important to recognise that pipeline systems operate within a wide range of system configurations and environmental contexts. This complexity makes achieving a truly comprehensive reporting or results unattainable. Furthermore, real systems often exhibit a higher degree of complexity compared to the scope considered in this paper. They frequently encompass multiple points with air valve protection and a diverse array of hydraulic devices. Additionally, there are inevitably uncertainties associated with the assumed numerical and behavioural models that are necessarily simplified in order to capture unsteady flow and hydraulic element behaviour. Nonetheless, the primary goal of this study is to elucidate general relevant causal relationships in the pump trip scenario, rather than presuming high-precision simulations are practical. Despite the relative simplicity of the general pipeline layout and tested conditions under examination, the inherent limitations of numerical models, and the limited variety of data acquired from the numerical simulations, valuable engineering insights naturally emerge from the mechanistic assessment approach employed in this paper.

The current study does not account for the potential influence of a hydropneumatic tank or other specialised surge protection equipment. Indeed, the primary objective of this study is to concentrate on the distinct impact of the air valve itself and its potential in mitigating the transient event.

The main set of simulations assumes a constant wave speed of $a = 1000$ m/s, a value appropriate for simulating hydraulic transients in many pipeline systems. In practice, there is some uncertainty regarding the wave speed arising from issues like temperature effects, bedding conditions, and the presence of air. Some simulations are undertaken involving lower a values, which are characteristic of viscoelastic pipes, though viscoelastic effects are not included.

The simulations assume that entrapped air remains localised near the air valve and disregard water level variations due to air pocket evolution, an assumption consistent with having relatively small air volumes. In the present study, the pipe lengths able to accommodate volumes corresponding to the maximum air pocket volumes are significantly shorter than the lengths of the pipeline segments themselves. This assumption is not universally valid [32].

The discrete vapour cavity model (DVCM) is employed for cavitation modelling. In this study, sensitivity is observed in cavitation modelling depending on the chosen size for the pipe reaches. Nonetheless, such sensitivity does not substantively alter the transient magnitudes. Moreover, the simulations do not consider wave speed reduction for pipe reaches adjacent to vapour cavities.

The pump check valves are assumed to close immediately upon reverse flow. In practice, however, the responsiveness of check valves varies based on factors such as type, age, model, and size. If the pumping system were to lack check valves, the transient behaviour could differ significantly from that presented here, and reverse flow through the pump(s) would become relevant. In rising water systems, check valves are nearly universal.

3.3. Evaluating Transient Severity

A common approach to assess the magnitude of a water hammer event involves analysing the extreme transient HG envelope obtained from a transient analysis. Agostinho et al. (2018) [33] provide an overview of 14 hydraulic transient indicators. Notably, Jung and Karney (2011) [34] introduced the negative and positive surge damage potential factors—denoted as $SDPF^{-}$ and $SDPF^{+}$, respectively—to facilitate the evaluation of water hammer protection strategies. These factors take into account the duration of extreme transient heads, addressing both the "how much" and "how long" aspects of the transient. For instance, Ramezani (2015) [24] utilised these factors to evaluate the role of air valves in the pump trip scenario.

This paper also addresses the important complementary aspects of "where" and "when" extreme transient heads occur. To this end, the water hammer phenomenon is divided into four sequential stages: depressurisation, air admission, air expulsion, and secondary wave. An investigation that considers the "where" and "when" aspects makes it easier to infer causal relationships. In this sense, regarding the pump trip scenario, the crucial aspects to be addressed in this paper are "where", "when", and "for what reason" the extreme transient heads occur.

The secondary waves resulting from the closure of the air valve at the end of the air expulsion stage are commonly discussed in the literature [15,18]. Indeed, these secondary waves should be controlled to prevent dangerous transients [35,36]. In fact, transient-critical air valves are typically situated at high points, making it rare for overall maximum transient heads to occur there. In actuality, pipeline segments situated at lower elevations are particularly susceptible to experiencing dangerous transient over-pressures.

For the pump trip scenario in a rising water pipeline, the highest transient HG occurs at the upstream section of the pipeline (i.e., at the lowest elevation region of the pipeline). The transient magnitude depends on several factors: pump(s) inertia, initial steady water velocity, pipeline layout, pipe material, diameter, and wall thickness, dynamic properties of check valve(s), air valve location, type, inflow and outflow capacities, and the occurrence of macro-cavitation. Given the intricate nature of the pump trip scenario, which involves air–water interactions within the pipe, air exchanges through the air valve, and macro-cavitation, the numerical simulation results for the maximum HG exhibit a certain degree of randomness and unpredictability.

To more comprehensively assess the extent of transient over-pressures in the system, a new parameter is introduced, providing an alternative to the maximum transient hydraulic grade line (HGL), the maximum transient HG in the system, and the magnitude of the secondary wave upon air valve closure. This parameter is referred to as the mean maximum hydraulic head and is defined as follows:

$$\overline{H}_{\max} = \overline{\mathrm{HG}}_{\max} - \overline{Z} = \frac{1}{N}\sum_{i=1}^{N}\mathrm{HG}_{\max_i} - \frac{1}{N}\sum_{i=1}^{N} Z_i \tag{1}$$

where $\overline{\mathrm{HG}}_{\max}$ is the mean maximum HG and $\overline{Z}$ is the mean pipeline elevation. These quantities are dependent on $\overline{\mathrm{HG}}_{\max_i}$ (the maximum HG at point i) and Z_i (the elevation at point i). Parameter N denotes the number of pipeline reaches.

4. Water Hammer Stages in the Pump Trip Scenario

For the pump trip scenario in a rising water pipeline, *depressurisation* starts when power to the pumping system is cut and ends when the under-pressure wave reaches the air valve. The *air admission* stage starts when the under-pressure wave reaches the air valve and ends when the maximum air pocket volume is achieved. The *air expulsion* stage starts when the maximum air pocket volume is achieved and ends when the air valve closes (i.e., when all the air has been expelled from the line). The *secondary wave* stage starts when the air valve closes and ends when a peak transient pressure occurs at the upstream end of the pipeline (in case of multiple transient cycles, such a peak precedes the re-occurrence of negative pressures at the air valve with consequent re-admission of air). Collectively, these four water hammer stages comprise a water hammer cycle. For large-outflow-capacity air valves, several water hammer cycles are expected before flow in the system is arrested. Nonetheless, the extreme transient HG envelope—minimum and maximum HGLs—is generally mostly determined during the first water hammer cycle.

4.1. *Stages for Reduced-Outflow-Capacity Air Valve*

Figure 2 depicts the water hammer stages in the first water hammer cycle in a pipeline with an air valve with a large-inflow-capacity orifice but reduced-outflow-capacity orifice considering $L_1 = 3000$ m and three values for the air valve elevation (20, 30, and 40 m). In each image in Figure 2, two sets of extreme transient HGLs are represented: the overall extreme HGLs (dotted red and blue lines) and the extreme HGLs up to the end of the water hammer stage shown in the image (continuous red and blue lines). This representation allows the assessment of the timing of each part of the overall extreme transient envelope. Furthermore, each image in Figure 2 displays the HGL at the conclusion of the water hammer stage shown in the image (continuous green line). For the air expulsion stage, specifically, the continuous green line refers to the HGL four time steps after the closure of the air valve in order to display the magnitude of the secondary transient waves.

Examples of systems experiencing depressurisation are shown in Figure 2a,e,i. During the depressurisation stage, the maximum HGL does not occur. Rather the maximum HGL at the end of this stage is determined by the initial steady-state HGL, whereas the global minimum HGL for the upstream branch is primarily formed during this stage. For the downstream branch, however, the minimum HGL at the end of this stage matches the initial steady-state HGL. The depressurisation stage is short lived, spanning between the inception of the transient and the time when the reduced-pressure wave reaches the air valve.

Simulation results associated with the air admission stages are shown in Figure 2b,f,j. The overall maximum HGL does not form during the air admission stage. However, for the downstream branch, the maximum HGL at the end of this stage differs from the initial steady-state HGL. For the downstream branch, the overall minimum HGL typically occurs during the air admission stage. For $h = 20$ m, the entire pipeline length experiences negative pressures by the end of this stage. In contrast, for $h = 40$ m, the lower portion of the downstream branch avoids negative pressures until the end of this stage. The closure of the check valves associated with the pumps takes place during the air admission stage. The check valves close at 9.5, 10.0, and 14.4 s for the air valve elevations of 40, 30, and 20 m, respectively.

Examples of air expulsion stages are shown in Figure 2c,g,k. For $h = 30$ m, the overall maximum HGL has been mostly formed by the end of the air expulsion stage. For $h = 40$ m, the maximum HGL at the end of this stage closely approximates the global maximum HGL. This is not the case, though, for $h = 20$ m. In Figure 2, given the reduced outflow capacity of the air valve, the secondary waves resulting from the air valve closure are mild and almost indistinguishable from the nearby HGL fluctuations. Moreover, the air pocket pressure as the air valve closes is considerably larger than atmospheric pressure (which is more pronounced for $h = 20$ m). This is a consequence of the pressurisation of the entrapped air pocket during the air expulsion stage.

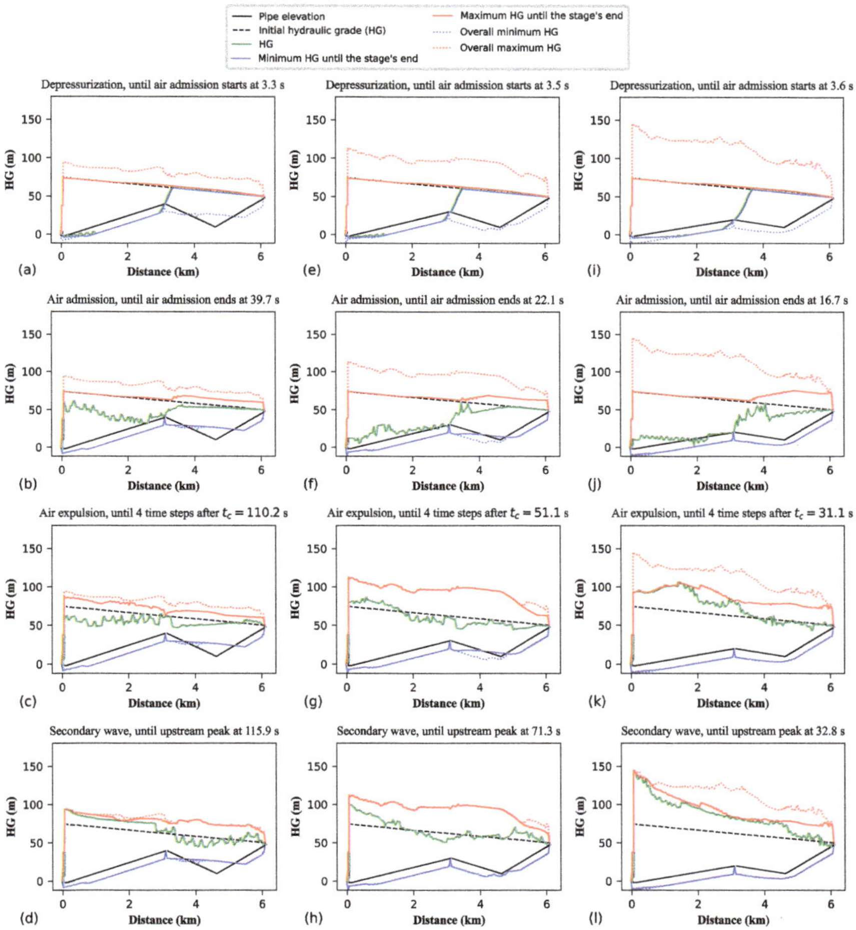

Figure 2. The four water hammer stages in the first water hammer cycle considering $L_1 = 3000$ m and air valve with reduced outflow capacity ($d/D = 0.04$): (**a**–**d**) $h = 40$ m; (**e**–**h**) $h = 30$ m; (**i**–**l**) $h = 20$ m.

Examples of secondary wave stages are shown in Figure 2d,h,l. For $h = 30$ m and $h = 40$ m, the overall maximum HGL forms near the end of the secondary wave stage. The HGL at the end of this stage contains the maximum transient HG in the system (HG_{max}) for $h = 20$ m. Nonetheless, for $h = 20$ m, the overall maximum HGL only occurs after $t_{HG_{max}}$. In contrast to the previous stages, in the secondary wave stage, the transient waves can move unimpeded across the whole pipeline as flow separation has been eliminated by the removal of the air.

Overall, the most severe HGL values occur for $h = 20$ m, while the least severe ones occur for $h = 40$ m. Note that for $h = 40$ m, the air valve is positioned at a higher elevation, rendering it less susceptible to high pressures. The maximum transient HG in the system, which occurs at the upstream section of the pipeline, is about 145 m for $h = 20$ m, 113 m for $h = 30$ m, and 94 m for $h = 40$ m. The maximum HGL tends to have higher values for points that are closer to the upstream reservoir.

The duration of the depressurisation stage is the same for the three air valve elevations. The duration of the other stages, however, is affected by h. The longest time for air admission and the largest admitted air pocket occur for $h = 40$ m. The air pocket collapse time for $h = 40$ m ($t_c = 110$ s) is more than double the value for $h = 30$ m ($t_c = 51$ s) and almost four times the value for $h = 20$ m ($t_c = 31$ s). In Figure 2, the air expulsion stage persists for a longer duration than the air admission stage. This difference becomes more pronounced as h increases. For $h = 40$ m, the air expulsion stage lasts 71 s, while the air admission stage lasts 36 s. On the other hand, for $h = 20$ m, the air expulsion stage lasts 14 s, while the air admission stage lasts 13 s.

4.2. Stages for Large-Outflow-Capacity Air Valve

Figure 3 shows the water hammer stages in the first water hammer cycle in a pipeline with an air valve with both large-inflow-capacity and large-outflow-capacity orifices considering $L_1 = 3000$ m and three values for the air valve elevation (20, 30, and 40 m). Like in Figure 2, each image in Figure 3 shows the overall extreme HGLs, the extreme HGLs until the conclusion of the corresponding stage, and the HGL at the end of the corresponding stage.

The depressurisation and air admission stages in Figure 3 are similar to those in Figure 2. In these two initial stages, the outflow function of the air valve does not influence the transient response. The timings of the depressurisation and air admission stages are equivalent for corresponding air valve elevations in Figures 2 and 3.

The air expulsion and secondary wave stages, however, are notably distinct between Figures 2 and 3 for corresponding air valve elevations. For example, in contrast to Figure 2c,g,k, in Figure 3c,g,k, the secondary waves formed as a consequence of air valve closure are much larger in comparison to the neighbouring HGL variations. In contrast to what is shown in Figure 2c,g,k, in Figure 3c,g,k, the air pocket pressure as the air valve closes is close to the atmospheric pressure.

For $h = 30$ m and $h = 40$ m, in Figure 3, similar to Figure 2, most of the overall maximum HGL has been formed by the end of the secondary wave stage. In contrast, for $h = 20$ m, in Figure 3, similar to Figure 2, the maximum HGL does not coincide with the end of the secondary wave stage. In Figure 3, similar to Figure 2, there is a notable tendency for larger HGL values to occur for points closer to the upstream reservoir. In contrast to Figure 2, in Figure 3, except for $h = 20$ m, there is a clear discontinuity of the maximum HGL at the air valve location.

In Figure 3, similar to Figure 2, the water hammer cycle is the longest for $h = 40$ m and the shortest for $h = 20$ m. However, for each h value, t_c is shorter in Figure 3 than in Figure 2. The air valve with a reduced-outflow-capacity orifice imposes significant resistance against air expulsion, while the air valve with a large-outflow-capacity orifice allows for the free exhaust of air. As a result, while the duration of the air expulsion stage is longer than the duration of the air admission stage in Figure 2, in Figure 3, such durations are equivalent. In Figure 3, for $h = 40$ m, the duration of the air expulsion stage is 34 s, while the duration of the air admission stage is 36 s.

In Figure 3, t_c is 80% longer for $h = 40$ m than for $h = 30$ m, and almost three times longer than the value found for $h = 20$ m. A larger sensitivity of t_c to h is found in Figure 2 in comparison to Figure 3. Indeed, for a small-outflow-capacity air valve, t_c increases substantially as h increases.

In Figure 3, the least intense HGL values occur for $h = 40$ m. The transient magnitudes for $h = 20$ m and $h = 30$ m, though, are quite similar—HG_{max} is about 168 m for $h = 20$ m, and about 175 m for $h = 30$ m. Notably, the least intense transient event shown in Figure 3 (for $h = 40$ m) has a magnitude comparable to the most intense event shown in Figure 2 (for $h = 20$ m).

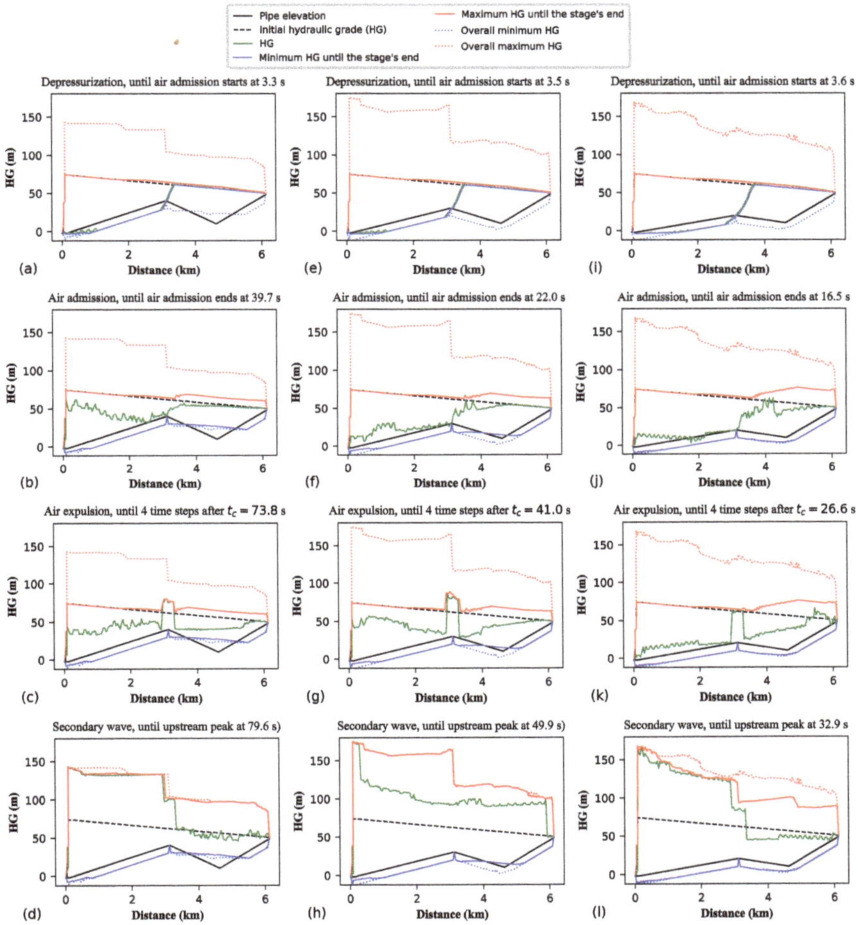

Figure 3. The four water hammer stages in the first water hammer cycle considering $L_1 = 3000$ m and air valve with large outflow capacity ($d/D = 0.10$): (**a–d**) $h = 40$ m; (**e–h**) $h = 30$ m; (**i–l**) $h = 20$ m.

5. Attenuated and Water-Hammer-Dominated Events

5.1. Effect of Air Valve Size on Transient Evolution

Figure 4 shows the evolution of the hydraulic head at the air valve considering $L_1 = 3000$ m, $h = 40$ m, $d_{\text{in}}/D = 0.20$, and d/D varying from 0.02 to 0.12. Two extreme transient behaviours can be distinguished: Figure 4a ($d/D = 0.02$) shows a transient with long-period, mild, and smooth head oscillations (such behaviour can be classified as attenuated or type 1 behaviour); Figure 4j ($d/D = 0.12$) shows a transient with short-period, intense, and sharp head oscillations (such behaviour can be classified as water-hammer-dominated or type 3 behaviour). In addition, the behaviour shown in Figure 4e ($d/D = 0.06$) can be classified as intermediary or type 2 behaviour. Note that distinguishing between these behaviours might become a little vague for intermediate d/D values.

For type 1 behaviour, the head evolution at the air valve includes at least one instance of relatively smooth head-versus-time variation with substantial compression of the air pocket before its collapse with a subsequent mitigated secondary wave (examples of type 1 behaviour are shown in Figure 4a–d). For type 2 behaviour, even though there is some compression of the air pocket before its collapse, the subsequent secondary wave is only partially mitigated (examples of type 2 behaviour are shown in Figure 4e,f). For type 3 behaviour, the air is essentially freely exhausted, resulting in a very sharp secondary wave upon air valve closure (an example of type 3 behaviour is shown in Figure 4j). The graphs in Figure 4g–i display the secondary head spike which typifies type 3 behaviour. However, note how in such cases the maximum hydraulic head increases as d/D increases. This indicates that for these cases, there is still a residual air cushioning effect.

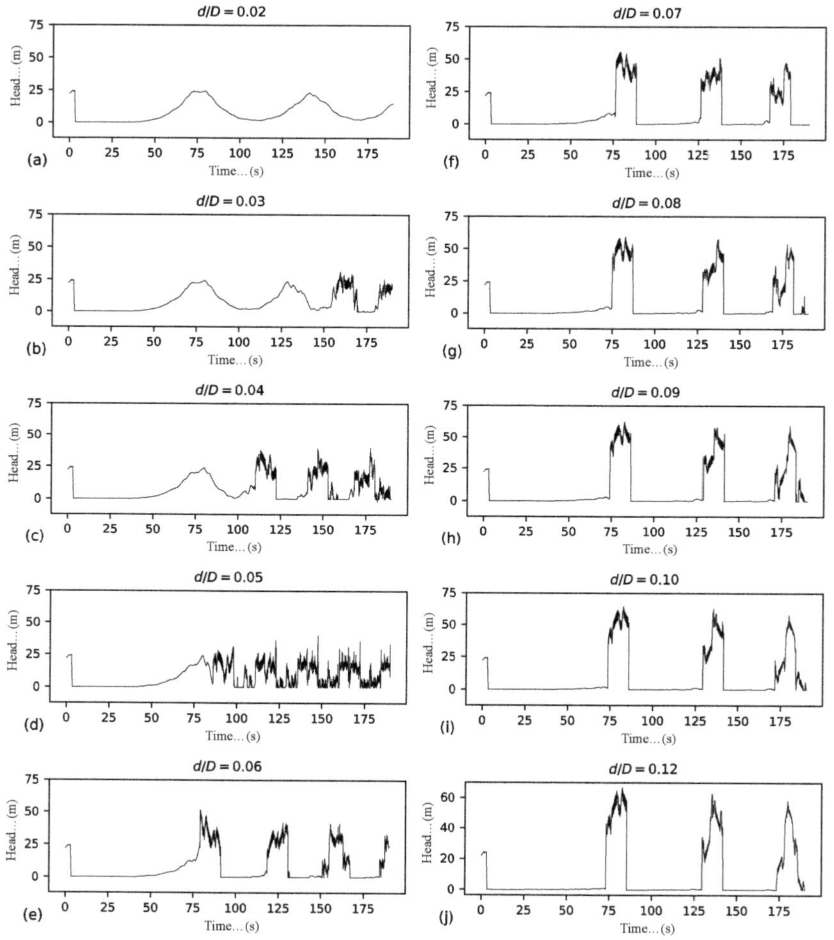

Figure 4. Hydraulic head at the air valve considering $L_1 = 3000$ m and $h = 40$ m: (**a**) $d/D = 0.02$; (**b**) $d/D = 0.03$; (**c**) $d/D = 0.04$; (**d**) $d/D = 0.05$; (**e**) $d/D = 0.06$; (**f**) $d/D = 0.07$; (**g**) $d/D = 0.08$; (**h**) $d/D = 0.09$; (**i**) $d/D = 0.10$; (**j**) $d/D = 0.12$.

In Figure 4a, the secondary wave stage is notably absent. For this case, $t_c = 423$ s, which is beyond the time range depicted in the figure. From Figure 4b onwards, however, the graphs include the secondary wave stage. The air pocket collapse time reduces as d/D increases—$t_c = 157$ s for $d/D = 0.03$, $t_c = 110$ s for $d/D = 0.04$, and $t_c = 86$ s for $d/D = 0.05$. However, for large d/D values, t_c is more stable—$t_c = 76$ s for $d/D = 0.07$, and $t_c = 73$ s for $d/D = 0.12$. In Figure 4, the oscillation period decreases from $d/D = 0.02$ to $d/D = 0.05$. However, from $d/D = 0.07$ onwards, the oscillation period becomes independent of d/D.

The mild and long-period head oscillations shown in Figure 4a occur due to the high resistance against air exhaust imposed by the small outflow orifice. In Figure 4j, however, air is exhausted under very reduced pressure differentials. For type 3 behaviour, sudden and large head rises are generated upon air valve closure (i.e., upon air pocket collapse). In general, for both type 1 and type 3 behaviours, the head oscillations while the air pocket is inside the pipeline are smooth, while the head oscillations after the closure of the air valve and before air re-admission are sharp and erratic. Figure 4 demonstrates the substantial variability in the fundamental nature of the transient response depending on air valve size.

The time evolution of the hydraulic conditions at the upstream section of the pipeline is also of interest. Figure 5 shows the time evolution of the HG at the upstream section of the pipeline (right downstream of the pumping station) considering $L_1 = 3000$, $h = 40$ m, $d_{\text{in}}/D = 0.20$, and d/D varying from 0.02 to 0.12. Type 1 behaviour is noticeable in Figure 5a, while type 3 behaviour is noticeable in Figure 5j. The upstream section (at −2 m elevation) experiences transient pressure fluctuations with a much larger amplitude than those at the air valve. For $d/D = 0.07$, the HG rise after the collapse of the air pocket reaches a value of about 128 m, whereas the corresponding head rise at the air valve reaches only about 55 m.

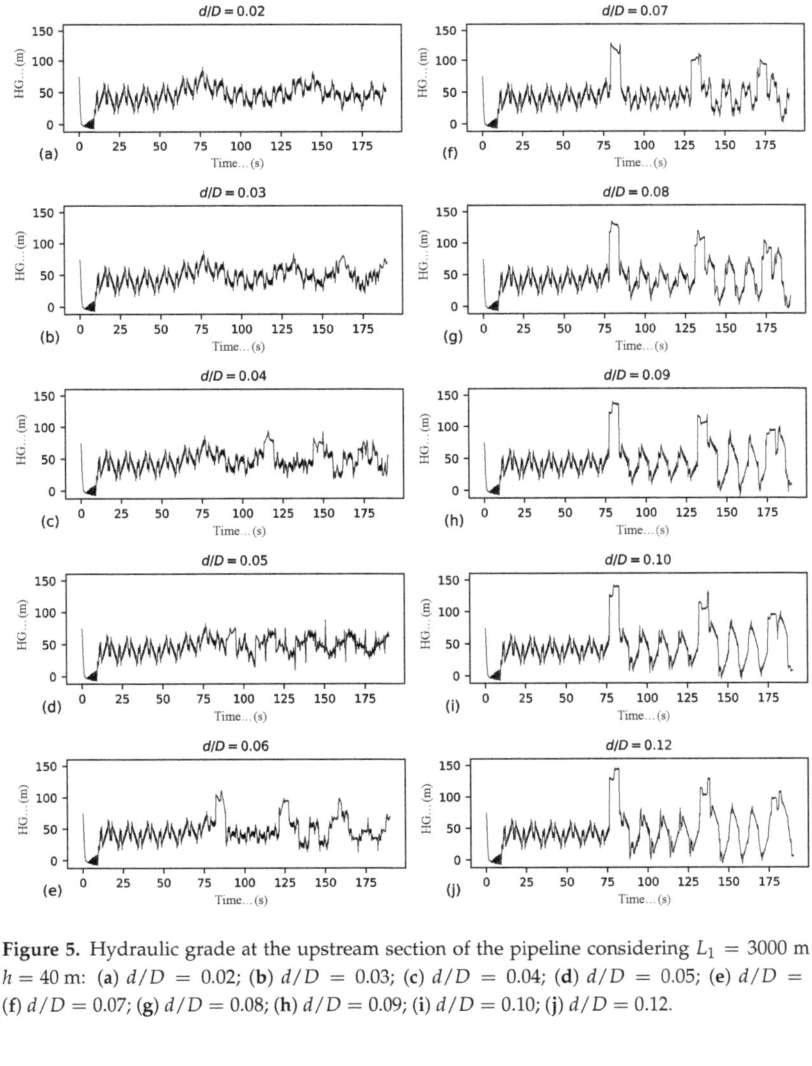

Figure 5. Hydraulic grade at the upstream section of the pipeline considering $L_1 = 3000$ m and $h = 40$ m: (**a**) $d/D = 0.02$; (**b**) $d/D = 0.03$; (**c**) $d/D = 0.04$; (**d**) $d/D = 0.05$; (**e**) $d/D = 0.06$; (**f**) $d/D = 0.07$; (**g**) $d/D = 0.08$; (**h**) $d/D = 0.09$; (**i**) $d/D = 0.10$; (**j**) $d/D = 0.12$.

For $d/D = 0.02$, the oscillation pattern is characterised by a high frequency, while for large d/D values, the oscillation pattern is divided into two consecutive periods: (i) an initial period with higher-frequency and lower-amplitude oscillations, and (ii) a posterior period with lower-frequency and higher-amplitude oscillations. In Figure 5, especially visible for the large d/D values, the sudden HG rises occur slightly after the sudden head rises in Figure 4. In fact, the transient oscillations at these two locations are causally connected. While the head rise in Figure 4f occurs at about 76 s, the associated HG rise in Figure 5f occurs at about 79 s (the timing difference corresponds to L_1/a, i.e., the time the transient wave takes to travel from the air valve to the upstream section of the pipeline).

5.2. Evolution of Key Transient Quantities

To complement the story told in Figures 4 and 5, Figures 6 and 7 show the evolution of key transient quantities for type 1 and type 3 behaviours. Figure 6 shows the evolution of the HG at both the air valve and the upstream section of the pipeline, accompanied by the water velocities upstream and downstream of the air valve.

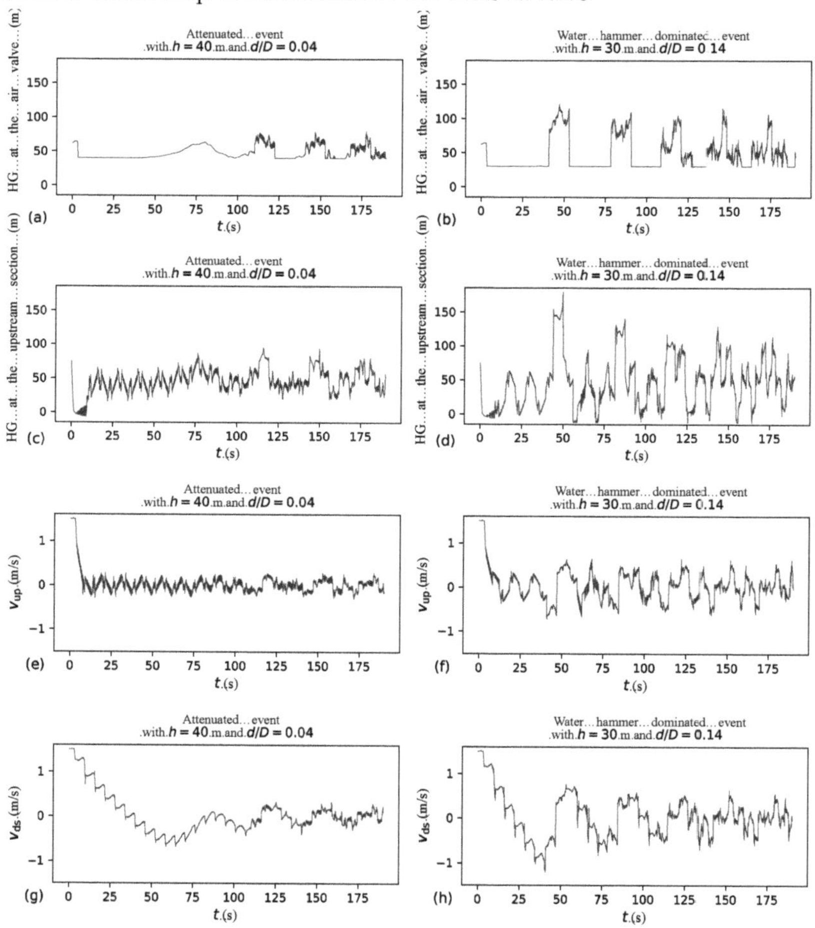

Figure 6. Hydraulic grade (HG) at the air valve, HG at the upstream section of the pipeline, and water velocity upstream (v_{up}) and downstream (v_{ds}) of the air valve considering $L_1 = 3000$ m: (**a**) HG at the air valve for type 1 behaviour (T1B); (**b**) HG at the air valve for type 3 behaviour (T3B); (**c**) HG at the upstream section for T1B; (**d**) HG at the upstream section for T3B; (**e**) v_{up} for T1B; (**f**) v_{up} for T3B; (**g**) v_{ds} for T1B; (**h**) v_{ds} for T3B.

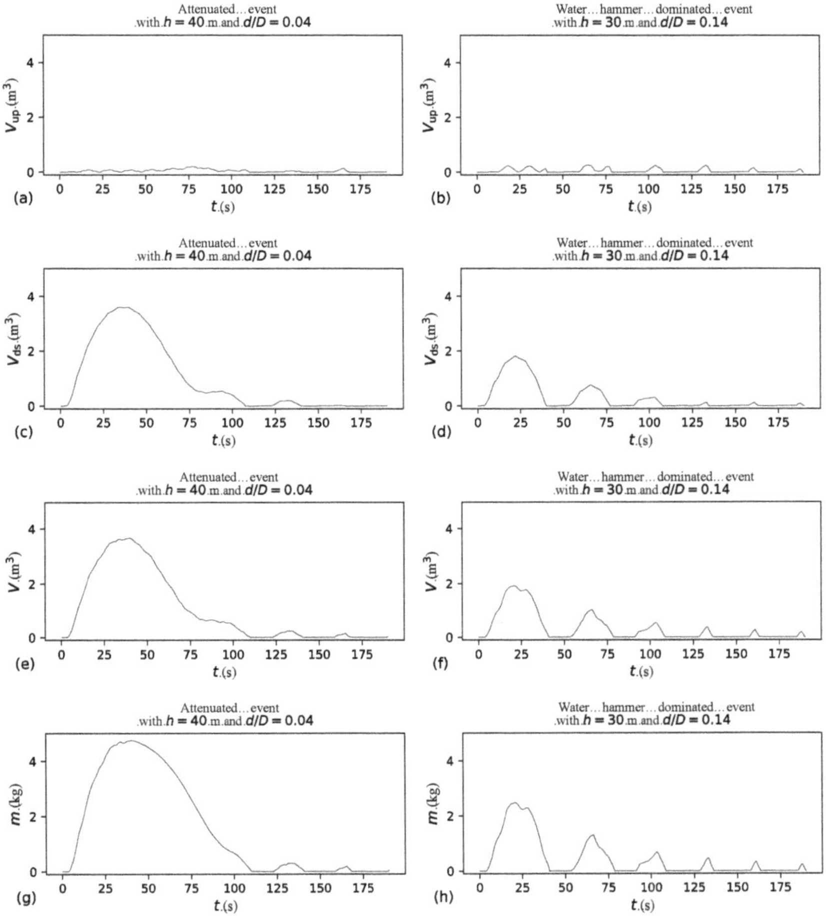

Figure 7. Air pocket volume upstream and downstream of the air valve, total air pocket volume, and air pocket mass considering L_1 = 3000 m: (**a**) upstream volume for type 1 behaviour (T1B); (**b**) upstream volume for type 3 behaviour (T3B); (**c**) downstream volume for T1B; (**d**) downstream volume for T3B; (**e**) total volume for T1B; (**f**) total volume for T3B; (**g**) mass for T1B; (**h**) mass for T3B.

If an air valve contains entrapped air, the hydraulic head evolution is relatively smooth (type 1 behaviour) or with negligible magnitude (type 3 behaviour). The type 1 behaviour in Figure 6c is characterised by a lower-amplitude HG evolution in comparison to the type 3 behaviour in Figure 6d. The HG evolution at the upstream section, though more complex than that at the air valve, also reveals essential aspects of the transient. This difference is of particular note since monitoring pressure signals close to a pumping station is generally more convenient than at remote locations, including at air valves.

The graphs in Figure 6e–h show that, after the transient inception, the water velocity in the upstream branch (v_{up}) varies around $v_{up} = 0$, while the water velocity downstream (v_{ds}) progressively decreases. When v_{ds} reaches $v_{ds} = 0$, the maximum air pocket volume is attained. Thereafter, v_{ds} becomes negative (i.e., the flow reverses). The reverse flow progressively intensifies until the air pocket is fully exhausted through the air valve. Such behaviour patterns for v_{up} and v_{ds} have also been identified by Ramezani and Karney (2017) [10], implying that the main contributor to the growth of the entrapped air pocket is the downstream branch. In fact, the physical process in the downstream branch during which v_{ds} remains positive and which is associated with air admission through the air valve constitutes a form of draining event. Given the 1.5 m/s initial steady water velocity, the air valve should be capable of accommodating air intake for a water draining velocity of a similar magnitude, especially during the initial part of the air admission stage. As this requirement significantly exceeds the conventional range of 0.3 to 0.6 m/s for controlled line draining [5], accounting for the pump trip scenario is essential when determining the size of the air valve's admission orifice.

As shown in Figure 6g,h, the water column in the downstream branch takes longer to be arrested for $h = 40$ m than for $h = 30$ m. Indeed, the velocity changes in Figure 6g are smaller than those in Figure 6h. Such a difference occurs because h is larger in Figure 6g than in Figure 6h. When the air valve is positioned at a higher elevation, a smaller portion of the initial under-pressure wave can traverse the high point and travel towards the downstream reservoir. Consequently, the initial reductions in head and water discharge in the downstream branch—along with subsequent changes in water discharge—are less pronounced for the higher-elevation air valve. Indeed, in Figure 6g,h, even when v_{ds} is still positive (i.e., there is no influence from the different d/D assumptions), the v_{ds} changes are smaller for the case with higher h. Note that a decrease in v_{ds} occurs with a periodicity of $2(L_2 + L_3)/a$. This period represents the time the wave requires to travel from the air valve to the downstream reservoir and back. Note that a reduction in v_{ds} occurs both when the wave reaches the downstream reservoir and when it returns to the air valve.

The dependence between the magnitude of the v_{ds} changes and h has also been identified by Ramezani and Karney (2017) [10]. However, by design, such work explicitly excluded the air cushioning effect that arises when a reduced-outflow-capacity air valve is used. This effect is marked by the water column's deceleration resulting from the compression of the entrapped air pocket. In Figure 6g, the outflow orifice is small and the air cushioning effect is relevant, whereas in Figure 6h, the outflow orifice is large and the entrapped air pocket presents no resistance against the moving water column during air expulsion. In Figure 6g, v_{ds} reaches zero at 34 s. It then progressively reduces, reaching -0.68 m/s at 64 s. Afterwards, v_{ds} starts to increase, reaching 0.10 m/s at 87 s. Following this, v_{ds} decreases again, reaching -0.43 m/s at 107 s. Once again, v_{ds} starts to increase, reaching -0.13 m/s when the air valve closes at 110 s. However, in Figure 6h, the behaviour of v_{ds} differs, with the air valve closing when v_{ds} reaches its minimum value. Additionally, no deceleration of the periodic v_{ds} reductions is observed in this case.

Figure 7 shows the air pocket evolution for type 1 and type 3 behaviours. The graphs in Figure 7a,b show the contribution of the upstream branch to the total air pocket volume, while the graphs in Figure 7c,d show the contribution of the downstream branch. Summing the volumes from the graphs in Figure 7a,c results in the total volume shown in Figure 7e, and summing the volumes from the graphs in Figure 7b,d results in the total volume shown in Figure 7f. The air pocket volume is mainly influenced by the movement of the water column in the downstream branch, while the upstream branch has a minor impact. Water discharge in the upstream branch oscillates around zero, leading to periodic expansion and contraction of the air pocket in this branch. Conversely, in the downstream branch, the water velocity is slower to reverse, creating an opportunity for the formation of a sizeable entrapped air pocket. Note that both the upstream and downstream air pocket volumes are integral components of the resulting volume of air that is expected to accumulate at or near the air valve.

6. Maximum Air Pocket Volume

The current study maintains $d_{in}/D = 0.20$ throughout all simulations (such a d_{in}/D value allows the free entry of air into the pipeline); however, it explores a wide range of relative d/D values. Given that the maximum air pocket volume (V_{max}) occurs at the end of the air admission stage, unaffected by the outflow orifice, V_{max} for a given air valve inflow capacity is expected to be influenced by the initial pipeline flow conditions and the air valve location. For instance, Tasca et al. (2021) [14] mention that an air valve positioned at a higher elevation leads to a larger V_{max} value compared to a lower elevation, assuming other factors remain constant.

Figure 8 shows the relationship between V_{max} and the air valve location (determined by L_1 and h). For each L_1 value, increasing h leads to smaller v_{ds} reductions that occur periodically each round trip across the downstream branch during the air admission stage. Consequently, with an increase in h, the duration of the air admission stage also extends. Similarly, for each h, decreasing L_1 increases the time taken for a wave round trip in the downstream branch. Hence, as h increases or L_1 decreases, the duration of the air admission stage increases. The range of the V_{max} values spans from 0.1 m^3 (for L_1 = 5000 m and h = 20 m) to 5.5 m^3 (for L_1 = 1000 m and h = 40 m), representing a substantial difference of nearly 50 times. For L_1 = 1000 m and h = 20 m, V_{max} = 2.2 m^3, while for L_1 = 5000 m and h = 40 m, V_{max} = 1.3 m^3.

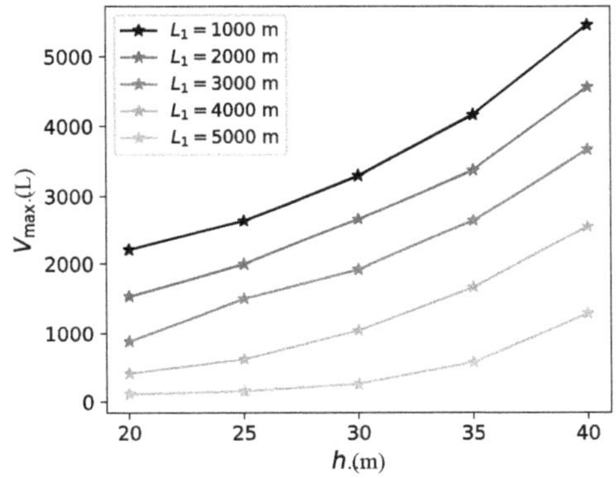

Figure 8. Dependence between the maximum air pocket volume and the air valve location.

The largest V_{max} value obtained in the simulations (approximately 5.5 m^3) corresponds to an air-occupied length of about 28 m, which is small relative to the pipeline length. Interestingly, the largest air pocket volume corresponds to the case with the largest descending pipe segment downstream of the air valve. Nonetheless, if the descending segment connected to the air valve were to have a mild slope, the air pocket could extend considerably farther. Since the V_{max} values found in the current study are small in relation to the length of the descending downstream segment, a metric that considered the ratio between air pocket volume and pipe volume was not used, though it might sometimes be of value.

7. Sensitivity Study

7.1. Assessing Extreme Transient Heads

The physical quantities most relevant to the study of extreme transient heads are as follows: pipeline length experiencing negative pressures (L_s), maximum HG in the system (HG_{max}), and mean maximum hydraulic head ($\overline{H}_{max}$). Variations in the following parameters are considered: five values for L_1 (1000, 2000, 3000, 4000, and 5000 m), five values for h (20, 25, 30, 35, and 40 m), and 17 values for d/D (ranging from zero to 0.40). Furthermore, the response without air valve protection is also included for comparison.

Air valves are typically thought of as devices responsible for mitigating negative pressures. However, as illustrated in Figures 2 and 3, even with a system equipped with a large-inflow-capacity air valve, momentary negative pressures can still occur during the pump trip scenario. Figure 9 shows how L_s/L is influenced by variations in L_1, h, and d/D. Generally, for the smallest h values (20 and 25 m), negative pressures are prevalent throughout the entire pipeline length, irrespective of d/D. Particularly for larger h values, smaller outflow orifices correspond to smaller L_s/L values. In fact, the smallest L_s/L values are those associated with h = 40 m and L_1 = 1000 m. Interestingly, even for the cases with the smallest L_s/L values, $L_s/L > 0.60$. As exemplified in Figures 2a–d and 3a–d, especially for the largest h values, the region of the pipeline neighbouring the low point downstream of the air valve is the least susceptible to negative pressures. Nonetheless, in the absence of air valve protection, the whole pipeline is generally subjected to negative pressures.

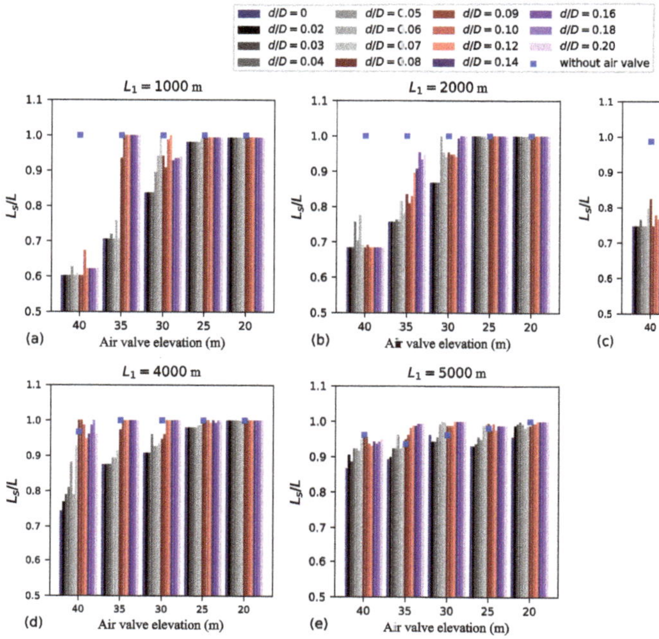

Figure 9. Ratio between the pipeline length that experiences negative pressures and the total pipeline length (L_s/L): (**a**) $L_1 = 1000$ m; (**b**) $L_1 = 2000$ m; (**c**) $L_1 = 3000$ m; (**d**) $L_1 = 4000$ m; (**e**) $L_1 = 5000$ m.

Having an air valve with a sufficiently small outflow orifice can help mitigate the intensity of secondary pressures upon air valve closure. Moreover, as shown in Figure 10, the presence of an air valve can also help mitigate the maximum transient HG at the upstream section of the pipeline. Figure 10 shows how HG_{max} is influenced by variations in L_1, h, and d/D. Figure 10 reveals that, except for $L_1 = 5000$ m and $h = 20$ m (case with the smallest V_{max} value), reducing d/D leads to greater attenuation of HG_{max} values. For every combination of L_1 and h (i.e., for every location of the air valve), the HG_{max} versus d/D data contains two distinct regions: a lower left region associated with smaller d/D values (indicating type 1 behaviour), and a higher right region associated with larger d/D values (indicating type 3 behaviour).

For $L_1 = 1000$ m and $L_1 = 2000$ m, with h ranging from 25 to 40 m, the transient magnitude tends to increase as h decreases. Similarly, for $L_1 = 3000$ m and $L_1 = 4000$ m, with h ranging from 30 to 40 m, the transient magnitude also tends to increase as h decreases. Interestingly, for the air valve location furthest from the upstream reservoir, the transient magnitude tends to decrease as h decreases. For $L_1 = 1000$ m and $L_1 = 2000$ m, HG_{max} is larger for the cases without an air valve than for the cases with it. Pipelines with smaller h values or those located farther from the upstream reservoir are more susceptible to macro-cavitation and tend to have smaller admitted air volumes. As mentioned earlier, the effectiveness of the air valve in reducing the transient magnitude is linked to the air cushioning effect, which is less pronounced in these situations.

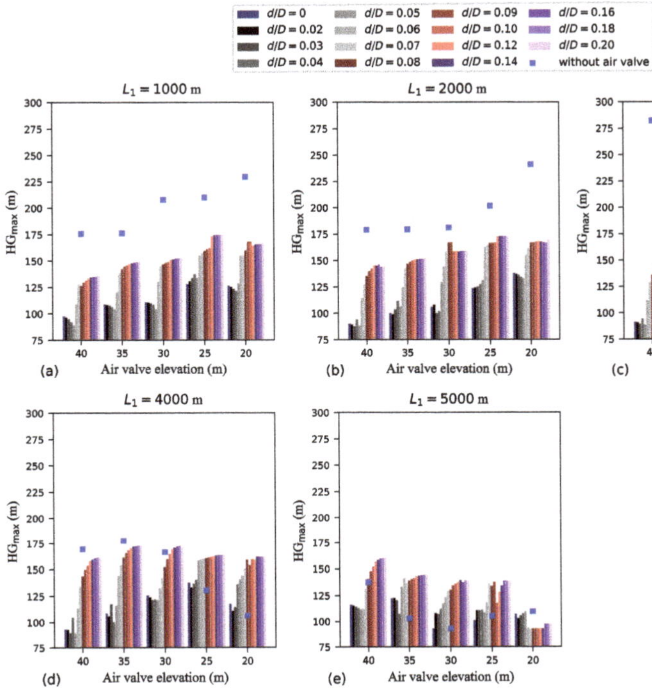

Figure 10. Maximum hydraulic grade in the system ($\mathrm{HG}_{\mathrm{max}}$): (**a**) $L_1 = 1000$ m; (**b**) $L_1 = 2000$ m; (**c**) $L_1 = 3000$ m; (**d**) $L_1 = 4000$ m; (**e**) $L_1 = 5000$ m.

Even though $\mathrm{HG}_{\mathrm{max}}$ is an important parameter for evaluating the magnitude of the transient (since the upstream section of the pipeline must withstand the head associated with $\mathrm{HG}_{\mathrm{max}}$), the mean maximum hydraulic head ($\overline{H}_{\mathrm{max}}$) offers a more comprehensive assessment of the entire transient event. Figure 11 shows how $\overline{H}_{\mathrm{max}}$ is influenced by variations in L_1, h, and d/D. The results in Figure 11 resemble those in Figure 10. However, the $\overline{H}_{\mathrm{max}}$ versus d/D data in Figure 11 appear smoother and show more distinct left and right plateaus. It is important to note that if the d/D ranges considered in Figures 10 and 11 were narrower, the presence of the left and right plateaus could shrink or even disappear. The effectiveness of a small-outflow-capacity air valve in mitigating transient events is evident in Figure 11. For $L_1 = 3000$ m and $h = 40$ m, $\overline{H}_{\mathrm{max}} = 51$ m for $d/D = 0.02$, whereas $\overline{H}_{\mathrm{max}} = 93$ m for $d/D = 0.10$ (a 45% difference). For $L_1 = 3000$ m and $h = 20$ m, $\overline{H}_{\mathrm{max}} = 90$ m for $d/D = 0.02$, whereas $\overline{H}_{\mathrm{max}} = 115$ m for $d/D = 0.10$ (a 22% difference). It is important to highlight the significance of air valve elevation: e.g., for $L_1 = 3000$ m, the water-hammer-dominated event for $h = 40$ m (with a lower driving head to induce reverse flow) displays a transient magnitude similar to that of the attenuated event for $h = 20$ m.

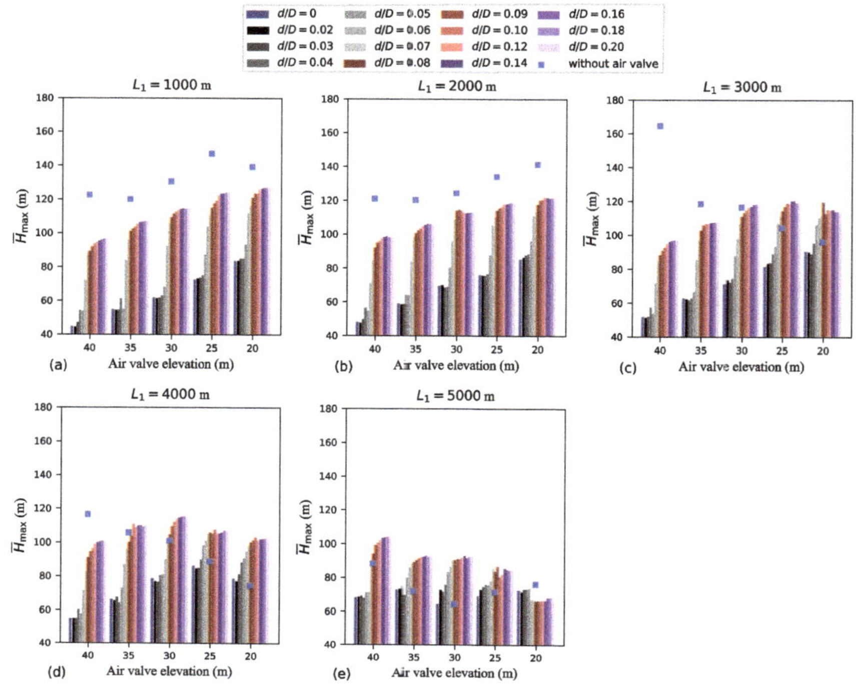

Figure 11. Mean maximum transient head ($\overline{H}_{max}$): (**a**) $L_1 = 1000$ m; (**b**) $L_1 = 2000$ m; (**c**) $L_1 = 3000$ m; (**d**) $L_1 = 4000$ m; (**e**) $L_1 = 5000$ m.

The data shown in Figure 11 are reinterpreted in Figure 12 using a dimensionless parameter denoted as $\overline{H}_{max}/\overline{H}_{steady}$, a parameter that relativises the mean maximum transient head ($\overline{H}_{max}$) relative to steady values ($\overline{H}_{steady}$). Note that there is a slight variation in the $\overline{H}_{steady}$ value depending on the location of the air valve. The highest $\overline{H}_{max}/\overline{H}_{steady}$ value, approximately 3, is observed for $L_1 = 1000$ m, the lowest air valve elevations, and type 3 behaviour. In contrast, the smallest $\overline{H}_{max}/\overline{H}_{steady}$ value, approximately 1.2, is observed for $L_1 = 1000$ m, $h = 40$ m, and type 1 behaviour. This implies that locating the air valve closer to the upstream reservoir results in more intense transient pressures for type 3 behaviour. However, locating the air valve near the upstream reservoir also offers the greatest potential for water hammer mitigation for type 1 behaviour.

Figures 10–12 show how the transient magnitude is significantly influenced by the air valve location. Figure 13 details the effect of h on $\overline{H}_{max}$ for three L_1 values (1000, 3000, and 5000 m) and two d/D values (0.05 and 0.20). For both $L_1 = 1000$ m and $L_1 = 3000$ m, $\overline{H}_{max}$ tends to increase as h decreases. For $d/D = 0.05$ and $L_1 = 5000$ m, the dependence between $\overline{H}_{max}$ and h is unclear, while for $d/D = 0.20$ and $L_1 = 5000$ m, $\overline{H}_{max}$ shows some tendency to decrease as h decreases. For each air valve location, $\overline{H}_{max}$ is generally smaller for $d/D = 0.05$ than for $d/D = 0.20$. Figure 13 reveals that, in the context of the pump trip scenario, it is the lower high points that result in more intense transient events.

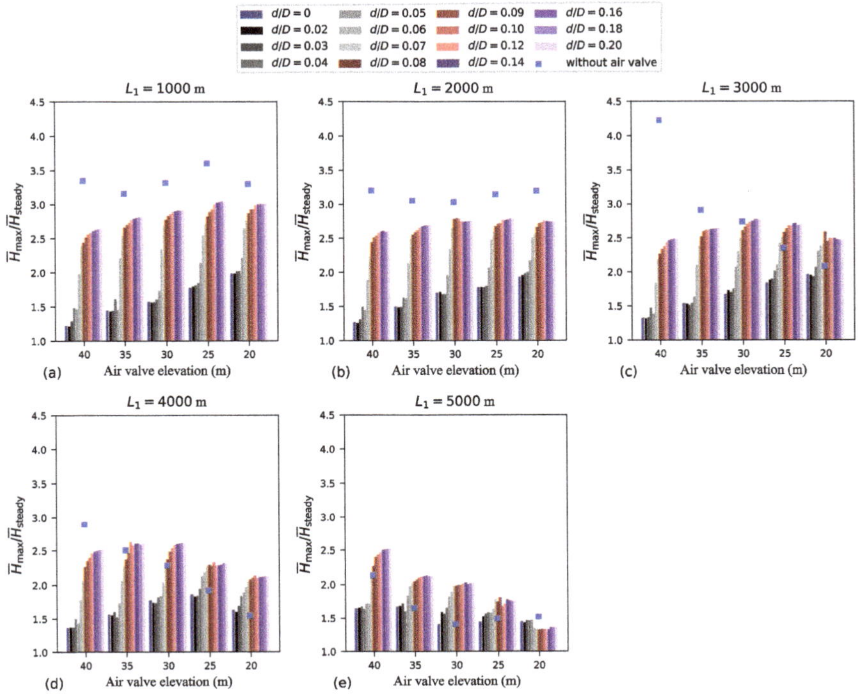

Figure 12. Mean maximum transient head ($\overline{H}_{max}$) divided by the mean head during steady flow ($\overline{H}_{steady}$): (**a**) $L_1 = 1000$ m; (**b**) $L_1 = 2000$ m; (**c**) $L_1 = 3000$ m; (**d**) $L_1 = 4000$ m; (**e**) $L_1 = 5000$ m.

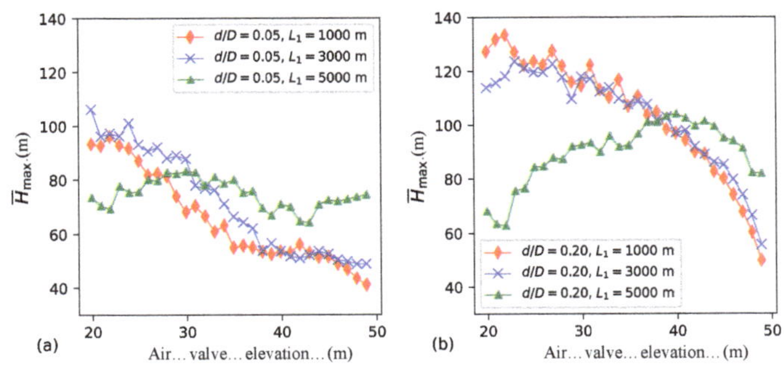

Figure 13. Influence of air valve elevation on $\overline{H}_{max}$ considering the h range from 20 to 49 m in detail: (**a**) $d/D = 0.05$; (**b**) $d/D = 0.20$.

7.2. Assessing the Timing of the Transient

As previously discussed, the duration of the initial water hammer cycle tends to extend with decreasing values of d/D and L_1, as well as increasing values of h. As illustrated in Figures 2 and 3, the initial duration of the water hammer cycle is typically slightly longer than t_c, making t_c a suitable proxy for this duration. Figure 14 shows how t_c is influenced by variations in L_1, h, and d/D. In this figure, the case with $d/D = 0$ was excluded because of the absence of air pocket collapse.

Figure 14 shows that, for each L_1 and d/D combination, as h decreases, so does t_c. For each h and d/D combination, as L_1 increases, t_c decreases. For each air valve location (i.e., for each L_1 and h combination), as d/D decreases, t_c increases exponentially, while as d/D increases, t_c tends to a constant value. Indeed, for each air valve location, the largest t_c

value occurs for $d/D = 0.02$. Notably, the t_c value for a small outflow orifice can be several times longer than that for a large outflow orifice. For $L_1 = 1000$ m and $h = 40$ m, $t_c = 423$ s for $d/D = 0.02$, whereas $t_c = 109$ s for $d/D = 0.20$ (a 74% reduction). For $L_1 = 1000$ m and $h = 20$ m, $t_c = 145$ s for $d/D = 0.02$, whereas $t_c = 43$ s for $d/D = 0.20$ (a 70% reduction). For $L_1 = 5000$ m, t_c is much smaller than for $L_1 = 1000$ m. For example, for $L_1 = 5000$ m, $h = 20$ m, and $d/D = 0.02$ (type 1 behaviour), $t_c = 14$ s—a value much smaller than that for type 3 behaviour considering $L_1 = 1000$ m, $h = 20$ m, and $d/D = 0.20$ ($t_c = 43$ s). A combined examination of Figures 11 and 14 reveals a consistent trend: event intensity attenuates with increasing t_c, as momentum considerations would indicate.

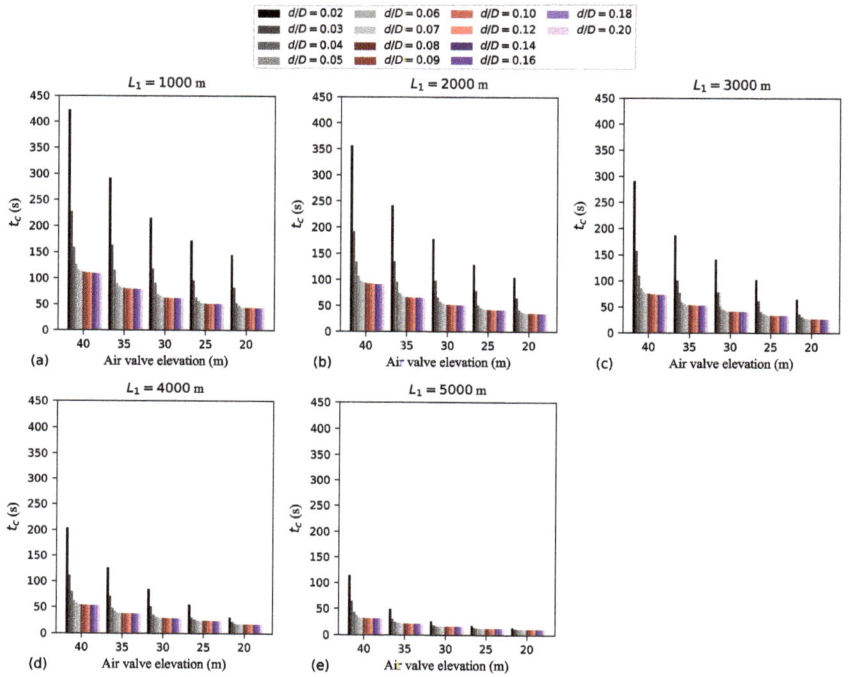

Figure 14. Air pocket collapse time (t_c): (**a**) $L_1 = 1000$ m; (**b**) $L_1 = 2000$ m; (**c**) $L_1 = 3000$ m; (**d**) $L_1 = 4000$ m; (**e**) $L_1 = 5000$ m.

An essential aspect regarding transient timing is the moment when the maximum hydraulic grade in the system (HG_{max}) occurs. The moment when HG_{max} occurs is referred to as $t_{HG_{max}}$. As discussed, a peak HG value occurs at the upstream section of the pipeline after the first instance of air pocket collapse. Yet, this peak HG does not always correspond to HG_{max}. In Figure 3, for example, the peak HG values at the upstream section of the pipeline at the conclusion of the secondary wave stage correspond to the HG_{max} values. However, since such correspondence is not always the case, one method to assess the timing of the maximum HG in the system is by comparing it to the air pocket collapse time. Figure 15 shows how $t_{HG_{max}}$ minus t_c is influenced by variations in L_1, h, and d/D. A negative bar in this figure indicates that HG_{max} occurs before t_c, whereas a positive bar indicates the opposite.

In Figure 15, three distinct $t_{HG_{max}} - t_c$ patterns can be identified, which are somewhat dependent on the air valve location and outflow capacity: (i) HG_{max} occurs much before t_c (typical for small d/D values); (ii) HG_{max} occurs much after t_c (sometimes observed for intermediate d/D values); (iii) HG_{max} occurs right after t_c (typical for large d/D values). For $L_1 = 5000$ m (especially for lower h values like 20, 25, and 30 m), the $t_{HG_{max}} - t_c$ patterns become less distinct compared to other L_1 values. For the smallest d/D values,

HG_{max} occurs much before t_c, while for the largest d/D values, HG_{max} usually occurs right after t_c. In the context of type 3 behaviour, the usually relatively small and positive value of $t_{HG_{max}} - t_c$ for large d/D values reinforces the connection between air valve closure and the occurrence of HG_{max}. However, for intermediate d/D values, $t_{HG_{max}} - t_c$ can be several times the pipeline's period.

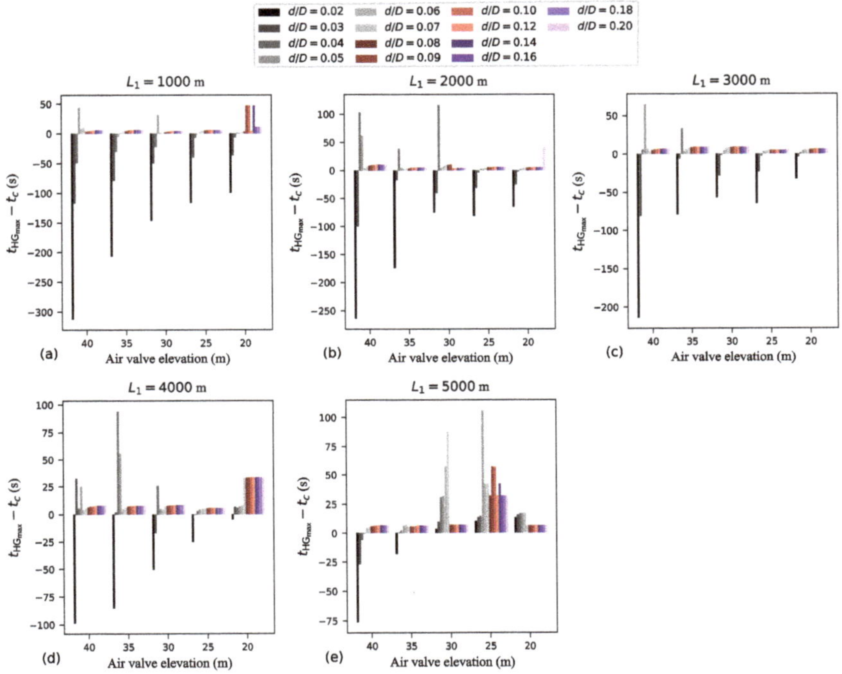

Figure 15. Time for the maximum hydraulic grade in the system ($t_{HG_{max}}$) minus the air pocket collapse time (t_c): (**a**) $L_1 = 1000$ m; (**b**) $L_1 = 2000$ m; (**c**) $L_1 = 3000$ m; (**d**) $L_1 = 4000$ m; (**e**) $L_1 = 5000$ m.

The striking difference in transient timing between type 1 and type 3 behaviours is evident in Figure 15. For $L_1 = 1000$ m and $h = 40$ m, $t_{HG_{max}} - t_c = -49$ s for $d/D = 0.04$ (which is actually a non-extreme case), which is a value comparatively much larger in absolute magnitude than $t_{HG_{max}} - t_c = 4$ s for $d/D = 0.10$. In Figure 15, the number of d/D cases with negative $t_{HG_{max}} - t_c$ values decreases as L_1 increases or h decreases. Negative $t_{HG_{max}} - t_c$ values indicate that HG_{max} occurs while the air pocket remains. Thus, in such cases, it makes sense to infer that $t_{HG_{max}}$ is associated with the air expulsion stage, i.e., HG_{max} occurs while the air pocket is being compressed by the reverse flow in the downstream branch. The transient timings associated with type 1 and type 2 behaviours occurring well before or much after t_c, respectively, do not necessarily rule out the possibility of a peak transient head occurring at the upstream section of the pipeline immediately after the air valve closure.

Figure 16 shows the evolution of the HG at the upstream section of the pipeline (i.e., close to the pumping station) for types 1, 2, and 3 behaviours. For $d/D = 0.02$ (Figure 16a), HG_{max} occurs much before t_c. For this case, t_c is about four times longer than $t_{HG_{max}}$. For $d/D = 0.05$ (Figure 16b), $t_{HG_{max}}$ is considerably longer than t_c. For this case, HG_{max} occurs after the second instance of air pocket collapse. In Figure 16b, note the similarity in the HG signal right after t_c and right before $t_{HG_{max}}$. This kind of HG variation is typical of the secondary wave stage. Also, note that in Figure 16b, the magnitude of the HG peak right after t_c is similar to the magnitude of HG_{max}. For $d/D = 0.20$ (Figure 16c),

HG_{max} occurs right after t_c. In general, in Figure 16, prior to t_c, the transient variations share the same frequency across all three cases.

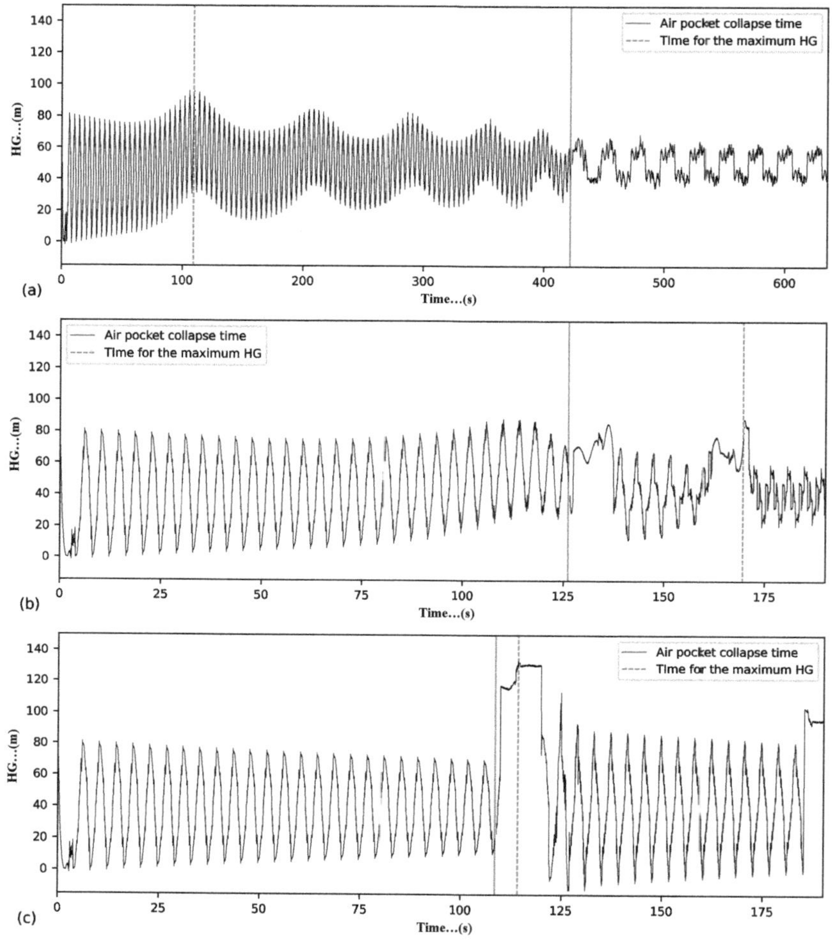

Figure 16. Hydraulic grade at the upstream section of the pipeline considering $L_1 = 1000$ m and $h = 40$ m: (**a**) $d/D = 0.02$; (**b**) $d/D = 0.05$; (**c**) $d/D = 0.20$.

Figure 17 shows the evolution of the HG at the upstream section of the pipeline for three examples of type 1 behaviour (i.e., for three L_1 values) considering $h = 35$ m and $d/D = 0.02$. Across all three cases, $t_{HG_{max}} < t_c$. Prior to t_c, the frequency of the transient oscillations decreases as L_1 increases. The time it takes for the transient wave to traverse the upstream branch is directly proportional to L_1. The absolute value of $t_c - t_{HG_{max}}$ decreases as L_1 increases. Unlike the wave pattern depicted in Figure 16b,c, the HG oscillation in Figure 17 is not momentarily disrupted after the collapse of the air pocket. Following t_c, the oscillation period stabilises at $4L/a$ for all three L_1 cases in the figure. For $L_1 = 5000$ m, with its relatively short t_c value, it becomes somewhat difficult to discern the period of the HG oscillation before t_c. In Figure 17a,b, the oscillation period before t_c is determined by $4L_1/a$ (as predicted by Ramezani and Karney (2017) [10]), resulting in 4 s for $L_1 = 1000$ m and 12 s for $L_1 = 3000$ m.

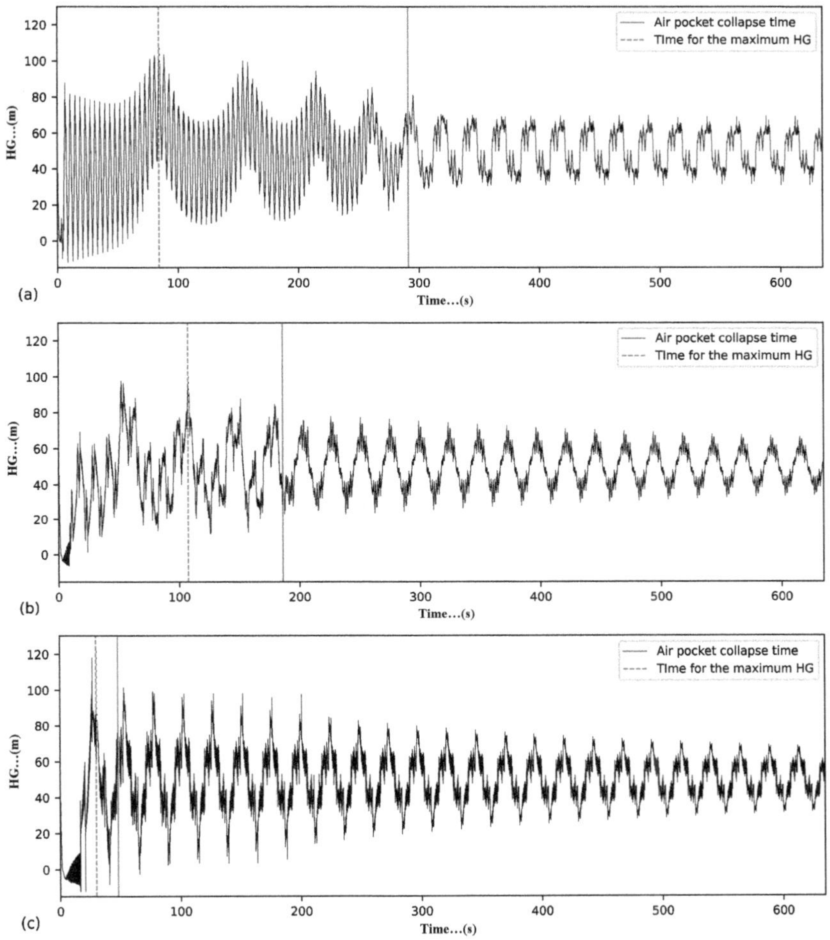

Figure 17. Hydraulic grade at the upstream section of the pipeline considering $h = 35$ m and reduced air valve outflow orifice ($d/D = 0.02$): (**a**) $L_1 = 1000$ m; (**b**) $L_1 = 3000$ m; (**c**) $L_1 = 5000$ m.

Figure 18 shows the evolution of the HG at the upstream section of the pipeline for three examples of type 3 behaviour (i.e., for three L_1 values) considering $h = 40$ m and $d/D = 0.20$. Across all three cases, $t_{HG_{max}} > t_c$. In contrast to Figure 17, in Figure 18, the $t_{HG_{max}} - t_c$ values across all three cases are similar. The frequency of the transient oscillations decreases as L_1 increases. Notably, the frequency of the oscillations remains the same both before and after the collapse of the air pocket. This behaviour pattern arises from the multiple instances of air admission and expulsion characteristic of type 3 behaviour, as shown in Figure 18, in contrast to the type 1 behaviour shown in Figure 17. This observation aligns with the data presented in Figure 7. The frequency of the transient oscillations in Figure 18 is determined by $4L_1/a$. In contrast to Figure 17, in Figure 18, the secondary wave stage can be easily discerned as a discontinuity in the HG oscillations. In Figure 18b, three instances of secondary wave stages can be observed, each showing progressively attenuated associated HG peaks.

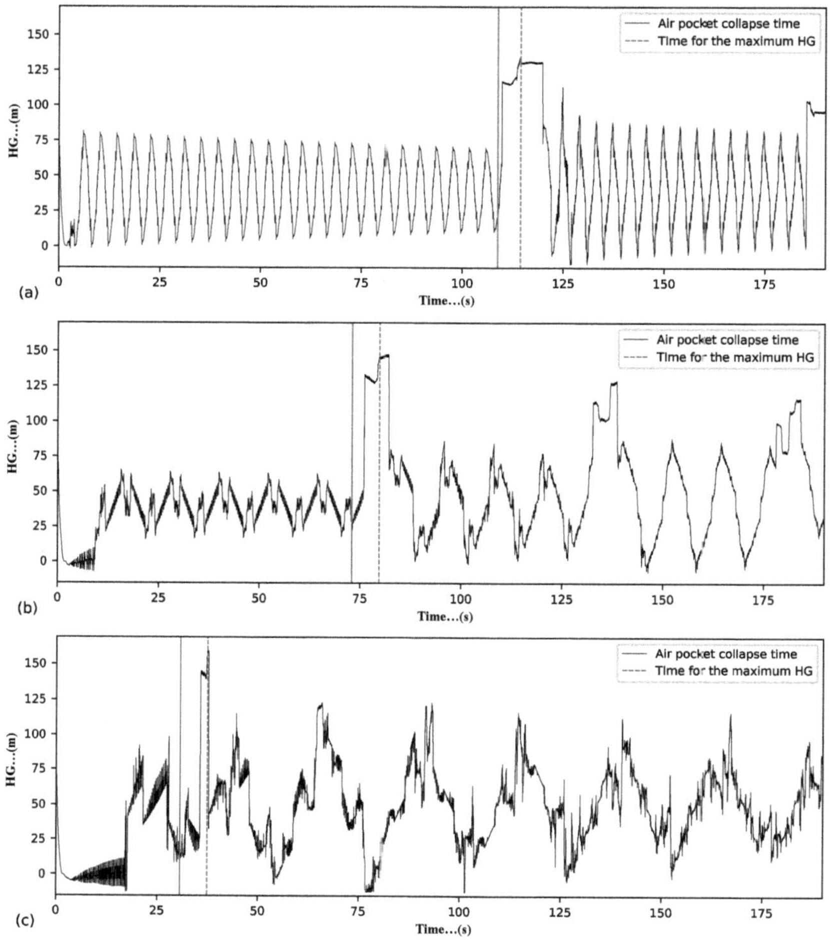

Figure 18. Hydraulic grade at the upstream section of the pipeline considering $h = 40$ m and a large air valve outflow orifice ($d/D = 0.20$): (**a**) $L_1 = 1000$ m; (**b**) $L_1 = 3000$ m; (**c**) $L_1 = 5000$ m.

7.3. Representing the Sensitivity Data

The air exchange apertures commonly examined in the literature are orifices and air valves. It is important to highlight the distinction in pipeline transient behaviour between these apertures. For the filling procedure in a pipeline with a downstream orifice, the absolute maximum transient head (often referred to as H^*_{max}) increases with increasing d/D until the maximum H^*_{max} value is reached. Then, H^*_{max} starts to decrease as d/D increases. The d/D value associated with the maximum H^*_{max} value is referred to as the critical d/D value. The pipeline transient behaviour associated with the occurrence of the maximum H^*_{max} value can be classified as water-hammer-dominated and is characterised by a sudden increase in head resulting from the impact of the filling water column against the downstream orifice. References providing data on the transient magnitude versus d/D for the pipeline filling procedure involving a downstream orifice include Zhou et al. (2002) [37], Lee (2005) [38], Zhou et al. (2019) [39], and Zhou et al. (2020) [3].

When filling a line with a downstream orifice, the orifice initially vents air to accommodate the incoming water. Nonetheless, as the water front reaches the orifice and in the subsequent moments, the orifice might expel not only air but also a mixture of air and water. In contrast, when filling involves a downstream air valve, only air is expected to be discharged. Even in cases where there might be a delay in air valve closure or if the last portion of the exhausted air contains moisture, the release of water or a mixture of air and water through the air valve is only momentary. Since air valves are designed to close upon contact with liquid water, the transient magnitude is anticipated to stabilise for large d/D values (i.e., to remain the same for d/D values for which the air cushioning effect is negligible). In fact, the air expulsion stage during the pump trip scenario represents a variation of the filling scenario, albeit within a more complex transient event. Notably, the transient magnitude during the pump trip scenario is directly influenced by the extent of the air cushioning effect during the air expulsion stage. As a result, both the pump trip and the pipeline filling scenarios share significant similarities, permitting a comparison between the physical processes involved in both situations.

Figure 19 allows a comparison between the transient responses associated with a pump trip and pipeline filling. Specifically, Figure 19a shows the relationship between $\overline{H}_{\text{max}}$ and d/D for the pump trip scenario involving air exchanges through an air valve. Figure 19b shows the relationship between H^*_{max}/H^*_r (ratio between the absolute maximum transient head and the absolute inlet head) and d/D for the pipeline filling scenario involving air exchanges through a simple orifice. In Figure 19a,b, a strong similarity in transient response is evident between the two cases for small d/D values. In both cases, if there is an increase in the maximum air pocket volume (achieved, for example, by decreasing L_1 in the pump trip scenario) or the initial air pocket volume, the intensity of the transient event decreases for d/D values consistent with type 1 behaviour. However, a significant difference in transient response arises if the d/D values are large.

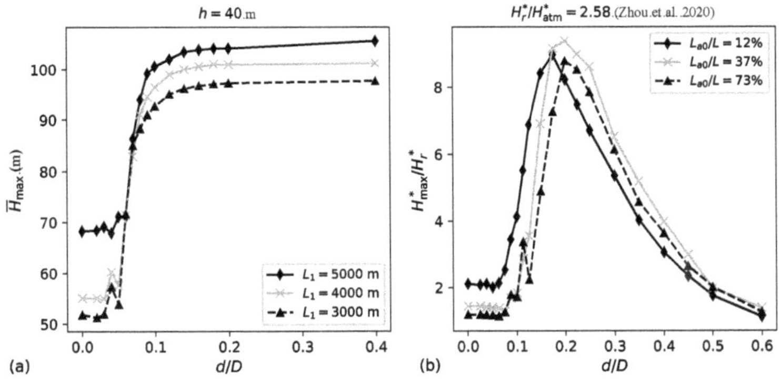

Figure 19. Influence of aperture size for air exchanges on system transient behaviour: (**a**) pipeline with distinct high point with air valve subjected to the pump trip scenario (data from the current study); (**b**) horizontal pipeline with downstream orifice subjected to the filling procedure (data from Zhou et al. (2020) [3]).

The critical aperture size for type 3 behaviour (referred to as d_{c3}) determines the threshold d/D value above which the transient can be classified as water-hammer-dominated. For the pump trip scenario, d_{c3} is defined here as the d/D value above which the transient magnitude stabilises. For the filling procedure, d_{c3} is defined here as the d/D value associated with the maximum transient magnitude. On the other hand, the critical aperture size for type 1 behaviour (referred to as d_{c1}) determines the d/D value below which the transient can be classified as attenuated. Figure 19 compares pump trip (Figure 19a) to filling (Figure 19b), showing that the transient magnitude curves experience a sharp incline immediately after the low left plateau. However, for small d/D values prior to this incline, the transient magnitude is relatively constant. The general shape of the curves in Figure 19a resembles that of a logistic curve. Each of these curves features a lower plateau to the left, followed by a steep (transitional) incline, ultimately reaching a higher plateau to the right. On the other hand, the general shape of the curves in Figure 19b resembles that of a right-skewed probability density function. Each of these curves features a lower plateau to the left, followed by a steep incline that reaches a maximum point, with a subsequent gradual descent.

Figure 20 shows the comprehensive $\overline{H}_{max}$ versus d/D results obtained from the main set of numerical simulations. The curves in Figure 20 exhibit a similar shape to those in Figure 19a, except for the case with $L_1 = 5000$ m and $h = 20$ m, which is associated with the smallest V_{max} value. As mentioned, this distinctive shape resembles the shape of a logistic curve. In Figure 20, for both $L_1 = 1000$ m and $L_1 = 2000$ m, regardless of whether it is type 1 or type 3 behaviour, the transient magnitude increases as h decreases. For L_1 values ranging from 1000 to 4000 m, this trend remains relatively consistent, particularly with respect to the low left plateaus. With the exception of $L_1 = 5000$ m, scenarios with $h = 40$ m consistently exhibit the mildest transient magnitudes for both type 1 and type 3 behaviours across different L_1 values.

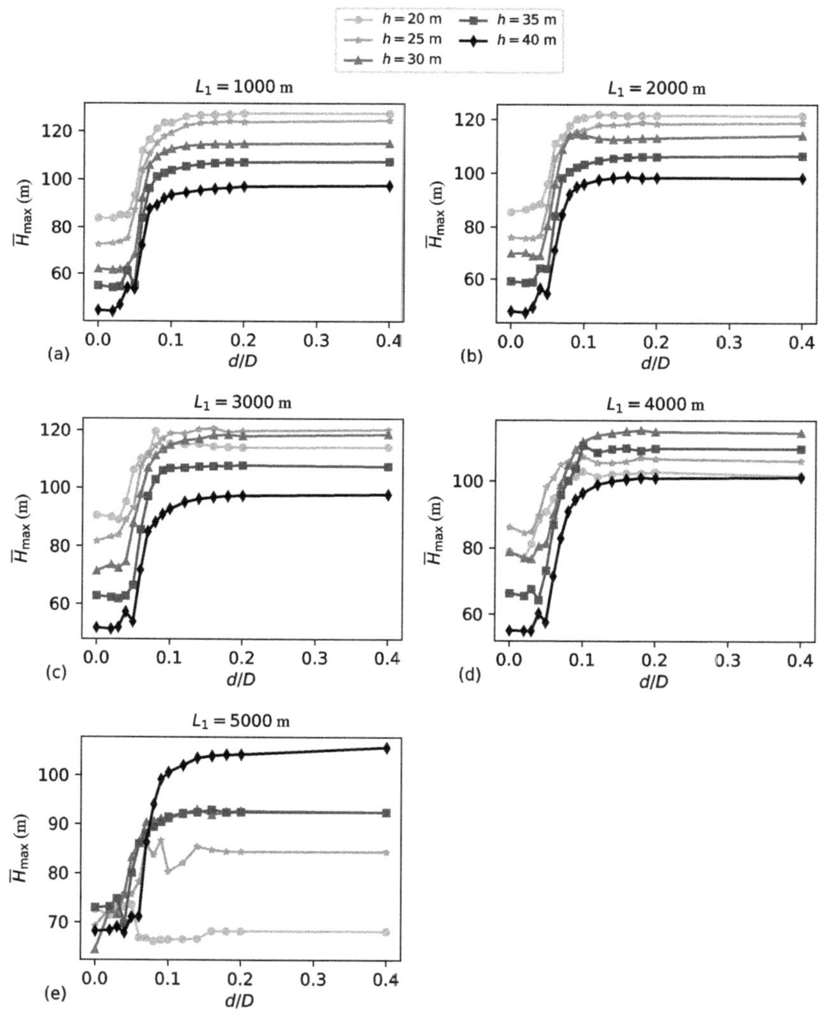

Figure 20. $\overline{H}_{max}$ versus d/D data: (**a**) $L_1 = 1000$ m; (**b**) $L_1 = 2000$ m; (**c**) $L_1 = 3000$ m; (**d**) $L_1 = 4000$ m; (**e**) $L_1 = 5000$ m.

Figure 21 validates the appropriateness of fitting logistic curves to the $\overline{H}_{\max}$ versus d/D results. In Figure 21a, $\overline{H}_{\max}$ remains relatively constant for large d/D values. In this figure, the length of the high right plateau is significantly longer than that of the low left plateau. In fact, a significant part of the d/D range in Figure 21a, specifically the largest d/D values, is of limited relevance for sizing. Air valves with d/D values exceeding 0.20 are typically not considered in practice. In Figure 21b, interpolated data is incorporated into the actual $\overline{H}_{\max}$ versus d/D results to establish an evenly spaced set of $\overline{H}_{\max}$ versus d/D pairs. These interpolated values are considered in the curve fitting process. The fitted curves shown in Figure 21b have the ability to predict the transient magnitude for both type 1 and type 3 behaviours.

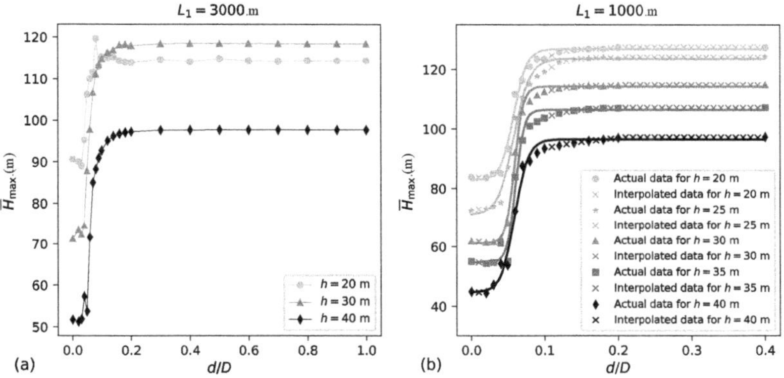

Figure 21. Detailed $\overline{H}_{\max}$ versus d/D data: (**a**) $0 \leq d/D \leq 1$ and $L_1 = 3000$ m; (**b**) actual data and interpolated data considering $0 \leq d/D \leq 0.40$ and $L_1 = 1000$ m.

Figure 22 shows logistic curves fitted to the dataset shown in Figure 20. In Figure 22, two points along each curve are emphasised: (d_{c1}/D, $\overline{H}_{\max_{c1}}$), which delimits type 1 behaviour; and (d_{c3}/D, $\overline{H}_{\max_{c3}}$), which delimits type 3 behaviour. The d_{c1}/D value is determined as the d/D value corresponding to the minimum $\overline{H}_{\max}$ plus 5% of the logistic curve's height, while d_{c3} is determined as the d/D value corresponding to the maximum $\overline{H}_{\max}$ minus 5% of the logistic curve's height. These definitions for the critical air valve orifice sizes, while simple, effectively capture the transitions between the transient behaviours. Using logistic curves instead of raw data enhances the ease of comparing various scenarios and determining the critical orifice sizes.

Table 1 summarises the d_{c1}/D and d_{c3}/D values for all the cases considered in the main set of simulations, as shown in Figure 22. Significant variation is found in both the d_{c1}/D and d_{c3}/D values, depending on the specific combination of L_1 and h. For example, for the range of h values considered in this paper, for $L_1 = 3000$ m, the range of d_{c1}/D spans from 0.024 to 0.042, while the range of d_{c3}/D spans from 0.064 to 0.092. To provide greater context, consider the d_{c1}/D ranges, which define type 1 behaviour, and the d_{c3}/D ranges, which define type 3 behaviour, based on the data reported in Zhou et al. (2019) [39] for vertical pipe filling and in Zhou et al. (2020) [3] for horizontal pipe filling with $H_r^*/H_{\text{atm}}^* = 2.58$ (ratio between the absolute inlet head and the absolute atmospheric head). When filling a line with a downstream orifice, the transient magnitude versus d/D data has a low left plateau for type 1 behaviour, but there is no high right plateau for type 3 behaviour, as shown in Figure 19. In a manner similar to the approach used in Figure 22 to define d_{c1}/D, for pipeline filling with a downstream orifice, the value of d_{c1}/D is the d/D value associated with the transient magnitude determined by adding 5% of the difference between the maximum value of the curve and the value of the low left plateau to the value of the low left plateau itself. Additionally, for this case, the value of d_{c3}/D is the d/D value associated with the maximum on the transient magnitude versus d/D curve.

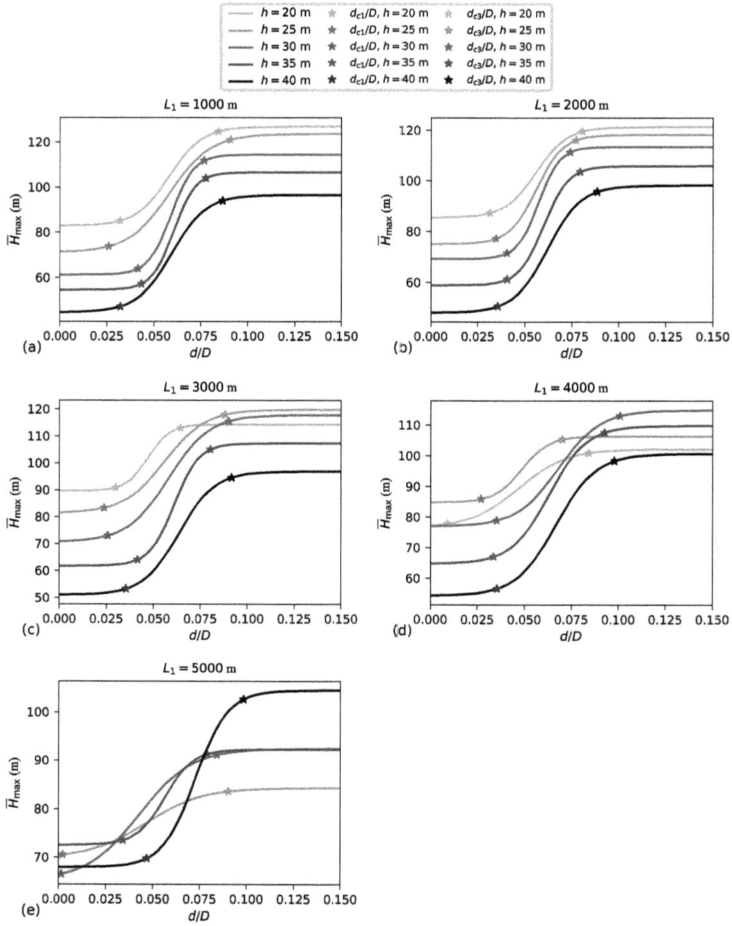

Figure 22. Logistic curves fitted to the $\overline{H}_{\max}$ versus d/D data together with the predicted points for the inception of type 1 (d_{c1}/D, $\overline{H}_{\max_{c1}}$) and type 3 (d_{c3}/D, $\overline{H}_{\max_{c3}}$) behaviours: (**a**) $L_1 = 1000$ m; (**b**) $L_1 = 2000$ m; (**c**) $L_1 = 3000$ m; (**d**) $L_1 = 4000$ m; (**e**) $L_1 = 5000$ m.

Table 1. Critical orifice ratios for the inception of both type 1 (d_{c1}/D) and type 3 (d_{c3}/D) behaviours.

	Length of the Upstream Pipeline Branch				
	1000 m	**2000 m**	**3000 m**	**4000 m**	**5000 m**
d_{c1}/D range	0.026–0.043	0.032–0.041	0.024–0.042	0.009–0.036	0.001–0.047
d_{c3}/D range	0.077–0.091	0.074–0.088	0.064–0.092	0.070–0.101	0.079–0.098
d_{c1}/D for $h = 20$ m	0.032	0.032	0.030	0.009	—
d_{c3}/D for $h = 20$ m	0.084	0.081	0.064	0.084	—
d_{c1}/D for $h = 25$ m	0.026	0.035	0.024	0.027	0.002
d_{c3}/D for $h = 25$ m	0.091	0.077	0.088	0.070	0.090
d_{c1}/D for $h = 30$ m	0.041	0.040	0.026	0.035	0.001
d_{c3}/D for $h = 30$ m	0.077	0.074	0.090	0.101	0.084
d_{c1}/D for $h = 35$ m	0.043	0.041	0.042	0.034	0.034
d_{c3}/D for $h = 35$ m	0.078	0.079	0.080	0.093	0.079
d_{c1}/D for $h = 40$ m	0.032	0.035	0.035	0.036	0.047
d_{c3}/D for $h = 40$ m	0.087	0.083	0.092	0.098	0.098

Figure 23 presents the data from Zhou et al. (2019) [39] and Zhou et al. (2020) [3] concerning the filling procedure with $H_r^*/H_{\text{atm}}^* = 2.58$. This figure also presents both the curves that represent the data (i.e., right-skewed probability density curves fitted to the data) and the d_{c1} and d_{c3} values associated with each curve. Only data within the range of $0 \leq d/D \leq 0.25$ was considered for two main reasons. Firstly, very large d/D values are generally not considered in practice. Secondly, these larger d/D values sometimes correspond to transient magnitude values lower than the low left plateau of the data. This poses challenges for fitting the representational curves, as the chosen function for fitting the data assumes identical vertical axis values for both the left and right low plateaus. Not imposing this limitation on the considered d/D range for curve fitting would compromise the predictive accuracy of the fitted curve for type 1 behaviour.

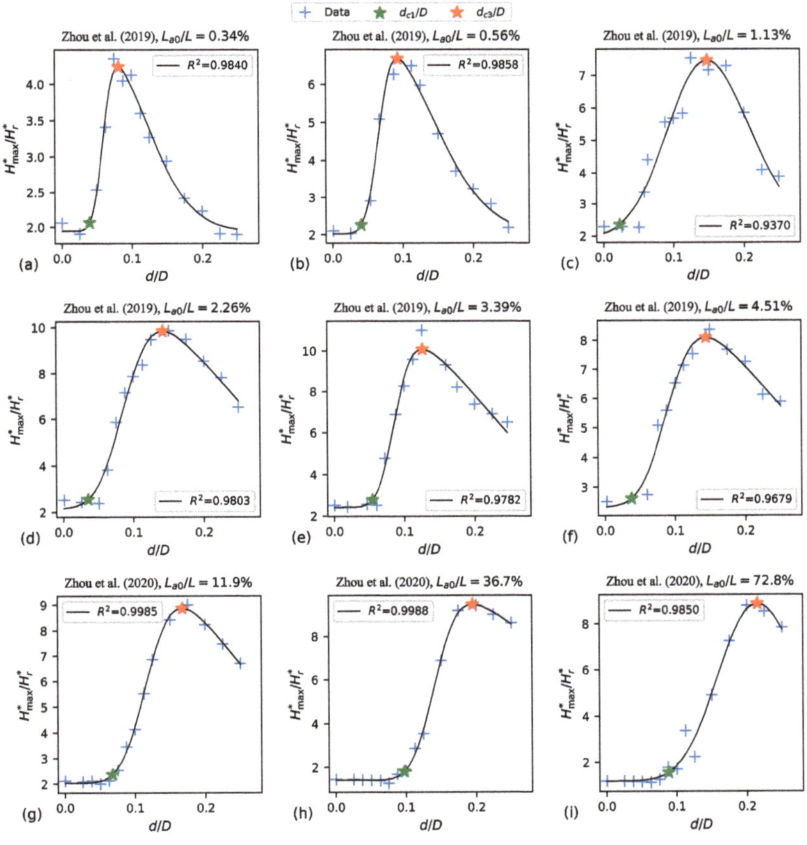

Figure 23. Critical orifice ratios for the pipeline filling procedure considering $H_r^*/H_{\text{atm}}^* = 2.58$: (**a–f**) vertical pipe filling (data from Zhou et al. (2019) [39]); (**g–i**) horizontal pipe filling (data from Zhou et al. (2020) [3]).

In Figure 23a–f (with data from Zhou et al. 2019 [39]), the d_{c1}/D values range from 0.023 to 0.054, while the d_{c3}/D values range from 0.081 to 0.148. In Figure 23g–i (with data from Zhou et al. 2020 [3]), the d_{c1}/D values range from 0.068 to 0.098, while the d_{c3}/D values range from 0.167 to 0.215. The critical d/D values identified in the current study (which are presented in Table 1) exhibit greater similarity to those reported in Zhou et al. (2019) [39] than to those reported in Zhou et al. (2020) [3]. This analysis indicates that the critical d/D values for both type 1 and type 3 behaviours are notably influenced by the specific characteristics of each system and the transient event. Note that the definitions used herein for type 1 and type 3 behaviours differ from those used in Zhou et al. (2019) [39] and Zhou et al. (2020) [3]. While not necessarily indicating a direct cause-and-effect relationship, similar to the current study, the experiments carried out by Zhou et al. (2019) [39] involved significantly reduced air fractions. Additionally, when compared to a horizontal filling procedure as examined in Zhou et al. (2020) [3], vertical filling procedures and the air expulsion stage with inclined pipe segments connected to the air valve are expected to exhibit a more predictable air–water interface behaviour.

7.4. Additional Intervening Parameters

The main set of numerical simulations considers $v_0 = 1.5$ m/s, $\Delta Z = 50$ m, $a = 1000$ m/s, and $C = 130$. However, such parameters themselves have an effect on the $\overline{H}_{max}$ versus d/D results. Such an effect is explored in Figure 24. The transient magnitude, especially for type 3 behaviour, is shown to increase with v_0, ΔZ, a, and C. It should be borne in mind that C is a conductance, so that increases in C correspond to decreased friction. In relation to v_0 and a, indeed, in the Joukowsky equation, the transient magnitude is proportional to the velocity variation that causes the transient and to a. In relation to ΔZ, having a downstream reservoir at a higher elevation, while maintaining the water discharge, means that the necessary head at the upstream section of the pipeline must be increased. In relation to C, it is worth noting that the attenuation of the transient during the pump trip scenario can be due to the air cushioning effect (for small d/D values) or due to friction (for any outflow orifice size). In scenarios with a high-outflow-capacity air valve, the transient attenuation is primarily due to friction. As the air admission and expulsion stages proceed, there is water movement in both the upstream and downstream branches. Such movement results in energy losses due to friction. Such energy dissipation in a system with water column separation due to air admission by an air valve has been investigated by Ramezani and Karney (2017) [10].

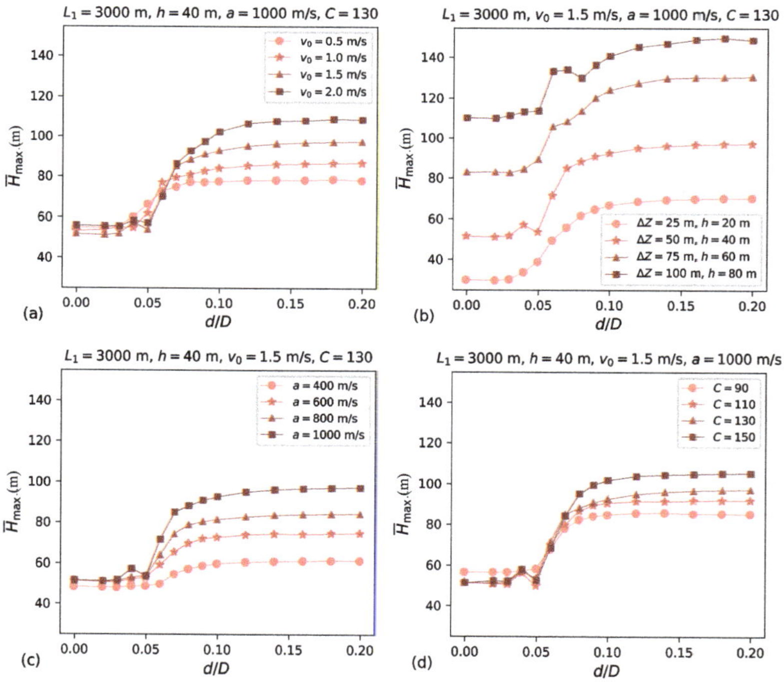

Figure 24. Considering additional variation in the initial parameters: (**a**) initial steady water velocity (v_0); (**b**) elevation difference between reservoirs (ΔZ); (**c**) wave speed (a); (**d**) friction coefficient according to the Hazen–Williams equation (C).

In Figure 25, logistic curves have been fitted to the $\overline{H}_{\max}$ versus d/D data shown in Figure 24. Notably, Figure 25a shows larger d_{c3}/D values for larger v_0 values—from 0.075 (for $v_0 = 0.5$ m/s) to 0.102 (for $v_0 = 2.0$ m/s). Similarly, in Figure 25d, d_{c3}/D ranges from 0.082 (for $C = 90$) to 0.091 (for $C = 150$). Figures 24 and 25 demonstrate that the patterns observed in the results from the main set of numerical simulations are also present in a broader spectrum of system configurations. Moreover, the outcomes of the current study corroborate and supplement the prior findings reported in Tasca et al. (2022) [11] and in Tasca et al. (2021) [14].

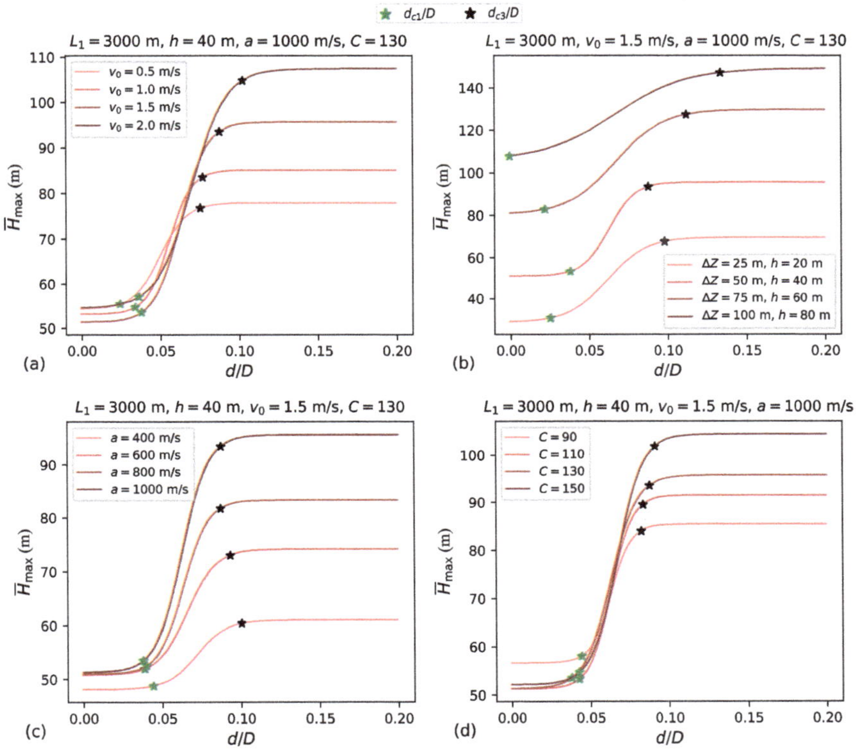

Figure 25. Logistic curves fitted to transient magnitude versus d/D data considering additional variation in the initial parameters: (**a**) initial steady water velocity (v_0); (**b**) elevation difference between reservoirs (ΔZ); (**c**) wave speed (a); (**d**) friction coefficient according to the Hazen–Williams equation (C).

7.5. Non-Slam Air Valves

To implement in practice the concept of enabling unrestricted inflow while restricting outflow, a non-slam air valve can be employed. A typical non-slam air valve contains two floats, which do not hinder air inflow. With the air valve under a reduced positive head, they do not obstruct the air flow through the large orifice of the valve. However, if this positive head exceeds the transition head, the upper float closes, permitting air flow only through a small orifice. Both floats move to the closing position when the water reaches the valve, effectively sealing it. The main design parameters of a non-slam air valve are the diameter of the inflow orifice, the diameter of the large outflow orifice, the diameter of the small outflow orifice, and the transition head. Schematics of non-slam air valves can be found in Tran (2017) [40] and Li et al. (2022) [18].

Figure 26 shows the transient response of a pipeline with a non-slam air valve during the pump trip scenario. The inflow orifice ratio is set as $d_{in}/D = 0.20$. The small outflow orifice ratio is set as $d/D = 0.04$, which is an orifice ratio found to be effective for transient attenuation. Figure 26 assesses the influence of the following parameters on the transient response: transition head ($\Delta H_{\text{transition}}$), which is the threshold head for transitioning from the large outflow orifice to the small outflow orifice; large outflow orifice ratio (d_{large}/D); and air valve elevation (h). The $\Delta H_{\text{transition}}$ values considered in this figure are in line with typical values found in product catalogues.

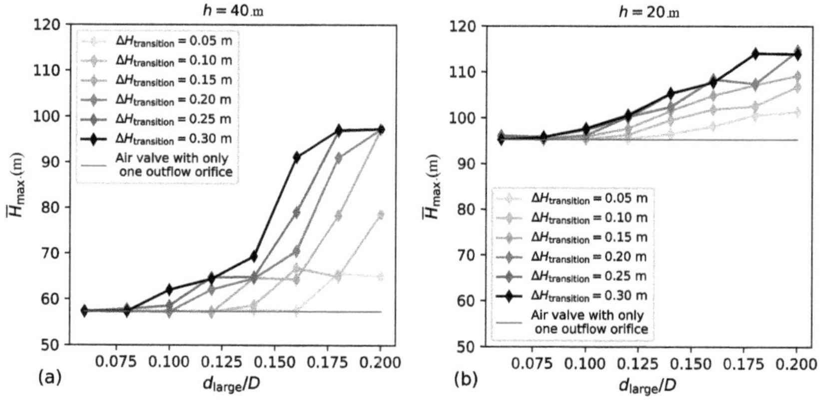

Figure 26. Influence of transition head ($\Delta H_{\text{transition}}$) and large outflow orifice (d_{large}) on the ability of the non-slam air valve to mitigate the transient event considering $L_1 = 3000$ m and $d/D = 0.04$: (**a**) $h = 40$ m; (**b**) $h = 20$ m.

As shown in Figure 26, in general, as d_{large}/D decreases, the behaviour of the non-slam air valve becomes more similar to that of the air valves with a single and small outflow orifice. This similarity becomes most pronounced when the $\Delta H_{\text{transition}}$ value falls below a specific threshold for each d_{large}/D value. This $\Delta H_{\text{transition}}$ threshold decreases as d_{large}/D increases. Figure 26 highlights the importance of carefully selecting both d_{large}/D and $\Delta H_{\text{transition}}$. In this sense, inadequate selection of these parameters may render the non-slam air valve ineffective in mitigating transients. This can happen if d_{large}/D is oversized so that $\Delta H_{\text{transition}}$ is never reached, or if $\Delta H_{\text{transition}}$ is too large for a given d_{large}/D value.

Figure 27 shows the evolution of both the HG at the air valve and the air pocket volume across four distinct situations: air valve with a single and small outflow orifice (Figure 27a,b), effective non-slam air valve (Figure 27c,d), ineffective non-slam air valve (Figure 27e,f), and air valve with a single and large outflow orifice (Figure 27g,h). For the non-slam air valve with $d_{\text{large}}/D = 0.10$ and $\Delta H_{\text{transition}} = 0.10$ m (Figure 27c,d), the transient response is similar to the case with the air valve with a single and small outflow orifice (Figure 27a,b). However, for the non-slam air valve with $d_{\text{large}}/D = 0.16$ and $\Delta H_{\text{transition}} = 0.30$ m (Figure 27e,f), the air valve fails to attenuate the transient event. In fact, the transient response for the case with the air valve with a single and large outflow orifice (Figure 27g,h) is similar to the case with the ineffective non-slam air valve (Figure 27e,f). Notably, having a small $\Delta H_{\text{transition}}$ value or a small d_{large}/D value might render the air valve unsuitable for pipeline filling, which usually requires a high air exhaust rate.

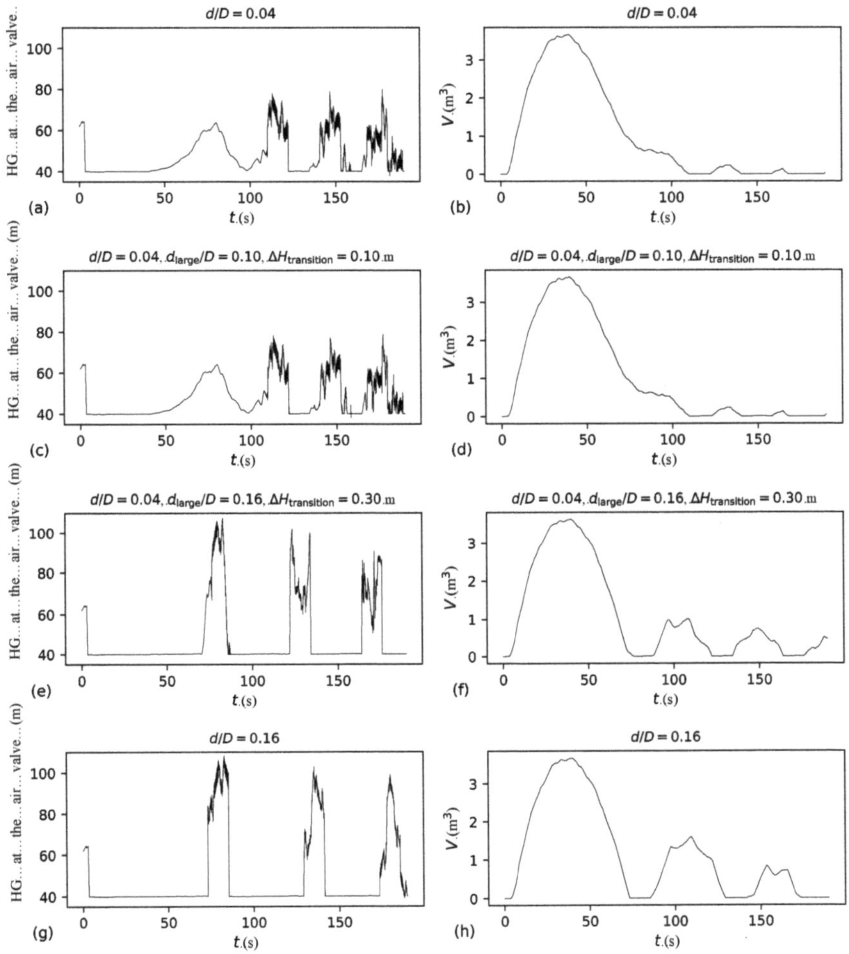

Figure 27. Regular air valve versus non-slam air valve considering $L_1 = 3000$ m, $h = 40$ m, and $d_{in}/D = 0.20$: (**a**,**b**) regular air valve with $d/D = 0.04$; (**c**,**d**) non-slam air valve with $d_{large}/D = 0.10$, $\Delta H_{transition} = 0.10$ m, and $d/D = 0.04$; (**e**,**f**) non-slam air valve with $d_{large}/D = 0.16$, $\Delta H_{transition} = 0.30$ m, and $d/D = 0.04$; (**g**,**h**) regular air valve with $d/D = 0.16$.

8. Conclusions

As future conditions can never be perfectly anticipated, and real-world systems invariably deviate from numerical idealisations, it becomes crucial for design and operational engineers to gauge the potential impact of real-world variability in system parameters on anticipated system performance. This paper seeks to at least partially address this concern, aiming to comprehend through extensive simulations coupled with a physical understanding how design choices and system variations can influence the critical transient pressures. Specifically, this paper considers a key transient event, namely, the pump trip event, but also extracts a number of insights relative to pipeline filling and draining. But significantly, the sensitivity of the system's transient response to varying conditions is shown to be remarkably diverse, demonstrating strong reactivity to changes in some cases with noticeable insensitivity in others.

In practice, the uses and specifications of air valves are vexing and problematic to many system designers and operators. This is because their behaviour is highly non-linear, sometimes showing high sensitivity to assumed conditions and sometimes displaying robust and stable results. This is at least partly because anything involving two phases is likely to be complex, but also because there is such a vast number of interacting components, including the many characteristics of the system (length, location, friction, initial state, pumping configuration), the specifics of the air valve (the absolute and relative sizes of both its admission and discharge orifices, its vertical and horizontal position, whether or not it possesses slow closing features, etc.) and of the transient itself (in this case, power failure). This work attempts the most comprehensive characterisation to date of the anticipated response of a realistic set of systems to a vast combination of characteristics. The immediate goal is to provide a systematic exploration and elucidation of the likely variation in transient response associated with a wide range of system and valve attributes. The long-term goal is to provide system owners, operators, designers, and managers a better understanding of air–system interactions so they can make better choices. This better understanding provides design engineers with a much clearer idea of what and where variations are likely to matter, and which ones do not.

Following pump trip, four stages constitute and complete a water hammer cycle: depressurisation, air admission, air expulsion, and the creation of a secondary wave. While most of the minimum hydraulic grade line (HGL) is formed during the initial two stages, the maximum HGL is often only completed after the first instance of air pocket collapse. But a key point is that each stage tends to be sensitive to different system attributes. So, for example, the smallest pipeline lengths experiencing negative pressures (i.e., smallest L_s values) are associated with small L_1 values but with large h values. L_s is found to be only slightly sensitive to d/D (ratio between the outflow orifice of the air valve and the pipe diameter). Especially for large h values, there is a tendency of having smaller L_s values for smaller d/D values. Moreover, if the air valve is located closer to the upstream reservoir rather than farther downstream, there is a tendency to observe milder transient magnitudes for higher h values.

Three main transient behaviours have been identified: attenuated (type 1 behaviour) for small d/D values, intermediate (type 2 behaviour) for moderate d/D values, and water-hammer-dominated (type 3 behaviour), which typically occurs for large d/D values. For type 1 behaviour, the head evolution at the air valve includes at least one instance of smooth head-versus-time variation with substantial compression of the air pocket prior to its collapse. However, for type 3 behaviour, the pressure level at the air valve remains close to the atmospheric pressure while there is still trapped air within the pipeline, i.e., there is no substantial compression of the air pocket prior to its collapse. For small d/D values, prior to air pocket collapse, the transient oscillation period at the air valve decreases as d/D increases. For large d/D values, however, the transient oscillation, which encompasses sequential instances of air pocket admission and collapse, becomes independent of d/D. Importantly, in cases exhibiting type 1 behaviour, the increase in head at the air valve immediately after the collapse of the air pocket is significantly reduced compared to cases exhibiting type 3 behaviour.

For type 3 behaviour, there is a causal connection between the closure of the air valve and a subsequent pressure peak at the upstream section of the pipeline. Notably, the hydraulic head magnitude is consistently larger at the upstream section of the pipeline than at the air valve. In fact, the maximum HGL tends to be higher at locations closer to the upstream reservoir. During the transient event, the water velocity in the upstream branch of the pipeline (v_{up}) varies around zero. In the downstream branch, though, the water velocity (v_{ds}) progressively decreases until reaching a minimum negative value, i.e., until a maximum reverse flow is reached. The periodic v_{ds} changes every round trip in the downstream branch increase as h decreases. Because of this, as h increases, both the time required for reverse flow to occur and the time for air pocket collapse (t_c) increase. While the variation in v_{ds} follows a linear pattern for type 3 behaviour, the variation in v_{ds} for type 1 behaviour initially appears linear but transitions to a non-linear pattern during the air expulsion stage. This non-linearity occurs due to the air cushioning effect.

Regarding air pocket evolution, the water velocity variations in the upstream branch have a minimal influence, while the water velocity variations in the downstream branch have a substantial impact. In cases exhibiting type 1 behaviour, a single large air pocket typically forms within the pipeline. Conversely, in cases exhibiting type 3 behaviour, a series of air pockets forms sequentially, each with progressively smaller volumes. Maximum air pocket volume increases with h but decreases with L_1. The difference in maximum air pocket volume between the least favourable location for air pocket entrapment ($L_1 = 5000$ m and $h = 20$) and the most favourable one ($L_1 = 1000$ m and $h = 40$ m) is approximately 50 times.

The smallest L_s values are found for type 1 behaviour, $L_1 = 1000$ m, and $h = 40$ m. For type 1 behaviour, there is a tendency of having smaller L_s values for larger h values. Nonetheless, in the current study, L_s/L is always greater than 60%. In fact, in case the pipeline does not contain an air valve, the entire pipeline experiences negative pressures. For $L_1 = 1000$ m and $L_1 = 2000$ m, not having an air valve leads to more pronounced transient pressures compared to having an air valve, regardless of the value of d/D. In the transient magnitude versus d/D data, for each air valve location, there is a lower left plateau (type 1 behaviour) and a higher right plateau (type 3 behaviour). Intermediate transient magnitudes correspond to type 2 behaviour. For example, for $L_1 = 3000$ m and $h = 40$ m, the mean maximum transient head for type 1 behaviour is about 50 m, while for type 3 behaviour, it is about 100 m. Especially for small L_1 values, there is a tendency for the transient magnitude to be milder as h increases.

For each air valve location, t_c increases exponentially as d/D decreases but stabilises for large d/D values. The collapse time of the air pocket is significantly affected by the location of the air valve. For example, for type 3 behaviour and $h = 40$ m, $t_c = 109$ s for $L_1 = 1000$ m, but $t_c = 31$ s for $L_1 = 5000$ m. The timing of the maximum HG in the system ($t_{\mathrm{HG}_{\max}}$) can be divided into three groups: well before t_c (typical of type 1 behaviour), right after t_c (typical of type 3 behaviour), and well after t_c (sometimes present for type 2 behaviour). Even when $t_{\mathrm{HG}_{\max}}$ occurs well after t_c, there is a pressure peak at the upstream section of the pipeline causally connected with the collapse of the air pocket. While the air pocket remains inside the pipeline, the frequency of the pressure oscillations at the upstream section of the pipeline increases as L_1 decreases. More specifically, the period of the pressure oscillations is given by $4L_1/a$.

Since the transient magnitude is determined during the air expulsion and secondary wave stages—with the air expulsion stage being a type of pipeline filling event—the comparison between the pump trip scenario and the filling scenario is meaningful. Previous studies concerning the filling scenario, which include a detailed assessment of transient magnitude in relation to aperture size for air exchanges, generally consider a system with a downstream end orifice instead of an air valve. For this reason, while the transient magnitude versus d/D results in the current work resemble a logistic curve, the results found in the literature regarding the pipeline filling scenario with a downstream orifice resemble a skewed-to-the-right normal probability density function. If the system contains an air valve, no water is expelled through the air exchange aperture, in contrast to the case with an orifice. For each air valve location, fitting a logistic curve to the transient response data allows for defining the d_{c1}/D and d_{c3}/D values for the inception of type 1 and type 3 behaviours, respectively. For each air valve location, if $d/D < d_{c1}/D$, then type 1 behaviour occurs; if $d/D > d_{c3}/D$, then type 3 behaviour occurs; and if $d_{c1}/D < d/D < d_{c3}$, then type 2 behaviour occurs. For example, for $L_1 = 3000$ m, for the range of h values from 20 to 40 m, the d_{c1}/D values range from 0.024 to 0.042, while the d_{c3}/D values range from 0.064 to 0.092. The d_{c1}/D and d_{c3}/D values are dependent on the air valve location and the initial flow conditions.

The current study examines the impact of varying certain key initial parameters that are initially locked-in for the main set of numerical simulations. Such parameters are the initial steady water velocity (v_0), elevation difference between the downstream and upstream reservoirs (ΔZ), wave speed (a), and friction coefficient (C). The influence of varying these parameters on the transient response is more evident in relation to type 3 behaviour compared to type 1 behaviour. Particularly in the case of type 3 behaviour, it is observed that $\overline{H}_{max}$ increases with v_0, a, ΔZ, and C. Of note, the same three transient behaviours found in the main set of transient simulations are also present when additional variations are considered for these initial parameters.

The current study assesses the impact of using a non-slam air valve with a single inflow orifice but two outflow orifices in comparison to an air valve with a single inflow orifice and a single outflow orifice. It is found that a well-sized non-slam air valve can mimic the behaviour of a large inflow orifice and small outflow orifice air valve during the pump trip scenario. For a non-slam air valve to be effective in mitigating the transient event, both the transition head ($\Delta H_{transition}$) and the large outflow orifice (d_{large}/D) should be sufficiently small. The transition head must be sufficiently small to enable the utilisation of the small outflow orifice for expulsion. It is found that the transition head threshold for ensuring the switching from the large outflow orifice to the small one during the transient event decreases as the size of the large outflow orifice increases.

Author Contributions: Conceptualisation, E.T.; methodology, E.T.; validation, E.T.; formal analysis, E.T.; investigation, E.T.; resources, E.L.J.; data curation, E.T.; writing—original draft preparation, E.T. and B.K.; writing—review and editing, E.T., M.B., H.M.R., and B.K.; supervision, E.L.J. and B.K. All authors have read and agreed to the published version of the manuscript.

Funding: This study was financed in part by the Coordenação de Aperfeiçoamento de Pessoal de Nível Superior—Brasil (CAPES)—Finance Code 001. This study was financed in part by the Pipeline Simulation Interest Group (PSIG) through the 2022 Orin Flanigan Scholarship awarded to Elias Tasca for his Ph.D. work.

Data Availability Statement: The data that support the findings of this study are available from the corresponding authors upon reasonable request.

Acknowledgments: The authors acknowledge the support from Bentley Systems that provided the simulation software HAMMER.

Conflicts of Interest: The authors declare no conflicts of interest.

Notation

a	Wave speed (m/s)
C	Friction coefficient according to the Hazen–Williams equation
D	Pipe diameter (m)
d	Diameter of the outflow orifice of the air valve (m)
d_{c1}	Critical orifice diameter below which an attenuated event occurs (m)
d_{c3}	Critical orifice diameter above which a water-hammer-dominated event occurs (m)
d_{in}	Diameter of the inflow orifice of the air valve (m)
d_{large}	Diameter of the large outflow orifice of the non-slam air valve (m)
d_{small}	Diameter of the small outflow orifice of the non-slam air valve (m)
HG	Hydraulic grade (m)
HG_{max}	Maximum hydraulic grade in the system (m)
$\overline{HG}_{max}$	Mean maximum hydraulic grade in the system (m)
$\overline{H}_{max}$	Mean maximum hydraulic head (m)
$\overline{H}_{steady}$	Mean hydraulic head during steady flow (m)
H^*_{atm}	Absolute atmospheric head (m)
H^*_{max}	Absolute maximum transient head (m)
H^*_r	Absolute inlet head (m)
h	Air valve elevation (m)
L	Pipeline length (m)
L_1	Length of the upstream branch of the pipeline in relation to the air valve (m)
L_2	Length of the descending pipe segment downstream of the air valve (m)
L_3	Length of the ascending pipe segment connected to the downstream reservoir (m)
L_s	Pipeline length experiencing negative pressures (m)
T1B	Type 1 behaviour (attenuated event)

T3B	Type 3 behaviour (water-hammer-dominated event)
t_c	Air pocket collapse time (s)
$t_{\mathrm{HG}_{\max}}$	Timing of the maximum HG in the system (s)
V	Air pocket volume (m^3)
V_{ds}	Air pocket volume at the downstream branch (m^3)
$V_{\max}$	Maximum air pocket volume (m^3)
V_{up}	Air pocket volume at the upstream branch (m^3)
v_0	Initial steady water velocity (m/s)
v_{ds}	Water velocity downstream of the air valve (m/s)
v_{up}	Water velocity upstream of the air valve (m/s)
Z	Pipe elevation (m)
$\overline{Z}$	Mean pipeline elevation (m)
$\Delta H_{\mathrm{transition}}$	Transition head of the non-slam air valve to switch from d_{large} to d_{small} (m)
ΔZ	Elevation difference between the downstream and upstream reservoirs (m)

References

1. Tasca, E.; Besharat, M.; Ramos, H.M.; Luvizotto, E., Jr.; Karney, B. Contribution of air management to the energy efficiency of water pipelines. *Sustainability* **2023**, *15*, 3875. [CrossRef]
2. Martins, S.C. Dinâmica da Pressurização de Sistemas Hidráulicos Com Ar Aprisionado. Ph.D. Thesis, Instituto Superior Técnico, Universidade Técnica de Lisboa, Lisbon, Portugal, 2013.
3. Zhou, L.; Cao, Y.; Karney, B.; Bergant, A.; Tijsseling, A.S.; Liu, D.; Wang, P. Expulsion of entrapped air in a rapidly filling horizontal pipe. *J. Hydraul. Eng.* **2020**, *146*, 04020047.
4. Boulos, P.; Karney, B.; Wood, D.; Lingireddy, S. Hydraulic transient guidelines for protecting water distribution systems. *J. Am. Water Work. Assoc.* **2005**, *97*, 111–124. [CrossRef]
5. American Water Works Association (AWWA). *Manual of Water Supply Practices M51–Air Valves: Air-Release, Air/Vacuum and Combination*, 2nd ed.; AWWA: Denver, CO, USA, 2016.
6. Ramos, H.M.; Fuertes-Miquel, V.S.; Tasca, E.; Coronado-Hernández, O.E.; Besharat, M.; Zhou, L.; Karney, B. Concerning dynamic effects in pipe systems with two-phase flows: Pressure surges, cavitation, and ventilation. *Water* **2022**, *14*, 2376. [CrossRef]
7. García-Todolí, S.; Iglesias-Rey, P.L.; Mora-Meliá, D.; Martínez-Solano, F.J.; Fuertes-Miquel, V.S. Computational determination of air valves capacity using CFD techniques. *Water* **2018**, *10*, 1433. [CrossRef]
8. Paternina-Verona, D.A.; Coronado-Hernández, O.E.; Espinoza-Román, H.G.; Fuertes-Miquel, V.S.; Ramos, H.M. Rapid filling analysis with an entrapped air pocket in water pipelines using a 3D CFD model. *Water* **2023**, *15*, 834. [CrossRef]
9. Tasca, E.; Karney, B. Improved air valve selection through better device characterization and modeling. *J. Hydraul. Eng.* **2023**, *149*, 04023019. [CrossRef]
10. Ramezani, L.; Karney, B. Water column separation and cavity collapse for pipelines protected with air vacuum valves: Understanding the essential wave processes. *J. Hydraul. Eng.* **2017**, *143*, 04016083. [CrossRef]
11. Tasca, E.; Karney, B.; Fuertes-Miquel, V.S.; Dalfré Filho, J.G.; Luvizotto, E., Jr. The crucial importance of air valve characterization to the transient response of pipeline systems. *Water* **2022**, *14*, 2590. [CrossRef]
12. Besharat, M.; Coronado-Hernández, O.E.; Fuertes-Miquel, V.S.; Viseu, M.T.; Ramos, H.M. Backflow air and pressure analysis in emptying a pipeline containing an entrapped air pocket. *Urban Water J.* **2018**, *15*, 769–779. [CrossRef]
13. Fuertes-Miquel, V.S.; Coronado-Hernández, O.E.; Mora-Meliá, D.; Iglesias-Rey, P.L. Hydraulic modeling during filling and emptying processes in pressurized pipelines: A literature review. *Urban Water J.* **2019**, *16*, 299–311. [CrossRef]
14. Tasca, E.; Karney, B.; Luvizotto, E., Jr. Performance similarity between different-sized air exchange valves. *J. Hydraul. Eng.* **2021**, *147*, 04021036. [CrossRef]
15. Li, X.; Yan, T.; Bi, X.; Fei, L. Influence of traditional and antislam air valve characteristics on transient pressure control. *J. Pipeline Syst. Eng. Pract.* **2022**, *13*, 04022022. [CrossRef]
16. Apollonio, C.; Balacco, G.; Fontana, N.; Giugni, M.; Marini, G.; Piccinni, A.F. Hydraulic transients caused by air expulsion during rapid filling of undulating pipelines. *Water* **2016**, *8*, 25. [CrossRef]
17. Zhou, L.; Liu, D.; Karney, B. Investigation of hydraulic transients of two entrapped air pockets in a water pipeline. *J. Hydraul. Eng.* **2013**, *139*, 949–959. [CrossRef]
18. Li, X.; Wang, T.; Zhang, Y.; Guo, P. Study on the factors influencing air valve protection against water hammer with column separation and rejoinder. *AQUA-Infrastruct. Ecosyst. Soc.* **2022**, *71*, 949–962. [CrossRef]
19. Catucci, D.; Briganti, R.; Heller, V. Numerical validation of novel scaling laws for air entrainment in water. *Proc. R. Soc. A* **2021**, *477*, 20210339. [CrossRef]
20. Bergant, A.; Kruisbrink, A.; Arregui, F. Dynamic behaviour of air valves in a large-scale pipeline apparatus. *Stroj. Vestn.-Mech. Eng.* **2012**, *58*, 225–237. [CrossRef]
21. He, J.; Hou, Q.; Lian, J.; Tijsseling, A.S.; Bozkus, Z.; Laanearu, J.; Lin, L. Three-dimensional CFD analysis of liquid slug acceleration and impact in a voided pipeline with end orifice. *Eng. Appl. Comput. Fluid Mech.* **2022**, *16*, 1444–1463. [CrossRef]

22. Zhou, L.; Lu, Y.; Karney, B.; Wu, G.; Elong, A.; Huang, K. Energy dissipation in a rapid filling vertical pipe with trapped air. *J. Hydraul. Res.* **2023**, *61*, 120–132. [CrossRef]
23. Pozos, O.; Sanchez, A.; Rodal, E.A.; Fairuzov, Y.V. Effects of water–air mixtures on hydraulic transients. *Can. J. Civ. Eng.* **2010**, *37*, 1189–1200. [CrossRef]
24. Ramezani, L. An Exploration of Transient Protection of Pressurized Pipelines using Air Valves. Ph.D. Thesis, University of Toronto, Toronto, ON, Canada, 2015.
25. Coronado-Hernández, O.E.; Fuertes-Miquel, V.S.; Besharat, M.; Ramos, H.M. A parametric sensitivity analysis of numerically modelled piston-type filling and emptying of an inclined pipeline with an air valve. In Proceedings of the 13th International Conference on Pressure Surges, Bordeaux, France, 14–16 November 2018.
26. Zhou, L.; Wang, H.; Karney, B.; Liu, D.; Wang, P.; Guo, S. Dynamic behavior of entrapped air pocket in a water filling pipeline. *J. Hydraul. Eng.* **2018**, *144*, 04018045. [CrossRef]
27. Chaudhry, M.H. *Applied Hydraulic Transients*, 3rd ed.; Springer: New York, NY, USA, 2014.
28. Abdeldayem, O.M.; Ferràs, D.; van der Zwan, S.; Kennedy, M. Analysis of unsteady friction models used in engineering software for water hammer analysis: Implementation case in WANDA. *Water* **2021**, *13*, 495. [CrossRef]
29. Bergant, A.; Simpson, A.R.; Tijsseling, A.S. Water hammer with column separation: A historical review. *J. Fluids Struct.* **2006**, *22*, 135–171. [CrossRef]
30. Wylie, E.B.; Streeter, V.L. *Fluid Transients*; FEB Press: Ann Arbor, MI, USA, 1983.
31. Fuertes-Miquel, V.S.; López-Jiménez, P.A.; Martínez-Solano, F.J.; López-Patiño, G. Numerical modelling of pipeline with air pockets and air valves. *Can. J. Civ. Eng.* **2016**, *43*, 1052–1061. [CrossRef]
32. Zhang, X.; Fan, C.; Yu, X.; Zhang, J.; Lv, J.; Xu, T. Study on the mathematical model of vacuum breaker valve for large air mass conditions. *Water* **2019**, *11*, 1358. [CrossRef]
33. Agostinho, M.; Fernandes, C.; Jung, B. Assessment of hydraulic transient indicators in water supply network. In Proceedings of the 1st International WDSA/CCWI Joint Conference, Kingston, ON, Canada, 23–25 July 2018.
34. Jung, B.; Karney, B. Application of fluid transients on pipeline optimization: Worst-case scenario search and systematic protection. In Proceedings of the Conference on Probabilistic Methodologies in Water and Wastewater Engineering, Toronto, ON, Canada, 23–24 September 2011.
35. Lingireddy, S.; Wood, D.J.; Zloczower, N. Pressure surges in pipeline systems resulting from air releases. *J. Am. Water Work. Assoc.* **2004**, 96, 88–94. [CrossRef]
36. Pothof, I.; Karney, B. Guidelines for transient analysis in water transmission and distribution systems. In *Water Supply System Analysis–Selected Topics*; Ostfeld, A., Ed.; InTech: Rijeka, Croatia, 2012; pp. 1–21.
37. Zhou, F.; Hicks, F.E.; Steffler, P.M. Transient flow in a rapidly filling horizontal pipe containing trapped air. *J. Hydraul. Eng.* **2002**, *128*, 625–634. [CrossRef]
38. Lee, N.H. Effect of Pressurization and Expulsion of Entrapped Air in Pipelines. Ph.D. Thesis, Georgia Institute of Technology, Atlanta, Georgia, 2005.
39. Zhou, L.; Pan, T.; Wang, H.; Liu, D.; Wang, P. Rapid air expulsion through an orifice in a vertical water pipe. *J. Hydraul. Res.* **2019**, *57*, 307–317. [CrossRef]
40. Tran, P.D. Pressure transients caused by air-valve closure while filling pipelines. *J. Hydraul. Eng.* **2017**, *143*, 04016082. [CrossRef]

Disclaimer/Publisher's Note: The statements, opinions and data contained in all publications are solely those of the individual author(s) and contributor(s) and not of MDPI and/or the editor(s). MDPI and/or the editor(s) disclaim responsibility for any injury to people or property resulting from any ideas, methods, instructions or products referred to in the content.

water

MDPI

Article

Dimensionless Pressure Response Analysis for Water Supply Pipeline Systems with or without Pumping Station

Sanghyun Kim

Department of Environmental Engineering, Pusan National University, Busan 46241, Republic of Korea; kimsangh@pusan.ac.kr; Tel.: +82-51-510-2479

Abstract: Dimensionless governing equations of unsteady flow and solutions for the pipeline systems having a surge tank were developed. Frequency domain pressure response expressions for two widely used water supply systems were analyzed in dimensionless frequency and time domains. One is the simple reservoir pipeline surge tank valve system and the other is the pipeline system with pump and check valve protected by surge tank. Two different dimensionless expressions for the surge tank were developed and the performance of two expressions was compared. The frequency response pattern of impedance at the downstream valve indicated that the system resonance was determined by the lengths of the main pipeline and the connector and the locations of the surge tank and check valve for the protection of the pumping station. The difference between the simple pipeline system and that of a pump with a check valve was expressed in terms of the phase difference in frequency response distribution. The integrated pressure response for the protected pipeline section was evaluated considering the impact of surge protection in the frequency domain. A better understanding of system behavior can be obtained in terms of specific component isolation both in the frequency and time domain pressure responses. The driven responses in the frequency and time domain can be an important basis for optimum design and operation conditions of water supply systems in dimensionless space.

Keywords: water hammer; pipelines with surge tank; dimensionless responses; pump with check valve

Citation: Kim, S. Dimensionless Pressure Response Analysis for Water Supply Pipeline Systems with or without Pumping Station. *Water* **2023**, *15*, 2934. https://doi.org/10.3390/w15162934

Academic Editor: Ahmad Shakibaeinia

Received: 21 July 2023
Revised: 10 August 2023
Accepted: 12 August 2023
Published: 14 August 2023

Copyright: © 2023 by the author. Licensee MDPI, Basel, Switzerland. This article is an open access article distributed under the terms and conditions of the Creative Commons Attribution (CC BY) license (https://creativecommons.org/licenses/by/4.0/).

1. Introduction

Water hammers in pipeline systems had been an important engineering problem due to burst, leakage, and water quality issues. Transient events can be generated from sudden valve maneuvers, and the abrupt action of pump and check valve reactions in pipeline systems. Pressure waves can originate from abrupt changes in the flow velocity, which introduce either overpressure or low pressure along the pipeline. Although high pressure can burst a weakened section of the pipeline, low pressure can generate column separation and cavitation, thus severely damaging the pipeline system. To protect pipeline structures from hydraulic transient events, surge protection devices such as surge tanks and air chambers have been widely used in front of control valves and pump stations. Therefore, the evaluation of water hammer events is a critical requirement for the proper design of pipeline surge arrest devices such as the surge tank and air chamber. The impact of surge pressure depends on various factors such as pipeline dimensions, layout, transient introducing conditions, and various hydraulic structures (e.g., valves and pumps). Substantial works have been explored to address many factors to consider each particular system feature.

To analyze surge events in pipeline systems, the characteristic method (MOC) has been [1–4]. Additionally, the size and location of surge protection devices have been determined to relax abrupt pressure variations, while considering the cost of surge protection [5–10]. Many studies showed that the control of water hammers or the design of surge tanks in pipeline systems can be performed based on an MOC basis [11,12].

Pressure wave generation and its propagation along the pipeline system can introduce pressure oscillations, which can be expressed by the surface water variation of the surge tank [13]. The length of the main pipeline and the location, cross-sectional area, and connector length of the surge tank in the pipeline system are important variables for determining the resonance characteristics of the pressure response. The application of the impedance method demonstrates the potential of the frequency domain approach in the context of resonance characterization for the design of hydraulic structures via transient analysis as well as its complementary time domain response [14]. Assuming that the layouts of pipeline systems with a surge tank are simple and similar to each other (e.g., the surge tank valve of a reservoir pipeline), the pressure wave propagation pattern of the pipeline system can be generalized via the dimensionless development of governing equations and its optimum solution in the time domain response [15]. Actually, the dimensionless approach can substantially generalize the characteristics of the pressure response pattern if the layout of the pipeline system is identical. In other words, if the general response pattern in dimensionless space is determined, then simple conversion into real dimension provides numerous representations of analysis as long as the structure of the pipeline system is identical.

Time domain responses of dimensionless solutions provided a general guideline for the design of surge tanks. However, a dimensionless analysis based on the pressure response must be performed in the frequency domain to configure the interaction of each pipeline segment and component, which allows one to comprehensively understand the system behavior and to conduct better management for various transient scenarios. Furthermore, a more general layout of pipeline systems, such as a pipeline system with pumping stations and check valves, must be developed and compared between systems with/without a specific component to provide a holistic evaluation of the pressure response in regard to impact identification for specific events. Actually, computational fluid dynamic analyses have been performed to improve the design of centrifugal pumps using multiple objective optimizations [16,17].

This study can be categorized into the following two aspects:

First, dimensionless transient governing equations for two widely adapted water-supply systems are developed: a reservoir pipeline surge tank valve reservoir system and a reservoir pump check valve pipeline surge tank valve reservoir system. Both a standard dimensionless expression for the surge tank and a simplified dimensionless lumped inertia are introduced, and solutions for combinations of various structures are developed.

Second, both frequency and time domain solutions for the two distinct systems are presented and compared. The effect of the surge tank on a specific section of the pipeline is investigated via the introduction of an integrated solution, and a comparison between different systems is performed to provide insights into the system response in terms of the contribution of isolation in the pressure response for a specific component.

Section 2 covered dimensionless governing equations and their solutions for two different water supply systems with surge tanks in the frequency domain. Section 3 provided frequency response functions and time domain responses for two systems in terms of point and integrated hydraulic impedance, respectively. Strengths and differences over other existing methods were discussed in Section 4. The summary of this study and future issues from this study were addressed in Section 5.

2. Materials and Methods

2.1. One-Dimensional Dimensionless Governing Equation with Steady Friction

The pressure head and flow rate variations in a pressurized pipeline system can be expressed using the partial differential equation (PDE) of continuity and momentum conservation as a function of two independent variables, i.e., time (t) and distance (x) [1]. Based on these PDFs, the dimensionless continuity and momentum equations can be developed as follows:

$$\frac{\partial V}{\partial t} + g\frac{\partial H}{\partial x} + \frac{fV|V|}{2DA^2} = 0 \tag{1}$$

$$\frac{a^2}{g}\frac{\partial V}{\partial x}+\frac{\partial H}{\partial t}=0, \tag{2}$$

where V is the mean velocity for the cross-sectional area A, H the piezometric head, a the wave propagation speed, g the gravitational acceleration, f the Darcy–Weisbach friction factor, and D the diameter.

This section may be divided into subheadings. It should provide a concise and precise description of the experimental results, their interpretation, as well as the experimental conclusions that can be drawn. By introducing the mean flow rate Q, which represents the mean velocity by multiplying the cross-sectional area, the dimensionless PDEs of momentum and mass conservation can be developed. The dimensionless independent variables for time and distance are defined as $\widehat{t}=at/L$ and $\widehat{x}=x/L$, respectively, where L denotes the length of the pipeline system. The dependent dimensionless variables are $\widehat{H}=gAH/(4aQ_o)$ for the pressure head and $\widehat{Q}=Q/Q_0$ for the flow rate, where Q_0 is the steady flow rate.

The dimensionless continuity and momentum equations can be derived as follows:

$$\frac{\partial\widehat{Q}}{\partial\widehat{t}}+\frac{\partial\widehat{H}}{\partial\widehat{x}}+\widehat{R}=0 \tag{3}$$

$$\frac{\partial\widehat{H}}{\partial\widehat{t}}+\frac{\partial\widehat{Q}}{\partial\widehat{x}}=0, \tag{4}$$

where $\widehat{R}$ is the dimensionless resistance, which can be estimated as $\widehat{R}=fLQ_o/(2DAa)$ under steady friction.

Applying the perturbation theory [1] to the dimensionless pressure head and flow rate expressed in Equations (3) and (4) yields the trigonometric relationship between the upstream and downstream dimensionless frequency ($\widehat{s}$) as follows:

$$\widehat{H_D}=\widehat{H_U}\cosh\widehat{\gamma}\widehat{x}-\widehat{Z_c}\widehat{Q_U}\sinh\widehat{\gamma}\widehat{x} \tag{5}$$

$$\widehat{Q_D}=-\frac{\widehat{H_U}}{\widehat{Z_c}}\sinh\widehat{\gamma}\widehat{x}+\widehat{Q_U}\cosh\widehat{\gamma}\widehat{x}, \tag{6}$$

where the dimensionless propagation constant, $\widehat{\gamma}$, can be expressed as

$$\widehat{\gamma}=\sqrt{\widehat{s}\left(\widehat{s}+\widehat{R}\right)}, \tag{7}$$

where $\widehat{s}$ is the dimensionless frequency.

2.2. Dimensionless Hydraulic Impedance from Surge Tank to Joining Point

The two-dimensional PDEs for the continuity and momentum in terms of time (t), axial distance (x), and radial distance (r) are expressed as follows:

$$\frac{\partial u}{\partial x}+\frac{g}{a^2}\frac{\partial h}{\partial t}=0 \tag{8}$$

$$\frac{\partial u}{\partial t}+g\frac{\partial h}{\partial x}+\frac{v}{r}\frac{\partial}{\partial r}\left(r\frac{\partial u}{\partial r}\right)=0 \tag{9}$$

where u and v are the axial velocity and dynamic viscosity, respectively.

By introducing dimensionless variables $\widehat{x} = x/L; \widehat{t} = at/L; \widehat{r} = r/R; \widehat{h} = gh/a^2$; and $\widehat{u} = u/a$, two-dimensional dimensionless equations for mass and momentum conservation can be derived as follows:

$$\frac{\partial \widehat{h}}{\partial \widehat{t}} + \frac{\partial \widehat{u}}{\partial \widehat{x}} = 0 \tag{10}$$

$$\frac{\partial^2 \widehat{u}}{\partial \widehat{r}^2} + \frac{1}{\widehat{r}}\frac{\partial \widehat{u}}{\partial \widehat{r}} - \frac{1}{S_a}\frac{\partial \widehat{u}}{\partial \widehat{t}} = \frac{1}{S_a}\frac{\partial \widehat{h}}{\partial \widehat{x}} \tag{11}$$

where $S_a = \nu L/aR^2$ is a dimensionless surge number. Based on the assumption of zero heat radiation, the integration of Equations (10) and (11) provides unsteady functions in the dimensionless frequency domain. The dimensionless pressure head and flow rate relationships between the upstream and downstream for Equations (10) and (11) can be expressed as follows:

$$\widehat{H_D}(\widehat{s}) = \widehat{H_U}(\widehat{s})\cosh\Gamma(\widehat{s}) - \widehat{Q_U}(\widehat{s})Z_c(\widehat{s})\sinh\Gamma(\widehat{s}) \tag{12}$$

$$\widehat{Q_D}(\widehat{s}) = -\frac{\widehat{H_U}(\widehat{s})}{Z_c(\widehat{s})}\sinh\Gamma(\widehat{s}) + \widehat{Q_U}(\widehat{s})\cosh\Gamma(\widehat{s}) \tag{13}$$

where $\widehat{H_D}(\widehat{s})$ and $\widehat{H_U}(\widehat{s})$ denote dimensionless complex heads; and $\widehat{Q_D}(\widehat{s})$ and $\widehat{Q_U}(\widehat{s})$ are dimensionless complex discharges for the downstream and upstream points, respectively.

The dimensionless characteristic impedance can be expressed as follows,

$$Z_c(\widehat{s}) = \sqrt{\frac{J_0(\sqrt{(\widehat{S}/S_a)}i)}{J_0(\sqrt{(\widehat{S}/S_a)}i) - 2/(\sqrt{(\widehat{S}/S_a)}i)J_1(\sqrt{(\widehat{S}/S_a)}i)}} \tag{14}$$

where J_0 and J_1 are Bessel functions of 0th and 1st order, respectively. The dimensionless propagation constant can be expressed as follows:

$$\Gamma(\widehat{s}) = \widehat{x}\widehat{s}\sqrt{\frac{J_0(\sqrt{(\widehat{S}/S_a)}i)}{J_0(\sqrt{(\widehat{S}/S_a)}i) - 2/(\sqrt{(\widehat{S}/S_a)}i)J_1(\sqrt{(\widehat{S}/S_a)}i)}} \tag{15}$$

2.3. Dimensionless Hydraulic Impedance from Surge Tank to Joining Point

The fluctuation in the flow rate from the surge tank to the main pipeline can be expressed as $q' = -A_S dh'/dt$, where A_S is the surge tank area. The disturbance of the pressure head (h') can be approximated as $h' = H_S e^{st}$, where H_S is the steady pressure head in the surge tank. By defining the dimensionless fluctuations of pressure head and flow rate as $\widehat{h'} = gAh'/(aQ_o)$ and $\widehat{q'} = q'/Q_o$, respectively.

The dimensionless pressure head variation at the surge tank entrance can be expressed as

$$\widehat{h'} = \widehat{H_S}e^{\widehat{ist}}, \tag{16}$$

where $\widehat{H_S} = \frac{gAH_S}{aQ_o}$.

The dimensionless flow rate fluctuation at the surge tank entrance can be expressed as

$$\widehat{q'} = \widehat{Q_S}e^{i\widehat{s}\widehat{t}}, \tag{17}$$

where $\widehat{Q_S} = -\frac{A_S a H_S}{LQ_o}i\widehat{s}$.

The hydraulic impedance at the surge tank outlet can be expressed as

$$\widehat{Z_S} = \frac{gAL}{A_S a^2 \widehat{s}} \tag{18}$$

If the surge tank is connected to the main pipeline by a connector, then the hydraulic impedance at the joining point of the main pipeline can be expressed as

$$\widehat{Z_J} = \widehat{H_j}(\widehat{s}) / \widehat{Q_J}(\widehat{s}) \tag{19}$$

The dimensionless pressure head and flow rate at the junction can be expressed, respectively, as follows:

$$\widehat{H_j}(\widehat{s}) = \widehat{H_S}(\widehat{s})\cosh\Gamma_c(\widehat{s}) - \widehat{Q_S}(\widehat{s})Z_{cc}(\widehat{s})\sinh\Gamma_c(\widehat{s}) \tag{20}$$

$$\widehat{Q_j}(\widehat{s}) = -\frac{\widehat{H_S}(\widehat{s})}{Z_{cc}(\widehat{s})}\sinh\Gamma_c(\widehat{s}) + \widehat{Q_S}(\widehat{s})\cosh\Gamma_c(\widehat{s}), \tag{21}$$

where $Z_{cc}(\widehat{s}) = \sqrt{\frac{J_0(\sqrt{(\widehat{S}/S_{ac})}i)}{J_0(\sqrt{(\widehat{S}/S_{ac})}i) - 2/(\sqrt{(\widehat{S}/S_{ac})}i)J_1(\sqrt{(\widehat{S}/S_{ac})}i)}}$, $S_{ac} = \nu L_c/aR_c{}^2$; L_c and R_c are the length and radius of the connector, respectively; and $\Gamma_c(\widehat{s}) = \widehat{x}\widehat{s}\sqrt{\frac{J_0(\sqrt{(\widehat{S}/S_{ac})}i)}{J_0(\sqrt{(\widehat{S}/S_{ac})}i) - 2/(\sqrt{(\widehat{S}/S_{ac})}i)J_1(\sqrt{(\widehat{S}/S_{ac})}i)}}$.

2.4. Dimensionless Lumped Inertia

Expression for accumulator in frequency domain had been approximated using the lumped inertia [1], which can be applied to surge tanks with short connectors. In this study, the existing lumped inertia expressions were extended into the dimensionless lumped inertia. The dimensionless lumped inertia model can widely address the effect of the surge tank. By introducing dimensionless variables for the lumped inertia, i.e., $\widehat{H} = \frac{gA}{aQ_o}H$, $\widehat{Q} = Q/Q_o$, $\widehat{l_c} = l_c/L$, the dimensionless relationship can be expressed as follows:

$$\widehat{H_D} = \widehat{H_U} - \widehat{l_c}i\widehat{s}\widehat{Q_U} \tag{22}$$

Assuming that the dimensionless upstream flow rate is identical to that of the downstream flow for a short connector, the following dimensionless relationship can be derived:

$$\widehat{Z_D} = \widehat{Z_U} - \widehat{l_c}i\widehat{s} \tag{23}$$

where $\widehat{Z_D}$ and $\widehat{Z_U}$ are the dimensionless hydraulic impedances downstream and upstream of the surge tank connector, respectively.

2.5. Development of Dimensionless Hydraulic Impedance for Two Different Systems

A surge tank is typically installed in a simple pipeline system, as shown in Figure 1. A hydraulic transient can be introduced through the abrupt closure of the downstream valve or the abrupt termination of the upstream pump. This can cause severe damage owing to either the overpressure or underpressure, which occasionally introduces cavitation.

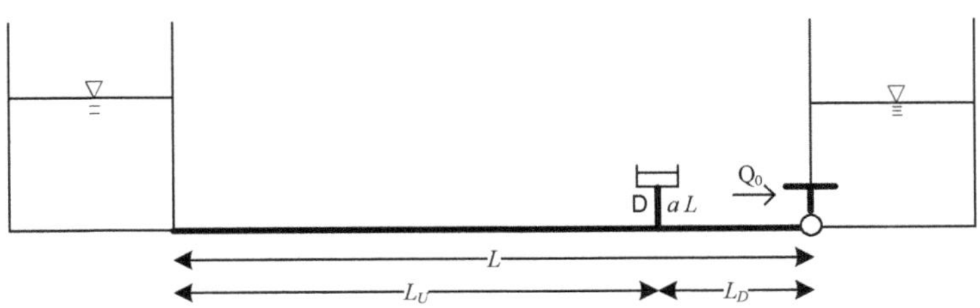

Figure 1. Schematic illustration of reservoir pipeline surge tank pipeline valve (R-P-ST-P-V) system.

The other widely used systems comprise a pump and check valve in the upstream section (see Figure 2), which allow the upstream pressure head to be increased when the pressure head of the upstream reservoir is low, or when the distribution of the pipeline elevation requires an elevating additional pressure head to secure flow conveyance downstream.

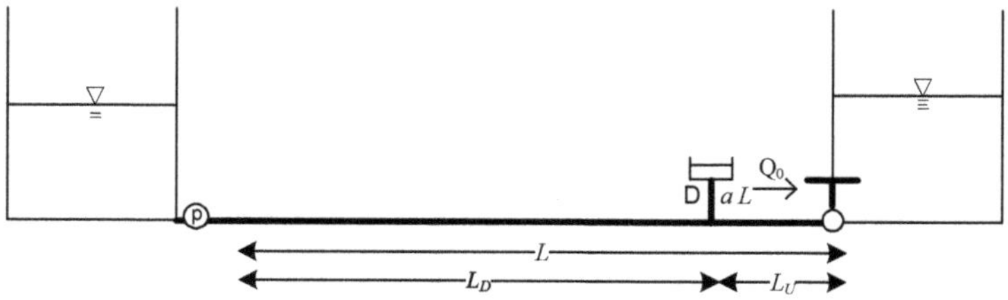

Figure 2. Schematic illustration of reservoir pump check valve pipeline surge tank pipeline valve (R-PP-CV-P-ST-P-V) system.

To generalize the system response feature in dimensionless space, the upstream length between the upstream reservoir and surge tank, L_U, can be converted into an upstream dimensionless length as $\widehat{L_U} = L_U/L$, and the dimensionless downstream length can be defined as $\widehat{L_D} = L_D/L$. If the driver of the hydraulic transient is the downstream valve in the pipeline system shown in Figure 1, then a dimensionless impedance from the upstream to downstream direction should be derived. The dimensionless hydraulic impedance upstream of the joining point is expressed as follows:

$$\widehat{Z}_{UJ} = -\widehat{Z_c}\tanh\widehat{\gamma}\widehat{L_U}, \tag{24}$$

where $\widehat{Z_c} = \widehat{\gamma}/\widehat{s}$ and $\widehat{\gamma} = \sqrt{\widehat{s}(\widehat{s}+\widehat{R})}$.

If the transient pressure head from downstream introduces a reversed flow in the upstream direction, then the check valve at the outlet of the pump can be instantly closed to protect the pump, and the system shown in Figure 2 can be changed in terms of the primary hydraulic impedance from the upstream boundary which can be expressed as follows:

$$\widehat{Z}_{UJ} = -\widehat{Z_c}\coth\widehat{\gamma}\widehat{L_U} \tag{25}$$

The dimensionless hydraulic impedance of the main pipeline downstream of the connecting element for the system shown in Figure 1 can be expressed as follows:

$$\widehat{Z_{DJ}} = \frac{-\widehat{Z_c}\tanh\widehat{\gamma}\widehat{L_U}}{1-\widehat{Z_c}\tanh\widehat{\gamma}\widehat{L_U}/\widehat{Z_J}} \tag{26}$$

The corresponding dimensionless hydraulic impedance downstream of the connecting element in Figure 2 can be expressed as follows:

$$\widehat{Z_{DJ}} = \frac{-\widehat{Z_c}\coth\widehat{\gamma}\widehat{L_U}}{1-\widehat{Z_c}\coth\widehat{\gamma}\widehat{L_U}/\widehat{Z_J}} \tag{27}$$

The dimensionless hydraulic impedance at the downstream valve can be expressed as follows:

$$\widehat{Z_{DV}} = \frac{\widehat{Z_{DJ}} - \widehat{Z_c}\tanh\widehat{\gamma}\widehat{L_D}}{1-\widehat{Z_{DJ}}/\widehat{Z_c}\tanh\widehat{\gamma}\widehat{L_D}} \tag{28}$$

The dimensionless pressure head response from the downstream flow rate variation, such as the valve maneuver between the downstream valve and the connecting point for the surge tank, can be derived as follows:

$$\frac{\widehat{H_{xD}}}{\widehat{Q_{DV}}} = \widehat{Z_{DV}}\cosh\widehat{\gamma}\widehat{x_D} + \widehat{Z_c}\sinh\widehat{\gamma}\widehat{x_D}, \tag{29}$$

where $\widehat{x_D}$ is the dimensionless distance from the downstream valve to an upstream point up to the connecting point.

The dimensionless pressure head response between the connecting point and upstream reservoir can be derived as follows:

$$\frac{\widehat{H_{xU}}}{\widehat{Q_{DV}}} = \widehat{Z_{Uc}}\cosh\widehat{\gamma}\widehat{x_U} + \widehat{Z_c}\sinh\widehat{\gamma}\widehat{x_U}, \quad (30)$$

where $\widehat{x_U}$ is the dimensionless distance from the connecting point to a point up to the upstream reservoir. The dimensionless hydraulic impedance upstream of the connection point can be estimated as

$$\widehat{Z_{Uc}} = \frac{\widehat{Z_J} - \widehat{Z_{Dc}}}{\widehat{Z_{Dc}}\widehat{Z_J}}, \quad (31)$$

The dimensionless hydraulic impedance downstream of the connecting point can be estimated as

$$\widehat{Z_{Dc}} = \frac{\widehat{Z_{DV}} + \widehat{Z_c}\tanh\widehat{\gamma}\widehat{L_D}}{1 + \widehat{Z_{DV}}/\widehat{Z_c}\tanh\widehat{\gamma}\widehat{L_D}} \quad (32)$$

If the hydraulic transient is introduced from the pump stoppage of upstream reservoir, then the derivation should be performed from the reservoir downstream to the pump upstream, and the dimensionless hydraulic impedance at the pump can be expressed as

$$\widehat{Z_p} = \frac{\widehat{Z_{PU}} + \widehat{Z_c}\tanh\widehat{\gamma}\widehat{L_U}}{1 + \widehat{Z_{PU}}/\widehat{Z_c}\tanh\widehat{\gamma}\widehat{L_U}}, \quad (33)$$

where $\widehat{Z_{PU}} = (\widehat{Z_J}\widehat{Z_c}\tanh\widehat{\gamma}\widehat{L_D})/(\widehat{Z_J} - \widehat{Z_c}\tanh\widehat{\gamma}\widehat{L_D})$. The dimensionless pressure head response from the abrupt pressure change in the pump between the pump and connecting point can be derived as follows:

$$\frac{\widehat{H_{xu}}}{\widehat{H_p}} = \cosh\widehat{\gamma}\widehat{x_u} - \frac{\widehat{Z_c}}{\widehat{Z_p}}\sinh\widehat{\gamma}\widehat{x_u}, \quad (34)$$

where $\widehat{x_u}$ is the dimensionless distance from the upstream to downstream connecting point. The dimensionless pressure head response from the sudden pressure change in the pump between the connecting point and downstream valve can be evaluated as follows:

$$\frac{\widehat{H_{xd}}}{\widehat{H_p}} = \cosh\widehat{\gamma}\widehat{x_d} - \frac{\widehat{Z_c}}{\widehat{Z_{cd}}}\sinh\widehat{\gamma}\widehat{x_d}, \quad (35)$$

where $\widehat{x_d}$ is the dimensionless distance from the connecting point downstream, $\widehat{Z_{cd}} = \widehat{Z_{cu}}\widehat{Z_J}/(\widehat{Z_{cu}} + \widehat{Z_J})$, and the dimensionless hydraulic impedance upstream of the connection point can be expressed as

$$\widehat{Z_{cu}} = \frac{\cosh\widehat{\gamma}\widehat{L_U} - \widehat{Z_c}/\widehat{Z_p}\sinh\widehat{\gamma}\widehat{L_U}}{-\sinh\widehat{\gamma\widehat{L_U}/\widehat{Z_c}} + \cosh\widehat{\gamma}\widehat{L_U}/\widehat{Z_p}} \quad (36)$$

The dimensionless impedance approach provides an integrated pressure response along a specific pipeline section. The total pressure response between the connecting point and the downstream valve owing to valve closure can be expressed as follows:

$$\int_0^{\widehat{L_D}} \widehat{H_{xD}}/\widehat{Q_{DV}}d\widehat{x} = \frac{\widehat{Z_D}}{\widehat{\gamma}}\sinh\widehat{\gamma}\widehat{L_D} + \frac{\widehat{Z_c}}{\widehat{\gamma}}(\cosh\widehat{\gamma}\widehat{L_D} - 1) \quad (37)$$

The total pressure response between the upstream pump and connecting point due to valve closure can be expressed as follows:

$$\int_0^{\widehat{L_U}} \widehat{H_{xU}}/\widehat{Q_{DV}} d\widehat{x} = \frac{\widehat{Z_U}}{\widehat{\gamma}} \sinh \widehat{\gamma} \widehat{L_U} + \frac{\widehat{Z_c}}{\widehat{\gamma}} (\cosh \widehat{\gamma} \widehat{L_U} - 1) \tag{38}$$

The total pressure response between the upstream pump and connecting point due to pump termination can be expressed as follows:

$$\int_0^{\widehat{L_U}} \widehat{H_{xu}}/\widehat{H_p} d\widehat{x} = \frac{1}{\widehat{\gamma}} \sinh \widehat{\gamma} \widehat{L_U} - \frac{\widehat{Z_c}}{\widehat{Z_p} \widehat{\gamma}} (\cosh \widehat{\gamma} \widehat{L_U} - 1) \tag{39}$$

The total pressure response between the connecting point and downstream valve owing to pump termination can be expressed as follows:

$$\int_0^{\widehat{L_D}} \widehat{H_{xd}}/\widehat{H_p} d\widehat{x} = \frac{1}{\widehat{\gamma}} \sinh \widehat{\gamma} \widehat{L_D} - \frac{\widehat{Z_c}}{\widehat{Z_{cd}} \widehat{\gamma}} (\cosh \widehat{\gamma} \widehat{L_D} - 1) \tag{40}$$

3. Results

In this study, the pipeline system used in a previous study was used to apply the proposed approach, as the widely used time domain modeling method (e.g., the method of characteristics) has been compared with the developed methods in multiple cases [16]. The total length of the pipeline was 150 m, and the diameters of the main and connected pipelines were 0.02 m. The pipeline lengths upstream and downstream from the surge tank were 145 and 5 m, respectively. The wave propagation speed in the pipeline was estimated at 1210.5 m/s The steady flow rate from the upstream reservoir or pump was 0.928×10^{-4} m^3/s. The diameter of the surge tank was assumed to be 2 m, and the length of the connector was 0.5 m. The abrupt valve closure and termination of the check valve might introduce hydraulic transients. The maximum frequency for frequency domain modeling was terminated in 3812 rad/s and the number of fast Fourier transform for the conversion of the frequency domain to time domain response function is 32,768.

3.1. Frequency Response Function

The amplitudes of the dimensionless hydraulic impedances at the downstream valves of the systems shown in Figures 1 and 2 are presented in Figure 3. The solid line in Figure 3, denoted as R-P-ST-P-V, is the dimensionless frequency response function of the hydraulic impedance at the downstream valve in Figure 1. The dotted line in Figure 3 represents the corresponding frequency response of the system (R-PP-CV-P-ST-P-V), as shown in Figure 2. The overall shapes of the frequency response functions of the two systems were similar, showing positive peak responses at 43 and 132 and negative peaks at 87 and 178 in the dimensionless frequency domain. The difference in the primary resonance response from the lower frequency to the higher frequency increased from 43 to 46 for the four peak responses. The resonance pattern of the dimensionless hydraulic impedance can vary depending on the location of the surge tank. The difference in the frequency response between R-P-ST-P-V and R-PP-CV-P-ST-P-V was the phase shift, which can be explained by the difference in the transient generation positions between the two systems. Whereas the hydraulic transient for R-P-ST-P-V was generated from the downstream valve, that for R-PP-CV-P-ST-P-V was initiated from the upstream check valve.

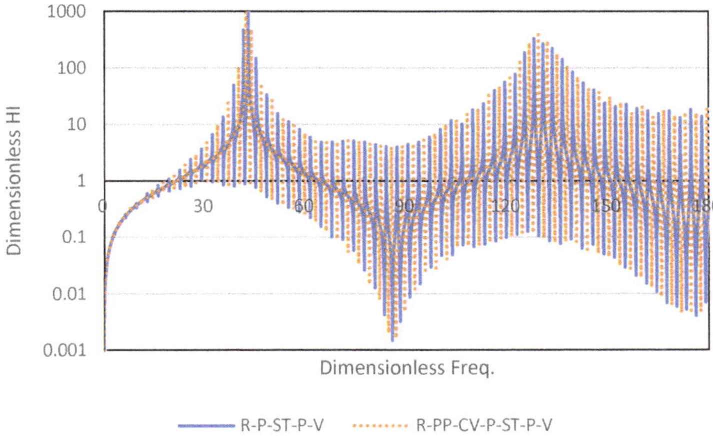

Figure 3. Amplitudes of dimensionless hydraulic impedance at downstream valve for reservoir pipeline surge tank pipeline-valve system (R-P-ST-P-V) in Figure 1 and that of reservoir pump check valve pipeline surge tank pipeline valve system (R-PP-CV-P-ST-P-V) shown in Figure 2.

Next, the frequency responses of the dimensionless expressions for the surge tank were analyzed. Figure 4 shows the amplitudes of the dimensionless hydraulic impedances at the downstream valve for the dimensionless surge tank expression (DLST) and dimensionless lumped inertia (DL LI) for the R-PP-CV-P-ST-P-V system. Even though slight differences in the amplitudes between the two distinct approaches were observed at several points, the distributions of the frequency responses were consistent with the development of the resonance pattern. This indicates that the dimensionless lumped inertia can be a reasonable approximation when the connector is short, which is applicable to most systems.

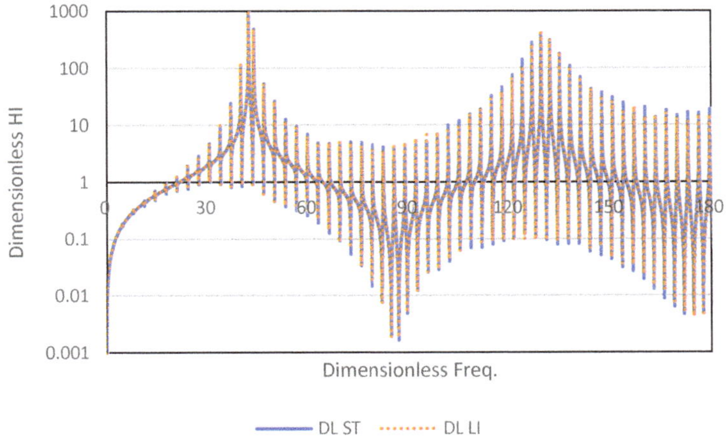

Figure 4. Amplitudes of dimensionless hydraulic impedance at downstream valve for reservoir pump check valve pipeline surge tank pipeline valve system (R-PP-CV-P-ST-P-V) using dimensionless surge tank (DLST) and dimensionless lumped inertia (DLLI).

The dimensionless frequency domain approach allows flexible analytical developments, such as obtaining integrated responses along any designated section of the pipeline system. This approach is particularly useful when the pipeline manager is required to estimate the potential pressure effect for a specific section and several designated points. Figure 5 shows the integrated dimensionless hydraulic impedances in the upstream section of the surge tank, which were protected by the surge tank from the generated hydraulic transient originating from the downstream valve. The distributions of the integrated frequency responses of the two systems were similar. The mitigated frequency responses compared with those in Figure 3 indicate that both systems were well protected by the surge tank. The phase discrepancy in the integrated frequency response between the two systems can be explained by the difference in the pressure wave reflections from the distinct upstream boundaries.

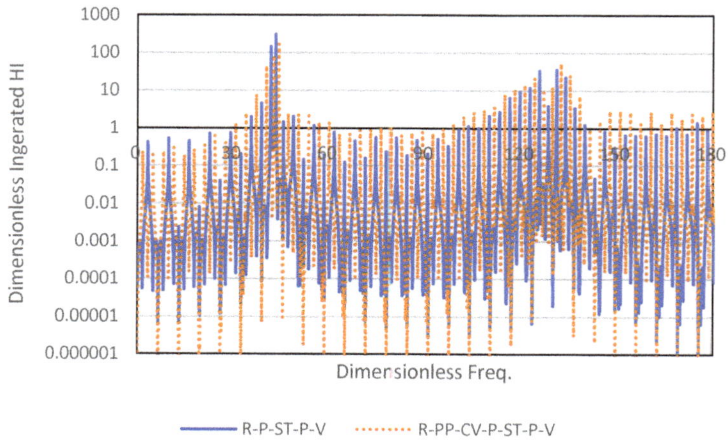

Figure 5. Amplitudes of integrated dimensionless hydraulic impedance along upstream section of surge tank for reservoir pipeline surge tank pipeline valve system (R-P-ST-P-V) in Figure 1 and that of reservoir pump check valve pipeline surge tank pipeline valve system (R-PP-CV-P-ST-P-V) shown in Figure 2.

3.2. *Time Domain Pressure Response*

The dimensionless frequency responses of the hydraulic impedance can be used to separate the corresponding time domain responses. Assuming that an abrupt flow rate change is introduced from the downstream valve via an instant valve closure, the flow rate variation can be convolved into the time domain response functions shown in Figures 3–5. Figure 6 shows the pressure responses at the downstream valve for R-P-ST-P-V and R-PP-CV-ST-P-V at the instant of the downstream valve closure. Figure 6 shows two distinct patterns in the pressure responses: a high-frequency component pattern between the downstream valve and surge tank, and a lower-frequency response pattern for all systems, which corresponds to the theoretical period of the pipeline system ($4L/a$). Both pressure responses showed perfectly matching results up to a dimensionless time of 2, which corresponded to a round travel time between different upstream boundaries and the downstream valve. The difference in the upstream boundary conditions between the two systems may have contributed to the phase difference in the pressure response from the dimensionless time 2, as shown in Figure 6.

As illustrated in Figure 4, the integrated pressure responses allow one to comprehensively evaluate the pipeline section. Figure 7 shows the normalized integrated pressure response for the upstream section of the system, as similarly shown in Figures 1 and 2.

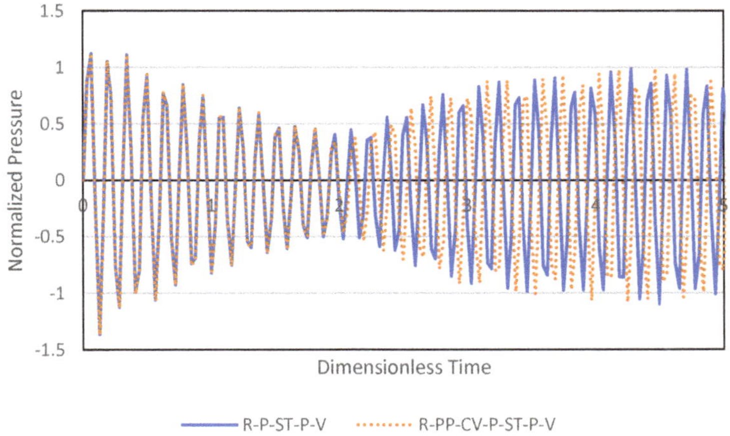

Figure 6. Normalized pressure responses at downstream valve due to instant valve closure of reservoir pipeline surge tank pipeline valve system (R-P-ST-P-V) in Figure 1 and that of reservoir pump check valve pipeline surge tank pipeline valve system (R-PP-CV-P-ST-P-V) in Figure 2.

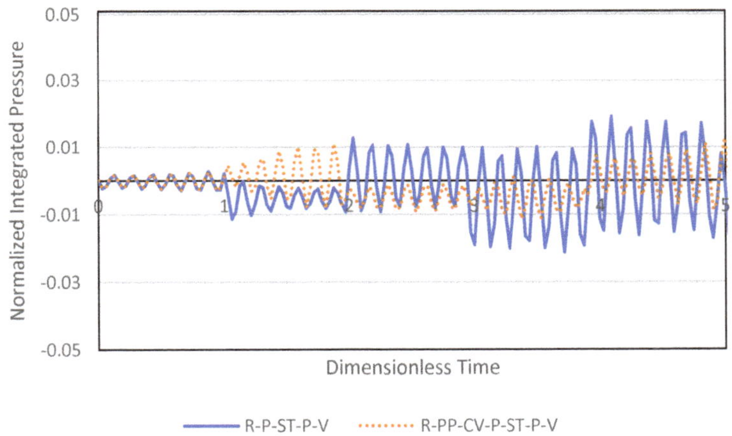

Figure 7. Normalized integrated pressure responses for upstream section of surge tank due to instant valve closure for reservoir pipeline surge tank pipeline valve system (R-P-ST-P-V) in Figure 1 and that of reservoir pump check valve pipeline surge tank pipeline valve system (R-PP-CV-P-ST-P-V) in Figure 2.

The scales of the integrated pressure response were substantially smaller than those of the pressure at the downstream valve, thereby indicating a significant surge-arresting effect of the surge tank. The integrated pressure responses for R-P-ST-P-V and R-PP-CV-P-ST-P-V matched perfectly with each other up to a dimensionless time of 1, which corresponded to the wave travel time from the downstream valve to the upstream boundary. As shown in Figure 7, the difference in the integrated responses was initiated from dimensionless time 1, and the pattern of R-P-ST-P-V reflected the mitigation from the upstream reservoir, whereas that of R-PP-CV-P-ST-P-V was doubled increasingly between dimensionless times 1 and 2 owing to the dead-end boundary from the check valve. The development of integrated pressure response patterns was evident at each dimensionless time interval, thus indicating that the dimensionless time defined in this study is a useful measure for characterizing pressure response patterns based on the boundary conditions and operational practices for pipeline systems.

4. Discussion

The solutions of a partial differential equation (PDE) for water hammer events have been explored through the conversion of PDE into an ordinary differential equation, namely MOC, in most applications. However, discretization approaches (e.g., MOC) have a critical restriction in representing the system. The requirement of Courant number related to wave speed and time–space interval in modeling introduces substantial computation costs and accurate representation of system dimensions. The location of the surge tank, check valve, pump and the dimensions of the surge tank should be fitted into the Courant number condition, which needs either assumptions or substantial computational costs. However, the introduction of a general solution in the frequency domain not only solved many existing issues in transient analysis but also provided the resonance response feature. The solutions of a hydraulic transient in the dimensionless space extend the applicability of the proposed method in general purpose if the system layout is identical. Furthermore, the evaluation of hydraulic impedance in dimensionless frequency provides a comprehensive understanding of system behavior as well as a universal solution in frequency response. As the dimensionless variables of system features, such as the location of the surge tank and dimensions, are changed, the frequency responses (Figures 3 and 5) provide an intuitive understanding of resonance response between pipeline and hydraulic structures. One additional strength of the frequency domain approach is the integrated expressions (Equations (37)–(40)). The direct integration in the frequency domain provides the evaluations of pressure or flow rate along a specific section. This means that the objective function for the optimization of hydraulic structure can be formulated not only from a point evaluation perspective but also through spatial integration, which can be useful for surge protection for vulnerable sections of the pipeline system.

5. Conclusions

Frequency domain responses due to the surge tank in pipeline systems with and without pumping stations and check valves were investigated. The frequency response of the hydraulic impedance at the downstream valve indicated that the resonance of the system can be explained by the pipeline length and the locations of the pipeline structures, such as the surge tanks, pumps, and check valves. The dimensionless approach provides response features in dimensionless time and amplitude, thus providing an intuitive explanation for the travel time, phase, and magnitude of the pressure response with the referenced scale. The difference in the frequency domain response was consistent with that in the time domain response, which depends on the system layout. The integrated response of the pressure along the pipeline segment allows one to comprehensively evaluate the pressure at a designated section. The difference in the time domain response between the two systems indicates that the amplification and mitigation originated from the boundary conditions with interaction with the surge tank. The dimensionless time and hydraulic impedance provide an intuitive understanding of the system response in the context of system dimensions, both in time and space. The characterization of the system behavior can be obtained through the evaluation of isolated hydraulic components in a dimensionless space. Future studies can be conducted to address the dynamic features of pumps and nonlinear valve behaviors. Additionally, the optimization of specific external disturbances (such as surges) can be considered in the design of surge protection devices.

Funding: This research was funded by the National Research Foundation of the Republic of Korea, grant number 2022R1A4A5028840.

Data Availability Statement: The data that support the findings of this study are available on request from the corresponding author, S. Kim.

Conflicts of Interest: The author declares no conflict of interest.

References

1. Wylie, E.B.; Streeter, V.L. *Fluid Transients in Systems*, 3rd ed.; Prentice-Hall International: London, UK, 1993; pp. 37–79.
2. Karney, B.W.; Simpson, A.R. In-line check valves for water hammer control. *J. Hydraul. Res.* **2007**, *45*, 547–554. [CrossRef]
3. Wan, W.; Zhang, B. Investigation of Water Hammer Protection in Water Supply Pipeline Systems Using an Intelligent Self-Controlled Surge Tank. *Energies* **2018**, *11*, 1450. [CrossRef]
4. Triki, A. Further invetogation on water-hammer control line strategy in water-supply systems. *J. Water Sup.* **2018**, *67*, jws2017073. [CrossRef]
5. Di Santo, A.R.; Fratino, U.; Iacobellis, U.V.; Piccinni, A.F. Effects of free outflow in rising mains with air chamber. *J. Hydraul. Eng.* **2002**, *128*, 992–1001. [CrossRef]
6. Jung, B.S.; Karney, B.W. Systematic Surge Protection for Worst-Case Transient Loadings in Water Distribution Systems. *J. Hydraul. Eng.* **2009**, *135*, 218–223. [CrossRef]
7. Duan, H.F.; Tung, Y.K.; Ghidaoui, M.S. Probabilistic Analysis of Transient Design for Water Supply Systems. *J. Water Resour. Plan. Manag. -ASCE* **2010**, *136*, 678–687. [CrossRef]
8. Martino, G.D.; Fontana, N. Simplified approach for the optimal sizing of throttled air chambers. *J. Hydraul. Eng.* **2012**, *138*, 1101–1109 [CrossRef]
9. Skulovich, O.; Bent, R.; Judi, D.; Perelman, L.S.; Ostfeld, A. Piece-wise mixed integer programming for optimal sizing of surge control devices in water distribution systems. *Water Resour. Res.* **2015**, *51*, 4391–4408. [CrossRef]
10. Bhattarai, K.P.; Zhou, J.X.; Palikhe, S.; Pandey, K.P.; Suwal, N. Numerical Modeling and Hydraulic Optimization of a Surge Tank Using Particle Swarm Optimization. *Water* **2019**, *11*, 715. [CrossRef]
11. Mahmoudi-Rad, M.; Najafzadeh, M. Effects of Surge Tank Geometry on the Water Hammer Phenomenon: Numerical Investigation. *Sustainability* **2023**, *15*, 2312. [CrossRef]
12. Wan, W.; Wang, Y.; Chen, X.; Zhan, H.; Wang, T.; Zhang, B. Investigation of partially expanded surge tank with self-adaptive auxiliary system controlling waterhammer in pipelines. *Eng. Sci. Technol. Int. J.* **2023**, *40*, 101379. [CrossRef]
13. Guo, J.; Woldeyesus, K.; Zhang, J.; Ju, X. Time evolution of water surface oscillations in surge tanks. *J. Hydraul. Res.* **2017**, *55*, 657–667. [CrossRef]
14. Kim, S.H. Design of surge tank for water supply systems using the impulse response method with the GA algorithm. *J. Mech. Sci. Technol.* **2010**, *24*, 629–636. [CrossRef]
15. Kim, S.H.; Choi, D. Dimensionless impedance method for general design of surge tank in simple pipeline systems. *Energies* **2022**, *15*, 3603. [CrossRef]
16. Wang, C.-N.; Yang, F.-C.; Nguyen, V.T.T.; Vo, N.T.M. CFD Analysis and Optimum Design for a Centrifugal Pump Using an Effectively Artificial Intelligent Algorithm. *Micromachines* **2022**, *13*, 1208. [CrossRef] [PubMed]
17. Nguyen, V.T.T.; Vo, N.T.M. Centrifugal Pump Design: An Optimization. *Eurasia Proc. Sci. Technol. Eng. Math.* **2022**, *17*, 136–151. [CrossRef]

Disclaimer/Publisher's Note: The statements, opinions and data contained in all publications are solely those of the individual author(s) and contributor(s) and not of MDPI and/or the editor(s). MDPI and/or the editor(s) disclaim responsibility for any injury to people or property resulting from any ideas, methods, instructions or products referred to in the content.

water

MDPI

Article

Finite Volume Method for Transient Pipe Flow with an Air Cushion Surge Chamber Considering Unsteady Friction and Experimental Validation

Yue Liu [1], Jianwei Lu [2], Jian Chen [3], Yong Xia [1], Daohua Liu [1], Yinying Hu [4], Ruilin Feng [4], Deyou Liu [4] and Ling Zhou [4,5,*]

1 Power China Chengdu Engineering Corporation Limited, Chengdu 610072, China
2 China Water Resources Beifang Investigation, Design & Research Co., Ltd., Tianjin 300223, China; ljw_hhu@163.com
3 Sichuan Huaneng Fujiang Hydropower Co., Ltd., Mianyang 622550, China; chenj@hnfjgs.com
4 College of Water Conservancy and Hydropower Engineering, Hohai University, Nanjing 210098, China; rl.feng@hhu.edu.cn (R.F.); liudeyouhhuc@163.com (D.L.)
5 Yangtze Institute for Conservation and Development, Nanjing 210098, China
* Correspondence: zlhhu@163.com

Abstract: In various water transmission systems such as long-distance water transfer projects and hydropower stations, accurate simulation of water hammer is extremely important for safe and stable operation and the realization of intelligent operations. Previous water hammer calculations usually consider only steady-state friction, underestimating the decay of transient pressure. A second-order Finite Volume Method (FVM) considering the effect of unsteady friction factor is developed to simulate the water hammer and the dynamic behavior of air cushion surge chamber in a water pipeline system, while an experimental pipe system is conducted to validate the proposed numerical model. Two unsteady friction models, Brunone and TVB models, were incorporated into the water hammer equations, which are solved by the MUSCL–Hancock method. One virtual boundary method was proposed to realize the FVM simulation of Air Cushion Surge Chamber. Comparisons with water hammer experimental results show that, while the steady friction model only accurately predicts the first pressure peak, it seriously underestimates pressure attenuation in later stages. Incorporating an unsteady friction factor can better predict the entire pressure attenuation process; in particular, the TVB unsteady friction model more accurately reproduces the pressure peaks and the whole pressure oscillation periods. For water pipeline systems with an air cushion surge chamber, energy attenuation of the elastic pipe water hammer is primarily due to pipe friction and the air cushion. The experimental results with the air cushion surge chamber demonstrate that the proposed FVM model with the TVB unsteady friction model and the air chamber polytropic exponent near 1.0 can well reproduce the experimental pressure oscillations.

Keywords: air cushion surge chamber; finite volume method; unsteady friction; water hammer

Citation: Liu, Y.; Lu, J.; Chen, J.; Xia, Y.; Liu, D.; Hu, Y.; Feng, R.; Liu, D.; Zhou, L. Finite Volume Method for Transient Pipe Flow with an Air Cushion Surge Chamber Considering Unsteady Friction and Experimental Validation. *Water* **2023**, *15*, 2742. https://doi.org/10.3390/w15152742

Academic Editor: Giuseppe Pezzinga

Received: 25 June 2023
Revised: 22 July 2023
Accepted: 27 July 2023
Published: 29 July 2023

Copyright: © 2023 by the authors. Licensee MDPI, Basel, Switzerland. This article is an open access article distributed under the terms and conditions of the Creative Commons Attribution (CC BY) license (https://creativecommons.org/licenses/by/4.0/).

1. Introduction

Water hammer often occurs in various water pipe systems, including long-distance water transfer projects and hydropower stations. An abnormal pressure surge may lead to a pipe burst, so many water hammer protection measures are introduced to reduce the water hammer intensity. The air cushion surge chamber is a closed chamber that is partially filled with water and compressed air [1]. As compared to an open-type pressure regulating chamber, it is rarely restricted by geological or topographical features, and offers numerous advantages, such as shorter construction supply, lower excavation volume, cost-effectiveness, and minimal ecological impact [2]. The air cushion surge chamber is widely used in hydroelectric power plants for water hammer protection, ensuring the safe hydraulic operation of the water pipe system. In order to realize the safe and stable operation of water systems and the realization of intelligent operations, it is extremely important to accurately simulate the transient flow of pipe systems with an air cushion surge chamber.

The Method of Characteristics (MOC) is currently a widely used simulation tool for modeling the hydraulic transition process of hydroelectric power plants. However, there are some complicated situations, such as short pipes, T-junctions, and series pipes, in actual water delivery systems. When using MOC, interpolation or wave speed regulation is required, which reduces computational efficiency and accuracy and introduces computational errors. Moreover, most existing water hammer calculation models only adopt a steady friction model, implying that the friction inside the pipeline remains the same as the steady-state during the transient process. However, in actual transient processes, the friction inside the pipeline is influenced by multiple factors, leading to significantly different calculation results from the actual results. Additionally, they cannot accurately describe the waveform distortion and peak attenuation of pressure waves [3].

The FVM discretized the calculation area in the pipeline into independent control units, and solved the differential equations in each unit separately. Godunov et al. [4] proposed a numerical scheme for solving nonlinear Riemannian problems. This scheme is very suitable for approximating smooth solutions and discontinuous solutions. Therefore, in recent years, a large number of researchers have gradually begun to construct the Godunov scheme to solve the hydraulic transient water hammer problem.

Yazdi et al. [5] pointed out in 2007 that when calculating hydraulic transients, if the Courant number condition is not met, the second-order Godunov scheme is more stable than the MOC method. Bi Sheng et al. [6] adopted the Godunov scheme to solve the two-dimensional flow-transport equation. This model can simultaneously solve the water flux and transport flux, which is highly efficient for simulating the dynamic characteristics of water flow in complex terrain, and effectively eliminates problems such as excessive numerical damping and unstable oscillation caused by the convection term in numerical calculation. Zheng Jieheng et al. [7] used the Godunov scheme to study the hydraulic transients in sequential transmission pipelines. Based on the finite volume method, Zhao Yue et al. [8] proposed a treatment method with double virtual boundary to numerically simulate the phenomenon of water hammer and water column separation in pipelines. Zhou et al. [9–11] proposed a method to simulate a liquid column separation-bridging water hammer using a second-order GODUNOV scheme. Hu et al. [12] proposed the application of a second-order GODUNOV scheme to simulate non-pressurized flow.

Currently, there are two main unsteady friction models extensively utilized. These models are the weighted function model represented by the Zielke [13] and the empirical correction model represented by the Brunone [14]. According to the Zielke unsteady friction model, the instantaneous shear stress in the pipe due to transient flow is composed of a constant term and an additional term. The additional term uses the weighted function to account for the impact of historical velocity and acceleration on the current flow state. However, this method has a long calculation time and requires a large storage space. Subsequently, Zielke's model was simplified by Trikha [15], Vardy [16], and other scholars, resulting in a weighted function class unsteady friction model with higher computational efficiency. The Brunone unsteady friction model links unsteady friction with instantaneous local acceleration and convective acceleration and proposes a new dynamic friction model. Vitkovsky [17] improved the Brunone model by predicting the direction of water flow and wave propagation, as well as the effects of specific acceleration and deceleration stages.

To simulate more accurately the hydraulic transient process of the pressurized delivery pipeline system with an air cushion surge chamber, this paper introduces the second-order Godunov format of FVM during the calculation process, incorporating the Trikha–Vardy–Brunone (TVB) and Brunone unsteady friction models. One virtual boundary method was proposed to realize the FVM simulation of an air cushion surge chamber. An experimental pipe system was designed and conducted to validate the proposed models in simulating the water hammer and dynamic behavior of an air cushion surge chamber.

The novelty of the paper is that a second-order FVM considering the effect of unsteady friction factor is developed to simulate the dynamic behavior of the air cushion surge chamber in a water pipeline system, while an experimental pipe system is conducted to validate the proposed numerical model. Pressure damping and energy dissipation of transient flow in a water pipeline system with air cushion surge chamber are carefully investigated and modeled, which have not been well considered in previous work.

2. Mathematical Models

2.1. Water Hammer Control Equations

Equations of motion and continuity for water hammer are [18]:

$$g\frac{\partial H}{\partial x}+V\frac{\partial V}{\partial x}+\frac{\partial V}{\partial t}+g(J_Q+J_u)=0 \tag{1}$$

$$V\frac{\partial H}{\partial x}+\frac{\partial H}{\partial t}+\frac{a^2}{g}\frac{\partial V}{\partial x}=0 \tag{2}$$

The matrix forms of the above two equations can be expressed as follows:

$$\frac{\partial \boldsymbol{U}}{\partial t}+\boldsymbol{A}\frac{\partial \boldsymbol{U}}{\partial x}=\boldsymbol{S} \tag{3}$$

where $\boldsymbol{U}=\binom{H}{V}$, $\boldsymbol{A}=\begin{pmatrix} V & a^2/g \\ g & V \end{pmatrix}$, $\boldsymbol{S}=\binom{0}{gS_0-J}$, where: x is the distance along the pipe axis coordinate. H is the piezometric head; V is the flow velocity in the pipe; a is the wave velocity of the water hammer; D is the inner diameter of the pipe; J_Q and J_u represent steady friction and unsteady friction J_u; and t is time.

If $V = 0$, the convection term can be ignored and the classical water hammer governing equation can be changed.

$$\frac{\partial \boldsymbol{U}}{\partial t}+\frac{\partial \boldsymbol{F}}{\partial x}=\boldsymbol{S} \tag{4}$$

where $\boldsymbol{F}=\overline{\boldsymbol{A}}\boldsymbol{U}, \overline{\boldsymbol{A}}=\begin{pmatrix} 0 & a^2/g \\ g & 0 \end{pmatrix}$.

Brunone used instantaneous local acceleration and convective acceleration to represent the unsteady frictional component of dynamic frictional resistance, resulting in an empirically modified model. Based on Brunone's work, Vitkovsky later added the identification of the flow direction of water and obtained an improved instantaneous acceleration model with higher calculation accuracy. The specific model forms are as follows:

$$J_u=\frac{k}{g}\left(\frac{\partial V}{\partial t}+a\cdot \mathrm{sign}(V)\left|\frac{\partial V}{\partial x}\right|\right) \tag{5}$$

in which the Brunone coefficient of friction $k=\sqrt{C^*}/2$; C^* is the shear attenuation constant and the value depends on the Reynolds number. When the water flow in the pipe is laminar flow, $C^* = 0.00476$; when the water flow in the pipe is turbulent flow, $C^*=7.41/\left[Re^{\log_{10}(14.3/Re^{0.05})}\right]$. if $V\geq 0$ then $\mathrm{sign}(V)=1$; if $V<0$ then $\mathrm{sign}(V)=-1$.

The Zielke model simplified by Vardy and Brown (TVB model) is as follows

$$J_u=\frac{16v}{gD^2}\sum_{i=8}^{1}\tau_u \tag{6}$$

$$\tau_u(t+\Delta t) = \frac{4\rho v}{D} \sum_{i=1}^{N} \Big[Y_{ai}(t) exp(-n_i v/R^2 \Delta t) + \frac{m_i D^2}{4 n_i v \Delta t}(1 - exp(-n_i v/R^2 \Delta t))(V(t+\Delta t) - V(t))\Big] \quad (7)$$

where τ_u is the unsteady shear stress; ρ is density of fluid; v is kinematic viscosity of the fluid; N is number of cells along the pipeline; $Y_{ai}(t)$ is weighting function; and R is pipe radius; When the water flow in the pipe is laminar flow, the values of the weighted function coefficients n_i and m_i are shown below n_i {i = 1, ..., 8} = {26.3744; 10^2; $10^{2.5}$; 10^3; 10^4; 10^5; 10^6;10^7} and m_i {i = 1, ..., 8} = {1; 2.1830; 2.7140; 7.5455; 39.0066; 106.8075; 359.0846; 1108.3666}.

When the water flow in the pipe is turbulent flow, the values of the weighted function coefficients n_i and m_i are shown below n_i^* ($n_i^* = n_i - B^*$) and m_i^* ($m_i^* = m_i/A^*$) n_i^*{ i = 1, ..., 13} = {10; $10^{1.5}$; 10^2; $10^{2.5}$; 10^3; $10^{3.5}$; 10^4; $10^{4.5}$; 10^5; $10^{5.5}$; 10^6; $10^{6.5}$; 10^7} and m_i^* { i = 1, ..., 13} = {9.06; −4.05; 12; 8.05; 22.7; 35.2; 65.9; 115; 206; 363; 664; 1070; 2620}.

The computational domain is discretized along both the x-axis and t-axis using the finite volume method. This results in the formation of multiple computational control volumes or cells, as shown in Figure 1. Calculations are then performed on these volumes.

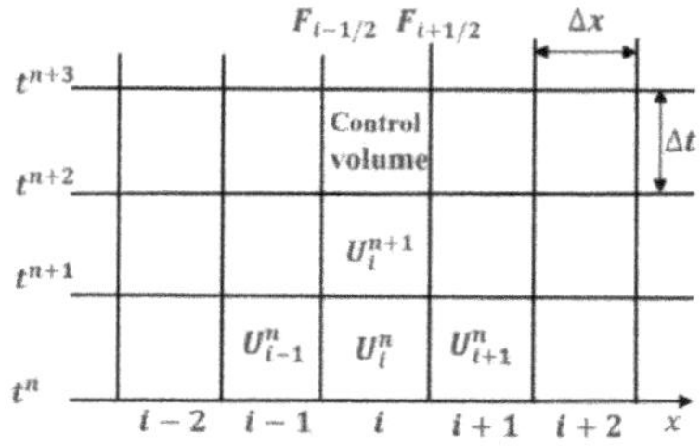

Figure 1. Grid of computational region.

Integrate Equation (4) from the control surface i − 1/2 to control surface i + 1/2. Additionally, since the control variable uniformly and continuously changes, substitute $\mathbf{U_i} = \frac{1}{\Delta x}\int_{i-1/2}^{i+1/2} udx$ to obtain the integration expression for control variable $\mathbf{U}$:

$$\mathbf{U}_i^{n+1} = \mathbf{U}_i^n - \frac{\Delta t}{\Delta x}\left(\boldsymbol{F}_{i-\frac{1}{2}} - \boldsymbol{F}_{i-\frac{1}{2}}\right) + \frac{\Delta t}{\Delta x}\int_{i-1/2}^{i+1/2} \mathbf{s}dx \quad (8)$$

where: $F_{i+1/2}$ and $F_{i-1/2}$ are the fluxes at i + 1/2 and i − 1/2, respectively; Δt is the time step; Δx is the length of control volume; the superscript n denotes the current time step; and n + 1 denotes the subsequent time step.

2.2. Control Equations of Air Cushion Surge Chamber

The continuity equation for node P at the bottom of the surge chamber is [18]

$$Q_T = Q + Q_S \quad (9)$$

where Q_T is the flow rate at the upstream pipe outlet of the surge chamber; Q is the flow rate of the downstream pipeline inlet; and Q_S is the flow into or out of the surge chamber (Figure 2).

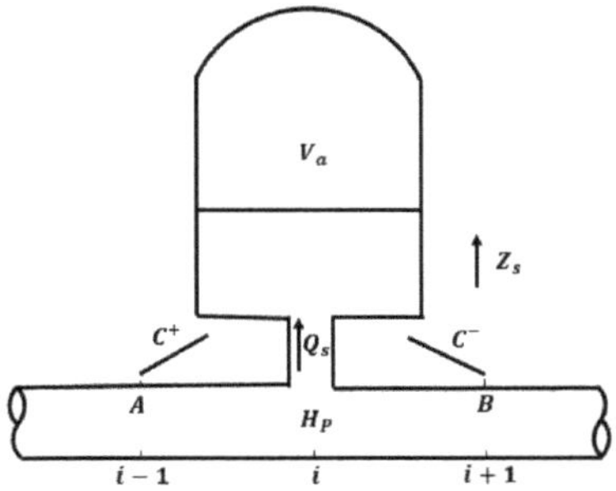

Figure 2. Schematic diagram of the air cushion surge chamber.

In order to determine the pressure tube head at the bottom point P of the surge chamber, it is necessary to use the feature compatibility equation of the last calculation section of the upstream pipeline and the first calculation section of the downstream pipeline in the surge chamber.

$$C^+ : H_P = C_P - B_P Q_T \tag{10}$$

$$C^- : H_P = C_M + B_M Q \tag{11}$$

where H_p is the piezometric head at the bottom of air cushion surge chamber.

Equations (10) and (11) are substituted into continuity Equation (9) at point P, and the variables Q_T and Q are eliminated

$$\left.\begin{array}{c} H_P = C_2 - C_1 Q_s \\ C_1 = \frac{1}{\frac{1}{B_{P1}} + \frac{1}{B_{M2}}}; C_2 = C_1\left(\frac{C_{P1}}{B_{P1}} + \frac{C_{M2}}{B_{M2}}\right) \end{array}\right\} \tag{12}$$

where C_{P1}, B_{P1}, C_{M2}, and B_{M2} are known variables at time t.

$$C_{P1} = H_{i-1} + BQ_{i-1} \tag{13}$$

$$C_{M2} = H_{i+1} - BQ_{i+1} \tag{14}$$

$$B_{P1} = B + R|Q_{i-1}| \tag{15}$$

$$B_{M2} = B + R|Q_{i+1}| \tag{16}$$

in which B is a function of the physical properties of the fluid and the pipeline, often called the pipeline characteristic impedance, and $B = a/gA$; R is the pipeline resistance coefficient $R = f\Delta x/(2gDA^2)$; f is the Darcy–Weisbach friction factor; D is pipe diameter; and A is the cross-section area.

Neglecting the water flow inertia and frictional head loss in the air cushion surge chamber, a relationship can be found between the pressure at the center point of the bottom and the water level in the air cushion surge chamber:

$$H_a = H_P - Z_s + H_{atm} - \left(R_S + \frac{1}{2gA_S^2}\right)|Q_S|Q_S \tag{17}$$

where H_a is the absolute head equal to the gauge plus barometric pressure heads; Z_s is the elevation of the air–water interface in air chamber; H_{atm} is the absolute barometric pressure head; R_s is the head loss coefficient of the impedance hole of the air chamber; and A_s is the cross-section area of the air chamber.

The air was assumed to follow the reversible polytropic relation [18]

$$PV_a^n = P_0V_0^n = Constant \tag{18}$$

where V_a is the volume of the air within the air chamber, and n is the polytropic exponent.

After combining Equations (12)–(18), the flow rate, pressure head, and water level at the air chamber can be obtained.

3. Second order Godunov Solution Scheme

3.1. Computation of Flux Term

The Riemann problem-solving method can be applied to obtain the flux values at the i + 1/2 and i + 1/2 boundary interfaces. The average value of the control variable U on the left side of the i + 1/2 interface is represented by $\mathbf{U}_L^n$, and the average value of the control variable U on the right side of the i + 1/2 interface is represented by $\mathbf{U}_R^n$.

$$U_x^n = \begin{cases} U_L^n & x < x_{i+1/2} \\ U_R^n & x > x_{i+1/2} \end{cases} \tag{19}$$

The flux values at i + 1/2 can be calculated by:

$$F_{i+1/2} = \overline{\mathbf{A}}_{i+1/2}u_{i+1/2} = \frac{1}{2}\overline{\mathbf{A}}_{i+1/2}\left\{\begin{pmatrix} 1 & a/g \\ g/a & 1 \end{pmatrix}\mathbf{U}_L^n - \begin{pmatrix} -1 & a/g \\ g/a & -1 \end{pmatrix}\mathbf{U}_R^n\right\} \tag{20}$$

The MUSCL-Hancock method for second-order linear reconstruction is used to realize the second-order accuracy of computation results.

First step: Data Reconstruction.

$$U_{i,L} = U_{i,n} - 0.5\Delta x \text{MINMOD}\left(\sigma_i^n, \sigma_{i-1}^n\right) \tag{21}$$

$$U_{i,R} = U_{i,n} - 0.5\Delta x \text{MINMOD}\left(\sigma_i^n, \sigma_{i-1}^n\right) \tag{22}$$

Second step: Advance time calculation

$$\overline{U}_i^L = U_i^L + \frac{1}{2}\frac{\Delta t}{\Delta x}\left(\overline{A}U_i^L - \overline{A}U_i^R\right) \tag{23}$$

$$\overline{U}_i^R = U_i^R + \frac{1}{2}\frac{\Delta t}{\Delta x}\left(\overline{A}U_i^L - \overline{A}U_i^R\right) \tag{24}$$

Third step: Solve the Riemann problem. To compute intercell flux $\mathbf{f}_{i+1/2}$, the conventional Riemann problem with data can be calculated by

$$\mathbf{U}_L^n \equiv \overline{\mathbf{U}}_i^R,\ \mathbf{U}_R^n \equiv \overline{\mathbf{U}}_{i+1}^L \tag{25}$$

By inserting the solved Equations (24) and (25) into the Equation (20), the fluxes in the second-order Godunov scheme at the boundary of each element can be obtained.

3.2. Time Integral

To advance the solution of the Godunov flux calculation at the n + 1 time step with second-order accuracy, time integration of Equation (8) is necessary. In the absence of friction, the following formula can be derived:

$$U_i^{n+1} = U_i^n - \frac{\Delta t}{\Delta x}\left(F_{i+(1/2)}^n - F_{i-(1/2)}^n\right) \tag{26}$$

The second-order Runge–Kutta method is used to obtain explicit results with a second-order calculation accuracy when pipe friction is taken into account. The specific calculation process is shown as follows:

First step:

$$\overline{U}_i^{n+1} = U_i^n - \frac{\Delta t}{\Delta x}\left(\mathbf{f}_{i+(1/2)}^n - \mathbf{f}_{i-(1/2)}^n\right) \tag{27}$$

Second step:

$$\overline{\overline{U}}_i^{n+1} = \overline{U}_i^{n+1} + \frac{\Delta t}{2} S\left(\overline{U}_i^{n+1}\right) \tag{28}$$

Last step:

$$U_i^{n+1} = \overline{U}_i^{n+1} + \Delta t S\left(\overline{\overline{U}}_i^{n+1}\right) \tag{29}$$

3.3. Boundary Condition

The virtual boundary method is used to process the boundary. To implement this method, virtual cells numbered 0, −1, N + 1, and N + 2 are added to the upstream and downstream boundaries, respectively. The condition that the upstream boundary is and the downstream boundary is $U_{-1} = U_0 = U_{1/2}, U_{N+1} = U_{N+2} = U_{N+1/2}$ enable the determination of the flux value at the boundary using the Riemann invariant equation. Solving the Riemann invariant equation at the upstream boundary of the reservoir yields [10]:

$$H_{1/2} - \frac{a}{g}V_{1/2} = H_1^n - \frac{a}{g}V_1^n \tag{30}$$

where V_1^n and H_1^n are the velocity and pressure head of the first control volume, and $V_{1/2}$ and $H_{1/2}$ are the velocity and pressure head at 1/2 interface at the upstream boundary.

Solving the Riemann invariant equation at the upstream boundary of the reservoir yields [10]:

$$H_N^n + \frac{a}{g}V_N^n = H_{N+1/2} + \frac{a}{g}V_{N+1/2} \tag{31}$$

where V_N^n and H_N^n are the velocity and pressure head of the last control volume, and $V_{N+1/2}$ and $H_{N+1/2}$ are the velocity and pressure head at N + 1/2 interface at the downstream boundary.

The physical variable values at each transient moment of the air cushion surge chamber can be determined by calculating the physical variable values of the virtual units at the upstream and downstream sides of the chamber based on the control equation. Thus, in combination with the Riemann invariant equation, C_{P1}, B_{P1}, C_{M2}, and B_{M2} are known variables at time t, which can conclude that:

$$C_{P1} = H_{Nu}^n + \frac{a}{g}V_{Nu}^n \tag{32}$$

$$B_{P1} = \frac{a}{gA_1} \tag{33}$$

$$C_{M2} = H_{1d}^n + \frac{a}{g}V_{1d}^n \tag{34}$$

$$B_{M2} = \frac{a}{gA_2} \tag{35}$$

where V_{Nu}^n and H_{Nu}^n represent the flow rate and the water head of the last control domain of the air cushion surge chamber on the upstream side pipeline, while V_{1d}^n and H_{1d}^n, respectively, represent the flow rate and water head of the first control domain of the air cushion surge chamber on the downstream side pipeline. A_1 and A_2 are the pipe cross-section areas at the upstream and downstream of the air chamber.

By substituting Equations (32)–(35) into Equation (8), the relation between the water level in the piezometric head at the bottom of the air cushion surge chamber and the flow rate into the air cushion surge chamber under virtual boundary conditions can be solved.

4. Results and Discussion

4.1. Experimental Setup

An experimental pipe system was designed and conducted to validate the proposed models in simulating the water hammer and dynamic behavior of the air cushion surge chamber. Figure 3 displays the experimental setup. The system consists of an upstream reservoir, variable frequency pump, constant pressure tank, upstream pipe, upstream electromagnetic flowmeter, and 1# ball valve. After the 1# ball valve, the pipeline serves as a return pipe. The water hammer experimental pipeline is 582 m long and is made of copper tubes, which have a wall thickness of 2 mm and an inner diameter of 21 mm. Along the pipeline, one 1/4 ball valve (1# ball valve) is arranged at the end, which is 582 m away from the upstream constant pressure tank. Additionally, five pressure sensors have been installed on the pipeline, with PT-1# measuring the pressure at the bottom of the constant pressure tank, PT-2# measuring 270 m downstream, PT-3# measuring 581.22 m downstream, PT-4# measuring 581.82 m downstream of the constant pressure tank, and PT-5# installed on the top of the air cushion surge chamber and used to measure the pressure of the gas in the gas chamber.

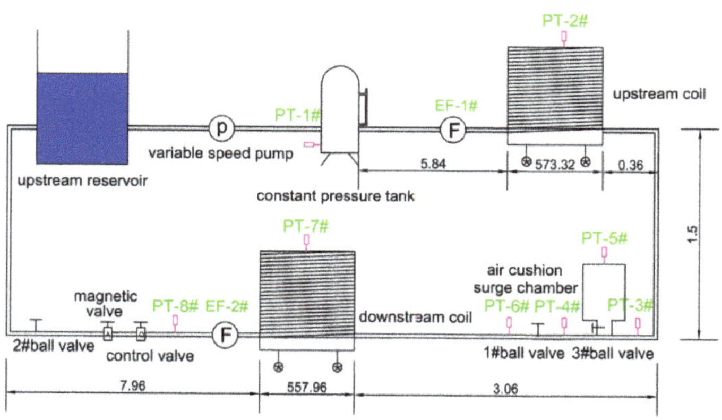

Figure 3. Experimental setup.

4.2. Water Hammer Problem in a Simple Reservoir–Pipe–Valve System

Four experimental conditions were conducted. The relevant parameters are shown in Table 1. H_0 is constant pressure head in the upstream pressure tank.

Table 1. Water hammer case.

Case	V_0 (m/s)	H_0 (m)
1	0.08	31
2	0.08	57
3	0.27	31
4	0.27	57

Cases 1 and 2 exhibit laminar flow with Re = 1284, while cases 3 and 4 demonstrate low Reynolds number turbulence with Re = 4334.

The wave velocity of the water hammer in this experiment is calculated according to the experimental data measured by the pressure sensor. According to the pressure experimental data, the time difference between any two adjacent wave peaks is recorded as $2T$, and then, according to the pipeline length L, water hammer wave $a = 2L/T$ can be obtained. In addition, many factors can affect the water hammer wave velocity. In order to eliminate the interference, repeated tests were carried out on different experimental conditions and many times. Finally, the average value of multiple tests was taken as the value of the water hammer wave velocity in the subsequent numerical simulation. The wave velocity of the water hammer measured by the experiment is between 1260 m/s and 1360 m/s. In the numerical simulations, the wave speed is a = 1290 m/s and the experimentally observed range of the average resistance coefficient for the pipeline system under stable flow conditions is 0.0312~0.0356. Valve closing times ranged between 0.012 s and 0.026 s. The experimental results show that the valve closing time is much less than half of the pressure fluctuation period. Therefore, it is believed that the valve is closed instantaneously.

As shown in Figure 4, Cases 1 and 2 can produce a similar trend for experimental pressure oscillations, although with different reference values. This is because Cases 1 and 2 have the same Reynolds number Re = 1284, but different driving pressure heads (H_0 = 31 and 57). When the initial steady condition is under laminar flow in Case 1 and Case 2, the steady friction water hammer model can only accurately predict the first pressure peak, but fail to calculate well the subsequent pressure oscillations. The reason should be that, during the fast transient flow event, the pressure damping is mainly caused by the dynamic shear force near the pipe wall. However, the traditional water hammer model assumes that the dynamic shear force coefficient is constant. In order to verify this, the unsteady friction factor is considered in the water hammer events. Figure 4a,b show that, compared to the steady friction water hammer model, the TVB and Brunone unsteady friction models can provide the much better simulation results. Meanwhile, it can be found that the TVB model shows the highest consistency with experimental results, while the Brunone unsteady friction model still produces the differences in simulating the later pressure peaks and the pressure oscillations. The main reason is that, as described above, the Brunone unsteady friction model is an empirically modified model, and the TVB model is a mechanism model which is derived from the physical equations.

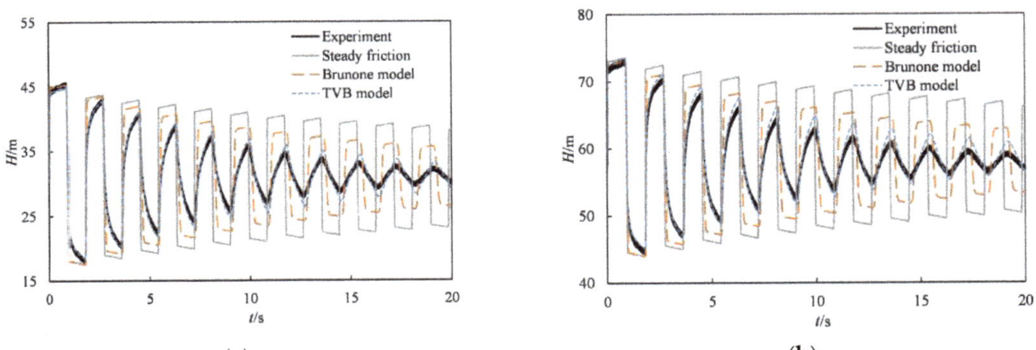

(a) (b)

Figure 4. Comparison of numerical simulation and experiment under laminar flow (PT−4#). (**a**) case 1; (**b**) case 2.

Figure 5 shows the calculated and experimental results under the turbulence flow condition in Case 3 and Case 4. Cases 3 and 4 still produce a similar trend for experimental pressure oscillations, although with reference values. Similar to laminar flow cases, the comparisons still demonstrate that, under turbulence flow condition, the steady friction model can accurately predict only the first pressure peak, and significantly underestimates the pressure damping in the later pressure oscillations. In contrast to the steady friction model, the TVB and Brunone unsteady friction models can accurately predict the entire pressure attenuation process. Among these models, the TVB model can most accurately reproduce the experimental results.

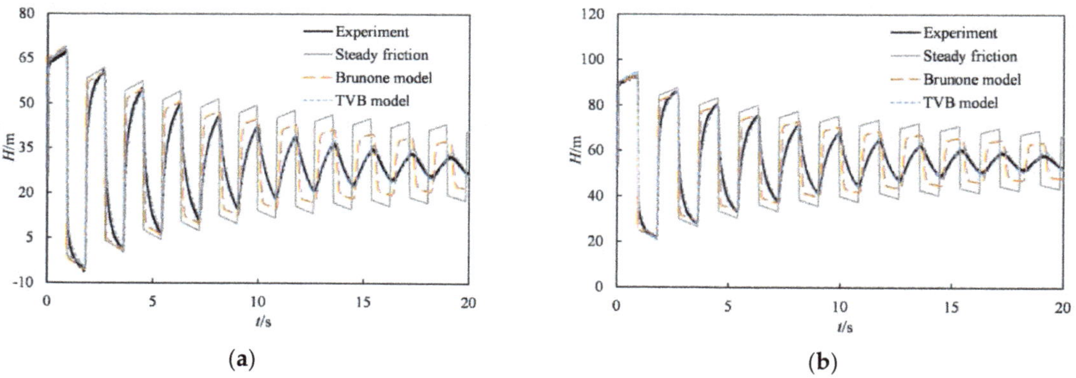

Figure 5. Comparison of numerical simulation and experiment under turbulence (PT−4#). (**a**) case 3; (**b**) case 4.

It can also be found from Figures 4 and 5 that, as the Reynolds number increases, the steady friction water hammer model can give better simulation results.

Overall, the TVB model can accurately reproduce the experimental results, regardless of the presence of laminar or turbulent flow, and is recommended to simulate the water hammer events.

4.3. Water Hammer Problem with Air Cushion Surge Chamber

Figure 1 demonstrates that, by opening ball valve #3 on the connection pipe of the air cushion surge chamber, the experimental device transforms into a pressurized pipeline water supply system equipped with an air cushion surge chamber. Furthermore, a pressure sensor PT-5# is placed on top of the air cushion chamber to monitor the gaseous pressure inside it, and the steady-state gas pressure of the cushion surge chamber is also measured by this pressure sensor. The velocity of pressure wave in the air cushion surge chamber experiment is consistent with that of the above water hammer experiment without air cushion surge chamber, i.e., the wave speed is a = 1290 m/s. Table 2 presents the experimental conditions of the air cushion surge chamber.

Table 2. Air cushion surge chamber case.

Case	V_0 (m/s)	H_0 (m)	Gas Volume (mL)
1	0.27	31	475 mL
2	0.08	43	275 mL
3	0.20	57	775 mL

The gas polytropic index, n, is varied to 1.0, 1.2 and 1.4, where n = 1.0 corresponds to an isothermal process, n = 1.4 corresponds to an adiabatic process, and n = 1.2 for the two in between, respectively, and the simulation results are depicted in Figure 6.

Figure 6a,b display the results calculated by steady friction model and experimental results in Case 1 (laminar flow) and Case 3 (turbulent flow). Results show that a value of n = 1.0 can simulate the pressure transient process of the gas in its first cycle more effectively than n = 1.2 and n = 1.4, although it still does not reproduce well the subsequent pressure oscillations.

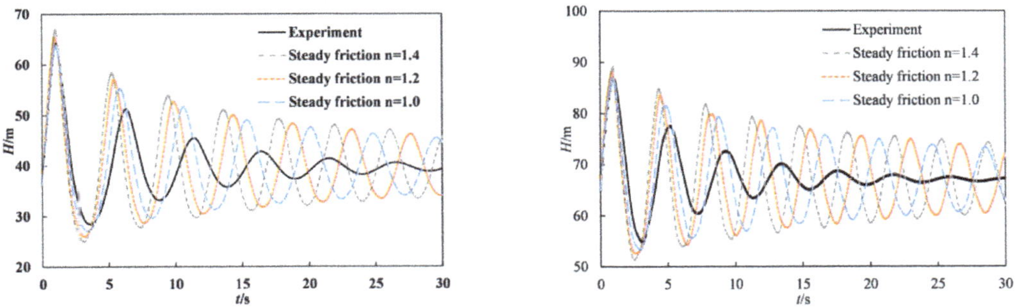

(**a**) (**b**)

Figure 6. Results calculated by steady friction model and experimental results. (**a**) Case 1 (PT−5#); (**b**) Case 3 (PT−5#).

Compared to $n = 1.2$ and $n = 1.4$, why is the result of $n = 1.0$ closer to the experimental data? The main reasons are: (1) during the transient flow, air in the chamber experiences an expansion and deceleration process, which involves complicated heat transfer and thermodynamics; (2) in the existing numerical models, the air was assumed to follow the reversible polytropic relation [18]; (3) the gas polytropic index, n, is varied to 1.0, 1.2 and 1.4, where $n = 1.0$ corresponds to an isothermal process, $n = 1.4$ corresponds to an adiabatic process, and $n = 1.2$ for middle process; (3) in this work, the heat transfer is serious, which makes the thermal process close to the isothermal process ($n = 1.0$).

Meanwhile, why, even for $n = 1.0$, do the calculated pressure oscillations attenuate lower than the measured ones? The reasons are that: (1) in the transient event in this work, the pressure damping (namely energy dissipation) is attributed to two aspects, in which one is energy loss due to heat transfer during air compression and expansion of air chamber, and another is hydraulic loss caused by the pipe friction; (2) results in Figure 6 only consider the effect of the steady friction factor, neglecting the effect of unsteady friction.

Therefore, in the following section, the effect of unsteady friction will be included in the numerical simulation to enhance the numerical accuracy, as well as to verify the above explanation.

Figure 7 gives the results calculated by numerical models (steady and unsteady friction water hammer model) and experimental results in Case 1 (laminar flow) and Case 3 (turbulent flow). According to Section 4.2, the TVB unsteady friction model performs better simulations for water hammer pressure of the pipeline in this experimental system. Therefore, the TVB unsteady friction model is adopted for this section. Additionally, examination of the simulation results from steady friction models indicated that $n = 1.0$ provides the optimal simulation effect. Consequently, when using unsteady friction simulation, n is set to 1.0.

As shown in Figure 7, the results of Case 1 (laminar flow) and Case 3 (turbulent flow) both show that introduction of the unsteady friction model causes the first peak pressure to slightly increase, compared to the steady friction model. This is because air in the chamber experiences an expansion and deceleration process when the valve is quickly closed, and the unsteady friction model suppresses its deceleration, which results in the first pressure peak value increasing. However, in comparison to the steady friction model, the peak and cycle of the first period align more closely with experimental data. The peak value and cycle of the pressure decay process also exhibits better agreement with experimental data, indicating that the use of the unsteady friction model can more accurately simulate the hydraulic transient process of the pressurized pipeline water supply system. The difference between the peak pressure of the numerical simulation and the experimental results becomes larger, which is caused by the absence of wall heat exchange in the mathematical model. In future work, we will also take into account the energy loss caused by the heat exchange of the tube wall.

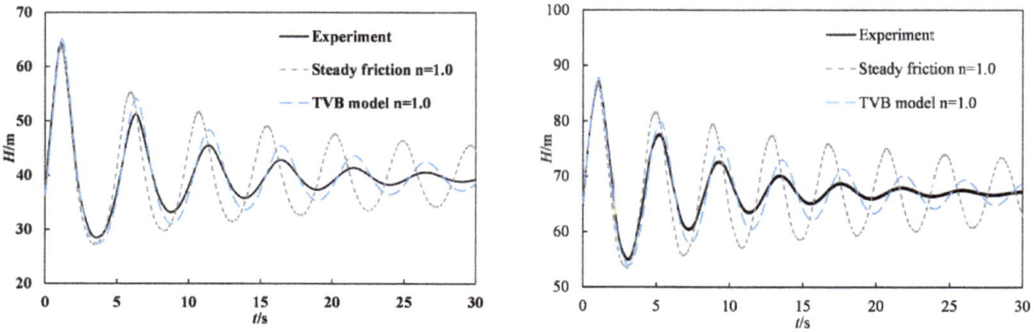

(**a**) (**b**)

Figure 7. Results calculated by numerical models and experimental results. (**a**) Case 1 (PT−5#); (**b**) Case 3 (PT−5#).

5. Conclusions

A second-order FVM considering the effect of unsteady friction factor is developed to simulate the water hammer and the dynamic behavior of air cushion surge chamber in a water pipeline system, while an experimental pipe system is conducted to validate the proposed numerical model. Two unsteady friction models, Brunone and TVB models, were incorporated into the water hammer equations, and the virtual boundary method was proposed to realize the FVM simulation of the air cushion surge chamber.

Comparisons with water hammer experimental results show that, while the steady friction model only accurately predicts the first pressure peak, it seriously underestimates pressure attenuation in later stages. Incorporating the unsteady friction factor can better predict the entire pressure attenuation process; in particular, the TVB unsteady friction model more accurately reproduces the pressure peaks and the whole pressure oscillation periods.

For water pipeline systems with an air cushion surge chamber, energy attenuation of the elastic pipe water hammer is primarily due to pipe friction and air cushion. The experimental results for the air cushion surge chamber demonstrate that the proposed FVM model with TVB unsteady friction model and the air chamber polytropic exponent near 1.0 can well reproduce the experimental pressure oscillations.

However, the difference between the peak pressures of the numerical simulation and the experimental results becomes larger, which is caused by the absence of wall heat exchange in the mathematical model. In the future, we will also take into account the energy loss caused by the heat exchange of the tube wall.

Author Contributions: Writing—original draft preparation, Y.L., J.L., Y.H. and D.L. (Deyou Liu); resources, J.C., Y.X., D.L. (Daohua Liu) and R.F.; writing, review and editing, L.Z. All authors have read and agreed to the published version of the manuscript.

Funding: This research was funded by National Natural Science Foundation of China, grant numbers 51839008 and 51679066.

Data Availability Statement: Some or all data, models, or code that support the findings of this study are available from the corresponding author upon reasonable request.

Conflicts of Interest: The authors declare no conflict of interest.

References

1. He, Y.; Yu, T. *Design of Air Cushion Surge Chamber for Hydropower Plant*; China Water & Power Press: Beijing, China, 2017. (In Chinese)
2. Liu, Q.; Peng, S. *Pressure Regulating Chamber of Hydropower Plant*; Water Resources and Hydropower Press: Beijing, China, 1995. (In Chinese)
3. Liu, J.; Zhou, L.; Cao, B. Second-order approximate solution of the weight-function unsteady friction model of transient flows. *J. Hydroelectr. Eng.* **2020**, *39*, 55–61.
4. Godunov, S.K. A difference method for numerical calculation of discontinuous equations of hydrodynamics. *Math. Sb* **1959**, *47*, 217.
5. Sabbagh-Yazdi, S.R.; Mastorakis, N.E. Water hammer modeling by Godunov type finite volume method. *Int. J. Math. Comput. Simul.* **2007**, *1*, 350–355.
6. Bi, S.; Zhou, J.; Chen, S. High precision two-dimensional fluid-transport coupling model in Godunov scheme. *Prog. Water Sci.* **2013**, *24*, 706–714. (In Chinese)
7. Zheng, J.; Jiang, M.; Guo, R. Finite volume method for hydraulic transient simulation of sequential transmission pipeline. *J. Comput. Mech.* **2015**, *32*, 418–422+428. (In Chinese)
8. Zhao, Y.; Zhou, L.; Liu, D. Water hammer model based on finite volume method and Godunov-type scheme. *Adv. Sci. Technol. Water Resour.* **2019**, *39*, 76–81. (In Chinese)
9. Zhou, L.; Li, W.; Liu, D.; Ou, C. Second-order Godunov mathematical model of water flow impinging trapped air mass in pipeline. *Hydropower Energy Sci.* **2021**, *39*, 95–99. (In Chinese)
10. Zhou, L.; Li, Y.; Karney, B. Godunov-type solutions for transient pipe flow implicitly incorporating Brunone unsteady friction. *J. Hydraul. Eng.* **2021**, *147*, 04021021. (In Chinese)
11. Zhou, L.; Elong, A.; Karney, B. Unsteady friction in rapid pipeline filling with trapped air. In Proceedings of the 2019 IAHR Congress, Panama City, Panama, 1–6 September 2019.
12. Hu, Y.; Zhou, L.; Pan, T. Godunov-type solutions for free surface transient flow in pipeline incorporating unsteady friction. *AQUA—Water Infrastruct. Ecosyst. Soc.* **2022**, *71*, 546–562. (In Chinese)
13. Zielke, W. Frequency-dependent friction in transient pipe flow. *J. Basic Eng.* **1968**, *90*, 109–115.
14. Brunone, B.; Golia, U.M.; Greco, M. Some remarks on the momentum equation for fast transients. In Proceedings of the International Meeting on Hydraulic Transients with Water Column Separation, 9th Round Table of the IAHR Group, Valencia, Spain, 4–6 September 1991.
15. Trikha, A.K. An efficient method for simulating frequency-dependent friction in transient liquid flow. *J. Fluids Eng.* **1975**, *97*, 97–105.
16. Vardy, A.E.; Brown, J. Transient turbulent friction in fully rough pipe flows. *J. Sound Vib.* **2004**, *270*, 233–257.
17. Vitkovsky, J.P.; Stephens, M.; Bergant, A. Efficient and accurate calculation of Zielke and Vardy-Brown unsteady friction in pipe transients. In Proceedings of the 9th International Conference on Pressure Surges, Cranfield, UK, 24–26 March 2004; Volume 2, pp. 405–419.
18. Wylie, E.B.; Streeter, V.L. *Fluid Transients in Systems*; Prentice Hall: New York, NY, USA, 1993.

Disclaimer/Publisher's Note: The statements, opinions and data contained in all publications are solely those of the individual author(s) and contributor(s) and not of MDPI and/or the editor(s). MDPI and/or the editor(s) disclaim responsibility for any injury to people or property resulting from any ideas, methods, instructions or products referred to in the content.

water

MDPI

Article

Gas Release and Solution as Possible Mechanism of Oscillation Damping in Water Hammer Flow

Giuseppe Pezzinga

Department of Civil Engineering and Architecture, University of Catania, Via Santa Sofia 64, 95123 Catania, Italy; giuseppe.pezzinga@unict.it

Abstract: Water hammer flow is examined, putting into evidence that unsteady friction cannot be fully responsible for observed oscillation damping. The measured piezometric head oscillations of water hammer flow experimental tests carried out for very long time (about 70 periods) are presented and compared with the numerical results of a quasi-two-dimensional (2D) flow model. The hypothesis is made that the energy dissipation could be partially due to the process of gas release and solution. An equation for the balance of gas mass is taken into account, already successfully used to improve the comparison between numerical and experimental head oscillations for transient gaseous cavitation. The models are based on a particular implementation of the method of characteristics (MOC-Z). The calibration of the empirical parameters of the models is carried out with a micro-genetic algorithm (micro-GA). The better performance of the proposed model is quantified with comparison of the mean absolute errors for three experimental tests at different Reynolds numbers, ranging from 5300 to 15,400. The corresponding ratios between the mean absolute errors of the models with and without gas release range between 47.3% and 17.7%. It is also shown that different turbulence models give very similar results. The results have some relevance in water hammer research, because sometimes dissipation that is not due to unsteady friction is attributed to it. However, the hypothesized mechanism has to be deepened and validated with further studies.

Keywords: water hammer; 2D models; energy dissipation; gas release; calibration; micro-GA

Citation: Pezzinga, G. Gas Release and Solution as Possible Mechanism of Oscillation Damping in Water Hammer Flow. *Water* **2023**, *15*, 1942. https://doi.org/10.3390/w15101942

Academic Editors: Helena M. Ramos and Kamil Urbanowicz

Received: 31 March 2023
Revised: 5 May 2023
Accepted: 16 May 2023
Published: 20 May 2023

Copyright: © 2023 by the author. Licensee MDPI, Basel, Switzerland. This article is an open access article distributed under the terms and conditions of the Creative Commons Attribution (CC BY) license (https://creativecommons.org/licenses/by/4.0/).

1. Introduction

Unsteady friction is one of the most important topics in water hammer flow. It is well known that analysis carried out using one-dimensional (1D) models with steady or quasi-steady resistance formulas gives rise to underestimation of friction forces and damping [1]. In 1D models, it is possible to adopt unsteady resistance, usually with dissipation terms to be added to quasi-steady resistance terms. However, in these models the evaluation of parameters is not general and rigorous [2]. The evaluation of energy dissipation due to friction can be carried out more properly with 2D models, in which the variation of the longitudinal component of velocity along the radial coordinate is considered. Different turbulent stress models were studied in 2D flow schematization [3–7], showing very similar results.

Vardy, in a recent review paper [8], examines the different possible mechanisms of dissipation and dispersion for water hammer flows, considering unsteady friction, fluid–structure interaction, viscoelasticity, bubbly flows, and porous pipe linings. He observes that it is important to distinguish between mechanisms with dissipative behaviour, giving rise to oscillation damping, and mechanisms with dispersive behaviour (in particular fluid–structure interaction), due to superposition of waves, that can lead to reduction of pressure amplitudes but also to pressure amplifications. Ferras et al. [9] deepened the mechanism of fluid–structure interaction, comparing the results obtained for four different experimental set-ups, showing the possibility of pressure reduction or amplification.

Many authors have proposed models in which the effects of free gas on transients are taken into account [10–16]. This aspect is mainly considered for the analysis of transient gaseous cavitation, and the reader can refer to the review paper by Bergant et al. [17] on water hammer with column separation for a more complete analysis of the literature. The reason for such attention is that water flowing through hydraulic plants usually contains free air or dissolved air very close to the theoretical saturation point [18]. In some of the models taking into account the effects of free gas on transients, the mass of free gas is considered constant for the sake of simplicity, whereas in others the process of dissolved gas release is taken into account. Models of the first kind reproduce the salient characteristics of the phenomenon, and in particular, the effect on the propagation of the increased compressibility of the liquid–gas mixture. Models of the second kind, assuming a release formulation that takes a relaxation process into account, can also explain the dissipation not due to friction. Although the bulk viscosity of pure liquid cannot be responsible for relevant dissipation, following Landau and Lifshitz [19] an equivalent bulk viscosity can be expressed, taking into account that when in transient phenomena the pressure variations are rapid with respect to the relaxation processes of restoration of equilibrium, and these processes, by nature irreversible and then characterized by energy dissipation, become important [19,20]. Only a few studies attempt to consider the combined effect of both unsteady friction and gas release and solution [20,21]. In particular the results of analysis of transient gaseous cavitation [21] showed that, although unsteady friction was taken into account by using a 2D model, there is a need to postulate other possible damping mechanisms to explain the observed pressure oscillation. Among the hypothesized mechanisms, that is, the thermic exchange between the gaseous phase and the surrounding liquid, and the dynamic of free gas due to dissolved gas release and solution, the first improves the simulation of experimental runs but does not always explain the observed dissipation, whereas the latter seems to explain the observed energy dissipation.

In the current study the results of long-duration experimental tests (about 70 periods) of water hammer without cavitation are presented. The experimental head oscillations are compared with the results of a 2D model [22]. The pipeline of the installation, on which the experimental tests were carried out, has long horizontal parts. Despite the presence of several air release valves, it is very difficult to completely eliminate the air from the circuit. Then the process of gas release and solution is considered as a possible reason for further oscillation damping for water hammer flow without cavitation. The process of gas release and solution is taken into account with a proper mass balance equation. The calibration of the model parameters is carried out with a micro-GA. A different turbulence model [23] is also considered for comparison.

2. Mathematical Models

2.1. Continuity Equation

The fundaments of modelling water hammer flow are well-established in the literature [24]. The assumptions here considered to obtain the presented model were already stated in a previous study [19], are well-established in the literature, appear very reasonable in the context of the considered phenomena, and are indirectly validated using a comparison with experimental results. Gas bubbles are distributed throughout the pipe and they are very small compared to pipe diameter; the difference in pressure due to surface tension across a bubble surface is neglected, as well as the momentum exchange between gas bubbles and surrounding liquid, so that gas bubbles and liquid have the same velocity. Under these hypotheses, the continuity equation can be written in the 1D form:

$$\frac{\partial}{\partial t}\left[\rho_l A\left(1-\frac{mRT}{p}\right)\right]+\rho_l A\left(1-\frac{mRT}{p}\right)\frac{\partial V}{\partial x}=0 \quad (1)$$

where m = mass of free gas per unit volume, R = gas constant, T = absolute temperature, p = absolute pressure, x = distance along the pipe, t = time, ρ_l = liquid density, A = cross-sectional area of pipe, and V = mean velocity.

Considering the mixture density as a function of time through pressure and, in general, also through mass of free gas, the continuity equation for a 1D two-phase flow in an elastic pipe, for small gas fraction, i.e., $mRT/p \ll 1$, can be written as:

$$\left(1+\frac{\rho_l a_0^2 mRT}{p^2}\right)\frac{\partial p}{\partial t}-\frac{\rho_l a_0^2 RT}{p}\frac{\partial m}{\partial t}+\rho_l a_0^2\frac{\partial V}{\partial x}=0 \tag{2}$$

with a_0 wave speed of pure liquid in an elastic pipe.

This form of the equation differs from that previously proposed [21] and is simpler, because of neglecting the density of gas with respect to the density of liquid.

Introducing the auxiliary variable φ defined as:

$$\varphi=\frac{p}{\rho_l g}-\frac{a_0^2 mRT}{gp} \tag{3}$$

where g is the gravitational acceleration, the continuity equation of the mixture can be written in the form [21]:

$$\frac{\partial \varphi}{\partial t}+\frac{a_0^2}{g}\frac{\partial V}{\partial x}=0 \tag{4}$$

A quasi-2D form of the continuity equation can be obtained by simply assuming a single instantaneous value of φ in each section by substituting the velocity component in the longitudinal direction u to the mean velocity V:

$$\frac{\partial \varphi}{\partial t}+\frac{a_0^2}{g}\frac{\partial u}{\partial x}=0 \tag{5}$$

where $u = u(x,r,t)$, r being the radial coordinate (distance from the pipe axis), while $\varphi = \varphi(x,t)$; in Equation (5), a_0 is then the value pertaining to the whole section.

After the auxiliary variable φ is known, the pressure can be computed as the positive solution of the quadratic equation derived from Equation (3)

$$p=\frac{\rho_l g\varphi+\sqrt{(\rho_l g\varphi)^2+4\rho_l a_0^2 mRT}}{2} \tag{6}$$

2.2. Gas Release and Solution Equation

As regards the gaseous phase, neglecting both the spatial derivative and the deformability of the area with respect to the term $\partial m/\partial t$, the continuity equation can be written as [22,25]:

$$\frac{\partial m}{\partial t}=\frac{1}{\theta_m}\frac{\beta}{RT}(p_s-p) \tag{7}$$

in which θ_m is a relaxation time, β the Henry's law constant, and p_s the gas saturation pressure computed as

$$p_s=p_0+\frac{RT}{\beta}(m_0-m) \tag{8}$$

where p_0 and m_0 are the steady state values of p and m, respectively. By substituting Equation (8) in Equation (7), the continuity equation for the gaseous phase becomes

$$\frac{\partial m}{\partial t}=\frac{1}{\theta_m}\left[m_0-m+\frac{\beta}{RT}(p_0-p)\right] \tag{9}$$

2.3. Momentum Equations

The momentum equation for a liquid–gas mixture in a pipe with circular cross section and axial-symmetric 2D flow, with the usual assumption of neglecting the convective term, can be expressed in cylindrical coordinates in the form [19]:

$$\frac{\partial u}{\partial t} + g\frac{\partial H}{\partial x} + \frac{1}{\rho r}\frac{\partial (r\tau)}{\partial r} = 0 \tag{10}$$

in which τ is the shear stress and H the piezometric head. In Equation (13) $\tau = \tau(x, r, t)$ while $H = H(x, t)$. The piezometric head is expressed with the following relation:

$$H = z + \frac{p + p_v - p_a}{\rho_l g} \tag{11}$$

where p_v is the vapour pressure, p_a is the atmospheric pressure, and z is the elevation head. The vapour pressure is added to the gas pressure because the gas bubbles also contain water vapour [24].

For turbulent flow, the shear stress depends on both viscosity and density. Here, the shear stress is expressed with the two-zone turbulence model of Santoro et al. [26].

In the 1D flow model, the following momentum equation is used instead of Equation (10):

$$\frac{\partial V}{\partial t} + g\frac{\partial H}{\partial x} + \frac{2\tau_0}{\rho r_0} = 0 \tag{12}$$

where τ_0 is the wall shear stress and r_0 is the pipe cross section radius.

In Equation (12), τ_0 is computed as $\tau_0 = f\rho V^2/8$, with the friction factor f evaluated as described by Santoro et al. [26].

2.4. Boundary Conditions

The same boundary conditions are used in all models; namely, with regard to the specific experimental installation that will be described below, a constant level reservoir with static head H_s at the downstream end ($x = L$, with L pipe length), and velocity linearly variable in time, from the initial steady state value to zero at the complete valve closure time T_c. For the latter condition, using pressure data from the literature for many other cases of transient flow, the author indirectly calibrated the law for the valve area by computing the coefficients of a third-order power function. In the examined cases of very fast closure, no differences existed, from a practical point of view, between the results obtained with a calibrated manoeuvre or with linear variation of discharge [26].

2.5. Method of Characteristics

By combining the continuity and momentum equations with standard treatment, the equations valid on the characteristic lines can be obtained. From Equations (5) and (10) for the 2D model one can write:

$$\frac{\partial u}{\partial t} + g\frac{\partial \varphi}{\partial x} + g\frac{\partial (H-\varphi)}{\partial x} + \frac{1}{\rho r}\frac{\partial (r\tau)}{\partial r} + \lambda\left(\frac{\partial \varphi}{\partial t} + \frac{a_0^2}{g}\frac{\partial u}{\partial x}\right) = 0 \tag{13}$$

where λ = auxiliary variable. In Equation (13) the term $g\partial\varphi/\partial x$ has been added and subtracted. Considering the conditions of the characteristic lines:

$$\frac{dx}{dt} = \frac{g}{\lambda} \tag{14}$$

$$\frac{dx}{dt} = \frac{\lambda a_0^2}{g} \tag{15}$$

Equation (13) can be written as:

$$\frac{d\varphi}{dt} \pm \frac{a_0}{g}\frac{du}{dt} \pm a_0\frac{\partial(H-\varphi)}{\partial x} \pm \frac{a_0}{\rho g r}\frac{\partial(r\tau)}{\partial r} = 0 \tag{16}$$

valid on the characteristic lines:

$$\frac{dx}{dt} = \pm a_0 \tag{17}$$

When the process of dissolved gas release and solution is taken into account, Equation (9) is resolved in addition.

2.6. Numerical Scheme

Equation (16) and, for variable mass, Equation (9), are solved on a cylindrical grid with constant step Δx in the longitudinal direction and constant area ΔA in the radial direction. Velocity is defined halfway in the radial direction, and shear stresses on the internal and external sides (Figure 1). All the calculations were carried out with 100 longitudinal steps and 50 steps in the radial mesh. Time step was $\Delta t = \Delta x / a_0$.

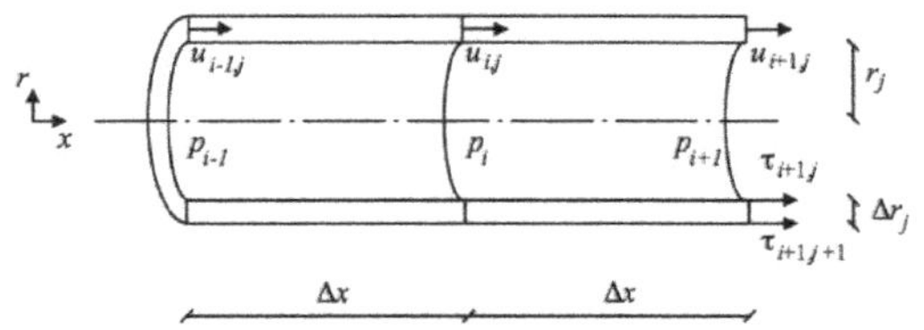

Figure 1. Grid for the 2D models.

The adopted numerical scheme is defined as Z-mirror scheme, due to the shape of the characteristic line coupled with the lines along which the $\partial(H-\varphi)/\partial x$ derivative is calculated (Figure 2). When the equation along the positive characteristic line is used, all the terms are computed on the segment AP, except the term containing the derivative with respect to x of $(H-\varphi)$, that is computed on the segment AM in the predictor step (p_+), and on the segment EP in the corrector step (c_+), respectively. When the equation along the negative characteristic line is used all the terms are computed on the segment BP, except the term containing the derivative with respect to x of $(H-\varphi)$, that is computed on the segment MB in the predictor step (p_-), and on the segment PF (c_-) in the corrector step, respectively. Segments EP, AP and AM form a "Z", while segments PF, BP, and MB form a "Z" as reflected in a mirror. For these reasons the scheme is defined as Z-mirror scheme and the associated model is called MOC-Z.

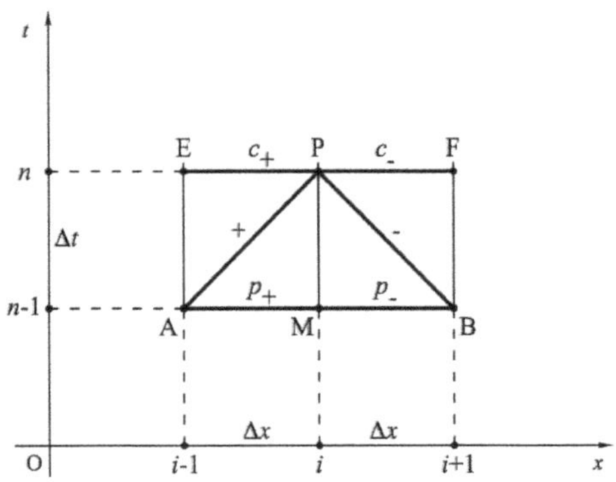

Figure 2. Lines for numerical resolution with the Z-mirror scheme.

In the predictor step, Equation (16) is solved as

$$\frac{\varphi_i^p-\varphi_{i-1}^{n-1}}{\Delta t}+\frac{a}{g}\frac{u_{i,j}^p-u_{i-1,j}^{n-1}}{\Delta t}+a\frac{H_i^{n-1}-\varphi_i^{n-1}-H_{i-1}^{n-1}+\varphi_{i-1}^{n-1}}{\Delta x}+\frac{2\pi a}{\rho g}\frac{r_{j+1}\tau_{j+1}^{p+}-r_j\tau_j^{p+}}{\Delta A}=0 \quad (18)$$

$$\frac{\varphi_i^p-\varphi_{i+1}^{n-1}}{\Delta t}-\frac{a}{g}\frac{u_{i,j}^p-u_{i+1,j}^{n-1}}{\Delta t}-a\frac{H_{i+1}^{n-1}-\varphi_{i+1}^{n-1}-H_i^{n-1}+\varphi_i^{n-1}}{\Delta x}-\frac{2\pi a}{\rho g}\frac{r_{j+1}\tau_{j+1}^{p-}-r_j\tau_j^{p-}}{\Delta A}=0 \quad (19)$$

where indices i, j, and n, refer, respectively, to directions x, r, and time t.

In the corrector step the corresponding set of equations is

$$\frac{\varphi_i^c-\varphi_{i-1}^{n-1}}{\Delta t}+\frac{a}{g}\frac{u_{i,j}^c-u_{i-1,j}^{n-1}}{\Delta t}+a\frac{H_i^p-\varphi_i^p-H_{i-1}^p+\varphi_{i-1}^p}{\Delta x}+\frac{2\pi a}{\rho g}\frac{r_{j+1}\tau_{j+1}^{c+}-r_j\tau_j^{c+}}{\Delta A}=0 \quad (20)$$

$$\frac{\varphi_i^c-\varphi_{i+1}^{n-1}}{\Delta t}-\frac{a}{g}\frac{u_{i,j}^c-u_{i+1,j}^{n-1}}{\Delta t}-a\frac{H_{i+1}^p-\varphi_{i+1}^p-H_i^p+\varphi_i^p}{\Delta x}-\frac{2\pi a}{\rho g}\frac{r_{j+1}\tau_{j+1}^{c-}-r_j\tau_j^{c-}}{\Delta A}=0 \quad (21)$$

In both steps, for the evaluation of the shear stress an implicit scheme is adopted [6]. At each step, velocity components u with quasi-2D models can be obtained by subtracting the negative characteristic equation from the positive one, without the need for knowing the "new" piezometric head nor the "new" auxiliary variable φ, as they cancel out. Then the variable φ is computed by adding Equations (18) and (19) (predictor step) or Equations (20) and (21) (corrector step). An analogous scheme is adopted for the 1D form of the equations.

It has been noted [22] that the MOC-Z models, not requiring interpolations, appear simpler and more straightforward if compared to the classic characteristics model based on the specified time intervals method [24].

2.7. Micro-GA

In order to compare numerical and experimental results, the calibration of m_0 and θ_m, considered as constants, has to be carried out. This was accomplished with a micro-GA. This optimization tool works with small populations, and it has the advantage, with respect to a genetic algorithm, of containing the calculation times. This is particularly useful when each evaluation of the fitness function requires the comparison of numerical results of a time-consuming mathematical model with experimental results.

The fitness of the micro-GA was evaluated as the inverse of the mean absolute error (MAE) function, as already explained by Pezzinga and Santoro [22],

$$MAE=\frac{\sum_{k=1}^{N}\left|H_c^k-H_m^k\right|}{N} \quad (22)$$

where H_c^k and H_m^k are, respectively, the computed and measured head, and N is the number of experimental values. Given that the considered duration of the experimental runs was 30 s and the experimental sampling frequency was 100 Hz, N is equal to 3001. A ten-bit binary coding (giving 2^{10} = 1024 possible values) of the parameters m_0 and θ_m, ranging, respectively, between 0 and 100 mg/m^3, and between 10 and 1000 s, was used. A population size N_P = 5 was used for the calibration of gas mass in the models with constant gass mass, where N_P = 9 was used for the calibration of initial gas mass and relaxation time in the models with variable gas mass [22].

3. Experimental Installation

The numerical results of the proposed mathematical model were compared with experimental results of water hammer flow tests, carried out on the installation sketched in Figure 3. Such installation is composed mainly of a pipe anchored to the wall. Furthermore, previous experimental tests showed pressure traces similar to that obtained for straight pipe by Ferras et al. [8]. These considerations lead to excluding any relevant effect due to structural damping [27] or to fluid–structure interaction. The pipe is made of zinc-plated steel (internal diameter 53.9 mm, thickness 3.2 mm, modulus of elasticity 2.06×10^{11} N/m^2, roughness 0.1 mm, length 144.3 m) and it is fed by a centrifugal pump. A pressure tank is located at the downstream end of the pipe. The line pressure was measured with strain gauge pressure transducers, having a range of 0 to 10 bar, with maximum error of ±0.5% of full-scale pressure. Discharge measurements were carried out with an electromagnetic flowmeter with adjustable full-scale velocity, with maximum errors of ±0.1% of full scale. Each experimental test started from steady-state conditions by manually closing the ball valve at the upstream end of the pipe. The valve closure was estimated to take 0.04 s. Temperature as measured in all the experimental tests was 24 °C. Each physical property of water and air was indirectly evaluated as a function of the measured temperature. In Table 1, for each experimental test, values of initial discharge Q_0 and static head H_s referred to the laboratory floor are shown.

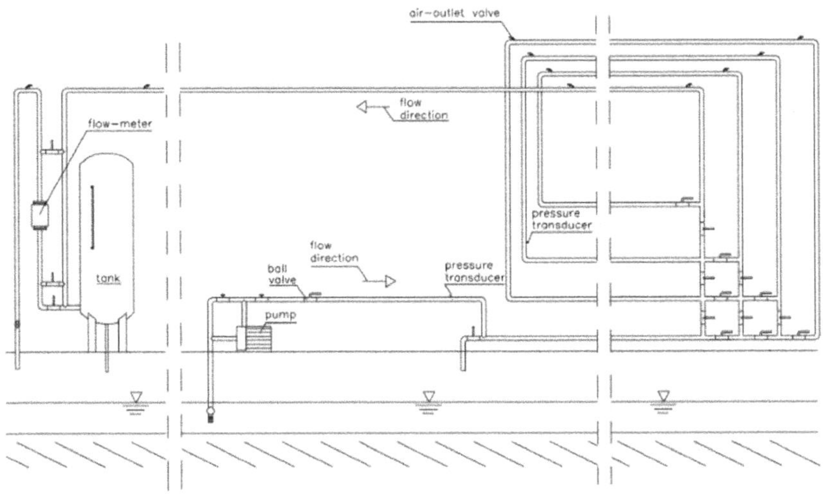

Figure 3. Experimental installation.

Table 1. Physical parameters of the experimental tests.

Test	Q_0 (L/s)	V_0 (m/s)	H_s (m)	Re_0
L1	0.207	0.091	68.12	5300
L2	0.409	0.179	66.87	10,500
L3	0.598	0.262	60.08	15,400

4. Analysis of Results

The comparison among results of all models is summarized in Figure 4, where MAEs for all tests are reported, respectively, for the 1D model without gas (1D $m = 0$), the 2D model without gas (2D $m = 0$), the 2D model with constant mass of gas (2D m const), and the 2D model with variable mass of gas (2D m var). It can be noted firstly the great decrease in MAE allowed with 2D models with respect to quasi-steady 1D models, confirming the importance of unsteady friction. Further reductions of MAE can be obtained by considering the presence of gas. When the mass of gas is considered as constant, the MAE reduction is due to the regulation of phase of the phenomenon, because the presence of gas allows the reduction of wave speed to match the experimental period and the computed one. The MAE for the models with variable mass of gas is the minimum for all the experimental tests, because the relaxation process taken into account in gas release and solution causes additional energy dissipation that allows a better reproduction of the experimental head oscillation. The ratios between the mean absolute errors of the models with and without gas release are, respectively, 47.3% for Test 1, 28.8% for Test 2, and 17.7% for Test 3.

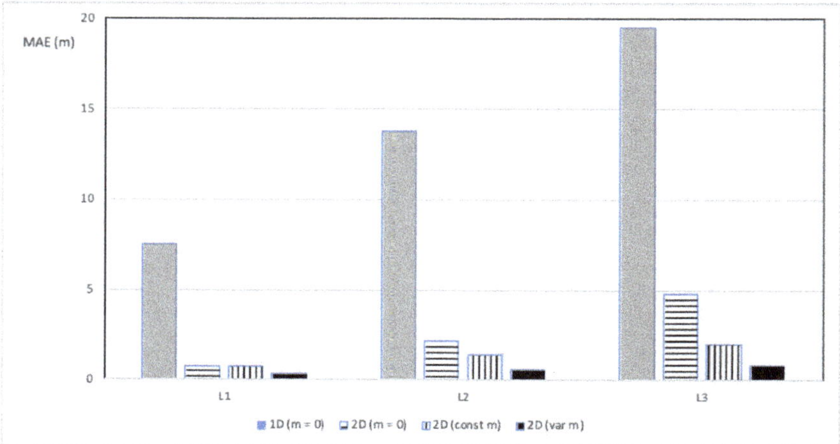

Figure 4. Mean absolute errors of different models with respect to experimental data.

These considerations are confirmed by observing the detailed comparisons reported in Figures 5 and 6 between the measured head oscillation Z with respect to the static head and the computed one with different models. The long duration of the test allows putting into evidence the great improvement due to the 2D flow models with respect to the 1D models (Figure 5). Furthermore, in Figure 6 the analogous results are reported for the 2D model with constant mass of gas, and with variable mass of gas. A similar behaviour to that of the model with constant mass of gas could also be obtained by reducing the wave speed of 0.5% with respect to the theoretical value (1361 instead of 1367), but with the same poor simulation of the amplitude of the oscillations. Instead, the comparison of the 2D model with variable mass of gas shows very good results in terms of phase and amplitude of the oscillations, due to cumulative effects of unsteady friction and relaxation due to gas release and solution. Longer durations were not considered due to the unreliability of smaller and smaller head oscillations.

To analyse the performance of different turbulence models, Figure 7 reports the comparison of the head oscillations computed with the Santoro et al. [26] turbulence model and the Lam and Bremhorst low Reynolds number k-ε turbulence model [23] with the experimental results. It is confirmed [7] that different turbulence models give almost the same results in terms of pressure head. Obviously more refined models can give more information on the turbulent variables, such as, for the k-ε models, the turbulent kinetic energy k and its dissipation rate ε.

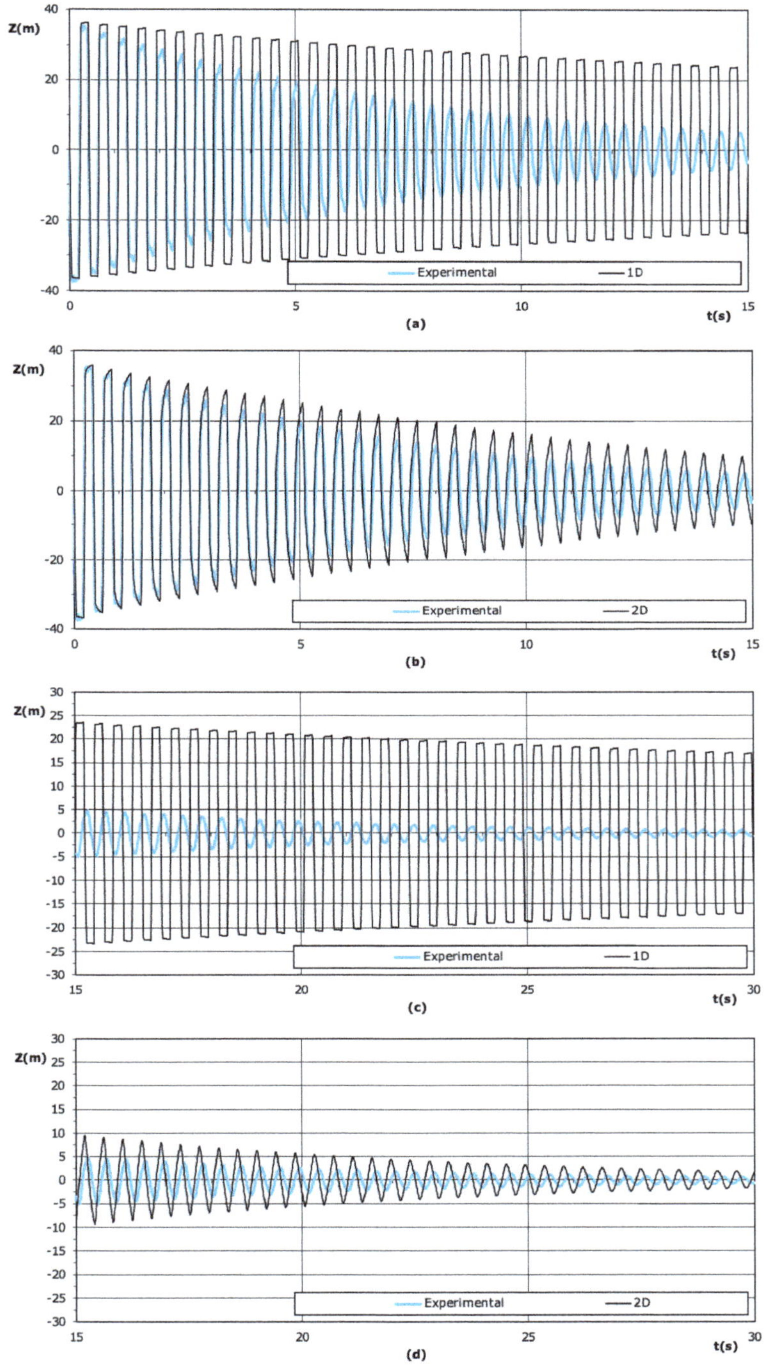

Figure 5. Comparison of head oscillations computed with the 1D or 2D model without free gas with the experimental results for Test L3: (**a**) 1D model, $t = 0 - 15$ s; (**b**) 2D model, $t = 0 - 15$ s; (**c**) 1D model, $t = 15 - 30$ s; (**d**) 2D model, $t = 15 - 30$ s.

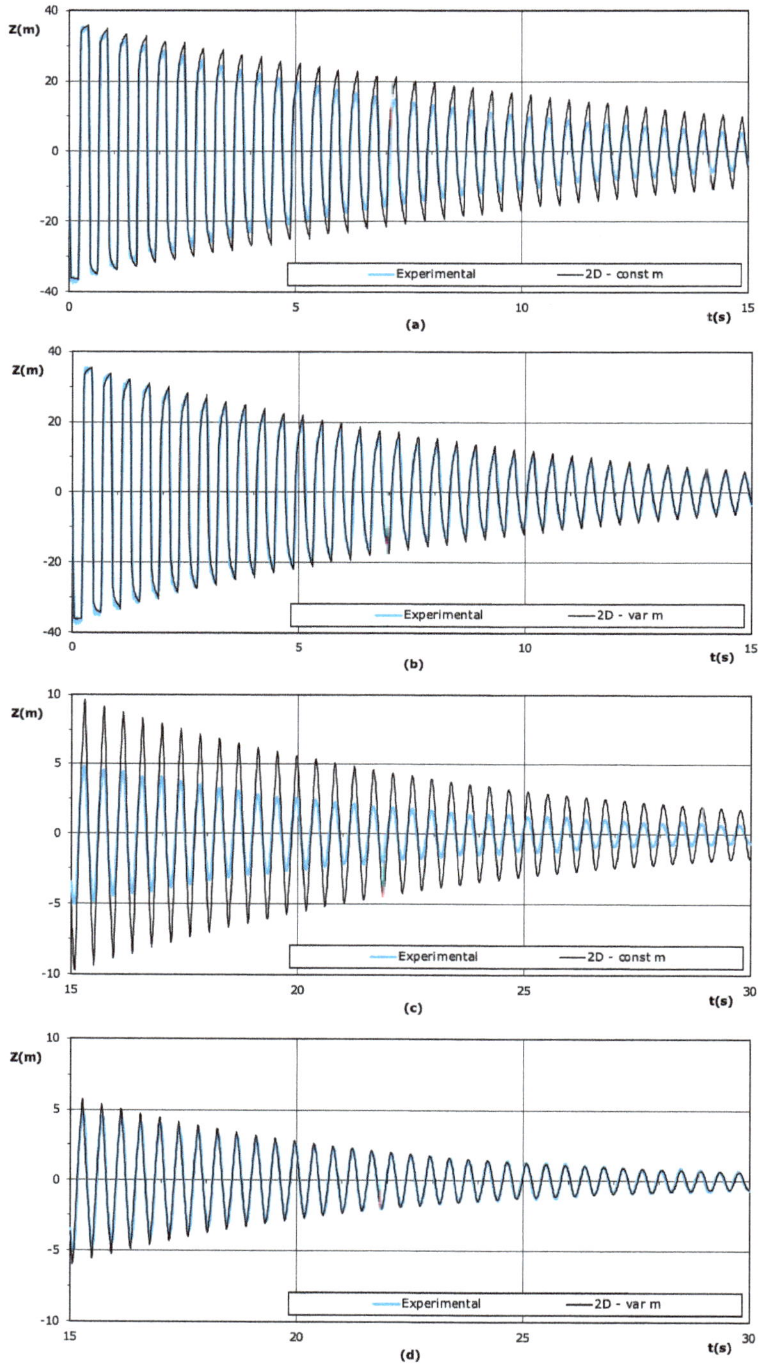

Figure 6. Comparison of head oscillations computed with the 2D model with constant or variable mass of free gas with the experimental results for Test L3: (**a**) 2D model with constant mass of free gas, $t = 0 - 15$ s; (**b**) 2D model with variable mass of free gas, $t = 0 - 15$ s; (**c**) 2D model with constant mass of free gas, $t = 15 - 30$ s; (**d**) 2D model with variable mass of free gas, $t = 15 - 30$ s.

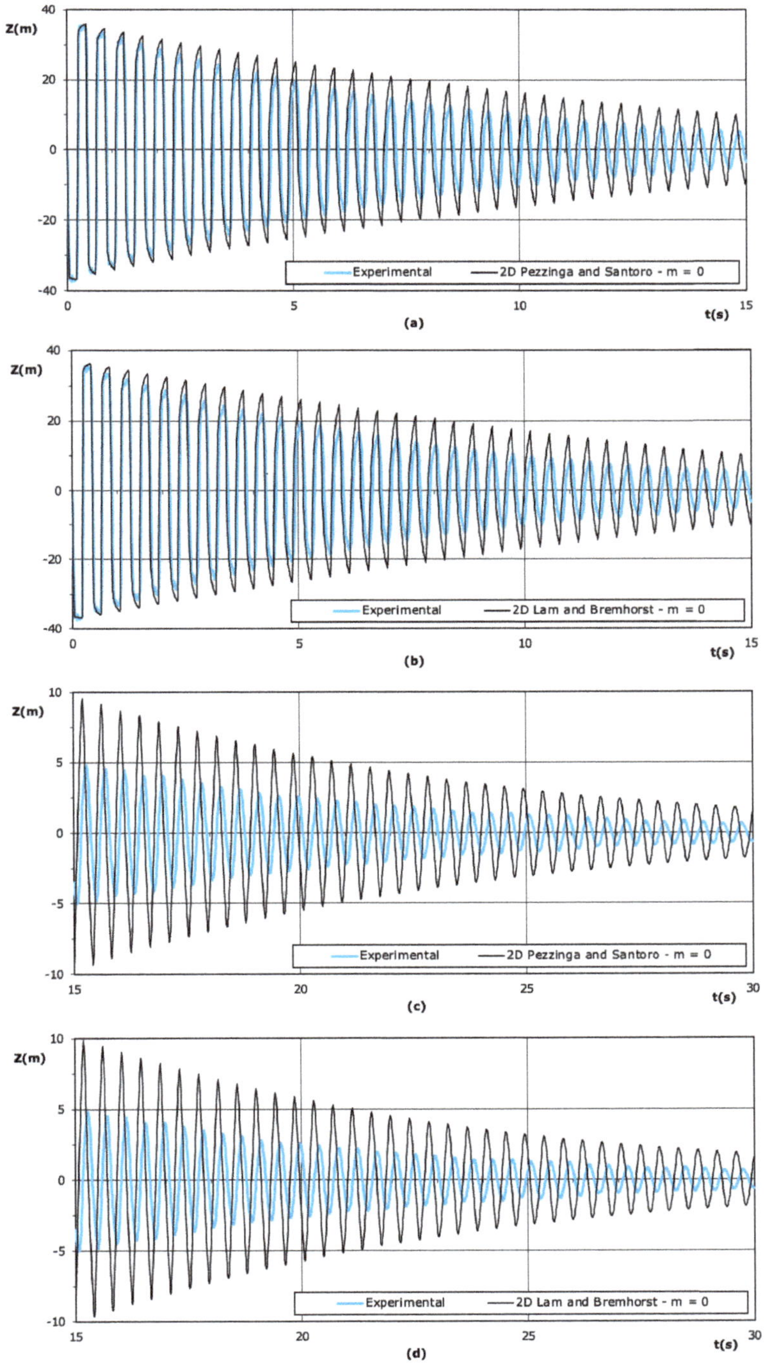

Figure 7. Comparison of head oscillations computed with the 2D present model or with the Lam and Bremhorst [23] model without free gas with the experimental results for Test L3: (**a**) 2D present model, $t = 0 - 15$ s; (**b**) 2D Lam and Bremhorst model, $t = 0 - 15$ s; (**c**) 2D present model, $t = 15 - 30$ s; (**d**) 2D Lam and Bremhorst model, $t = 15 - 30$ s.

The calibrated vales of the parameters for the 3 considered tests are reported in Table 2. An increase in the values of the mass of gas for increasing discharge already observed for transient gaseous cavitation can be noted [22]. However, this effect is probably due to particular conditions of the installation and of the experiments. With regard to the relaxation time, it can be observed that the calibrated values are of the same order of magnitude of the calibrated values for tests with transient gaseous cavitation [22].

Table 2. Values of calibrated parameters.

Test	2D—Constant Mass	2D—Variable Mass	
	m_0 (mg/m^3)	m_0 (mg/m^3)	θ_m (s)
L1	1.96	0.00	753.2
L2	17.30	6.16	754.2
L3	28.64	14.96	815.2

5. Conclusions

In the present study, the use of both 1D and 2D models was considered, as well as both constant and variable gaseous mass for water hammer flow. The use of such models has the aim to examine the damping of head oscillations for tests of very long duration (about 70 periods). Main conclusions follow.

From the methodological point of view, a recently developed form of the MOC, called MOC-Z, was used, operating without the need FOR interpolation as the standard MOC for flow with liquid and gas. The calibration of the parameters was carried out with a micro-GA, to obtain results with contained computer times.

The main improvement was in the modelling of head oscillation damping results from the 2D flow schematization with respect to the 1D one with quasi-steady friction. Taking into account the mass of gas, considered as constant, reduces the MAE because it allows to phase the computed oscillations, approaching it to the observed one. If the mass of free gas is considered as a variable, taking into account a gas release and solution process, the oscillation damping is caught altogether, provided that a proper calibration of the parameters of the model is made.

In conclusion, the oscillation damping observed in water hammer flow is mainly due to unsteady friction, but other mechanisms of dissipation exist. The attribution of further energy dissipation to the gas release and solution process, here used as a hypothesis, can explain the observed oscillation damping, with values of calibrated parameters similar to other ones previously obtained. Neglected effects, in particular the thermic exchange between bubbles and surrounding liquid, could influence the values of the calibrated parameters, but, as already obtained previously [21], they do not seem capable of fully reproducing the observed pressure traces. The obtained results have some relevance in the field of water hammer research, mainly because sometimes all the energy dissipation is attributed to unsteady friction, for example calibrating the coefficients of 1D unsteady friction models using comparison of numerical and experimental pressure oscillations. The use in this study of a 2D flow model and the comparison with very long duration experimental tests allows correctly putting into evidence the role of unsteady friction and the need of other possible mechanisms of dissipation. However, the proposed hypothesized mechanism of gas release and solution has to be deepened and validated with further experimental and theoretical studies. In particular, more complex models of gas release and solution could be considered in future studies.

Funding: This research received no external funding.

Data Availability Statement: The experimental and numerical results generated or used during the study are available from the author by request.

Conflicts of Interest: The author declares no conflict of interest.

References

1. Ghidaoui, M.S.; Zhao, M.; McInnis, D.A.; Axworthy, D.H. A review of water hammer theory and practice. *Appl. Mech. Rev.* **2005**, *58*, 49–76. [CrossRef]
2. Pezzinga, G. Evaluation of unsteady flow resistances by quasi-2D or 1D models. *J. Hydraul. Eng.* **2000**, *126*, 778–785. [CrossRef]
3. Vardy, A.E.; Hwang, K.L. A characteristics model of transient friction in pipes. *J. Hydraul. Res.* **1991**, *29*, 669–684. [CrossRef]
4. Eichinger, P.; Lein, G. The Influence of Friction on Unsteady Pipe Flow. In *Proceedings of the International Conference on Unsteady Flow and Fluid Transients, Durham, UK, 29 September–1 October 1992*; IAHR: Durham, UK, 1992; pp. 41–50.
5. Silva-Araya, W.F.; Chaudhry, M.H. Computation of energy dissipation in transient flow. *J. Hydraul. Eng.* **1997**, *123*, 108–115. [CrossRef]
6. Pezzinga, G. Quasi-2D Model for Unsteady Flow in Pipe Networks. *J. Hydraul. Eng.* **1999**, *125*, 676–685. [CrossRef]
7. Pezzinga, G.; Brunone, B. Turbulence, friction and energy dissipation in transient pipe flow. In *Vorticity and Turbulence Effects in Fluid Structures Interactions*; Brocchini, M., Trivellato, F., Eds.; WIT Press: Southampton, UK, 2006; pp. 213–236.
8. Vardy, A.E. On Sources of Damping in Water-Hammer. *Water* **2023**, *15*, 385. [CrossRef]
9. Ferras, D.; Manso, P.A.; Schleiss, A.J.; Covas, D.I. Experimental distinction of damping mechanisms during hydraulic transients in pipe flow. *J. Fluids Struct.* **2016**, *66*, 424–446. [CrossRef]
10. Kranenburg, C. Gas release during transient cavitation in pipes. *J. Hydraul. Div.* **1974**, *100*, 1383–1398. [CrossRef]
11. Wiggert, D.C.; Sundquist, M.J. The Effect of Gaseous Cavitation on Fluid Transients. *J. Fluids Eng.* **1979**, *101*, 79–86. [CrossRef]
12. Wylie, E.B. Low void fraction two-component two-phase flow. In *Unsteady Flow and Fluid Transients*; Bettess, R., Watts, J., Eds.; Balkema Rotterdam, The Netherlands, 1992; pp. 3–9.
13. Huygens, M.; Verhoeven, R.; Van Pocke, L. Air entrainment in water hammer phenomena. *WIT Trans. Eng. Sci.* **1998**, *18*, 10.
14. Hadj-Taieb, E.; Lili, T. Transient flow of homogeneous gas-liquid mixtures in pipelines. *Int. J. Numer. Methods Heat Fluid Flow* **1998**, *8*, 350–368. [CrossRef]
15. Lee, T.S.; Low, H.T.; Nguyen, D.T. Effects of air entrainment on fluid transients in pumping systems. *J. Appl. Fluid Mech.* **2007**, *1*, 55–61.
16. Lee, T.S.; Low, H.T.; Huang, W.D. Numerical study of fluid transient in pipes with air entrainment. *Int. J. Comput. Fluid Dyn.* **2004**, *18*, 381–391. [CrossRef]
17. Bergant, A.; Simpson, A.R.; Tijsseling, A.S. Water hammer with column separation: A historical review. *J. Fluids Struct.* **2006**, *22*, 135–171. [CrossRef]
18. Fanelli, M. *Hydraulic Transients with Water Column Separation*; IAHR Working Group 1971–1991 Synthesis Report; IAHR: Delft, The Netherlands; ENEL-CRIS: Milan, Italy, 2000.
19. Landau, L.D.; Lifshitz, E.M. *Fluid Mechanics: Course of Theoretical Physics*; Pergamon Press: London, UK, 1959; Volume 6.
20. Pezzinga, G. Second viscosity in transient cavitating pipe flows. *J. Hydraul. Res.* **2003**, *41*, 656–665. [CrossRef]
21. Cannizzaro, D.; Pezzinga, G. Energy dissipation in transient gaseous cavitation. *J. Hydraul. Eng.* **2005**, *131*, 724–732. [CrossRef]
22. Pezzinga, G.; Santoro, V.C. MOC-Z Models for Transient Gaseous Cavitation in Pipe Flow. *J. Hydraul. Eng.* **2020**, *146*, 04020076. [CrossRef]
23. Lam, C.K.G.; Bremhorst, K.A. Modified Form of the k-ε Model for Predicting Wall Turbulence. *J. Fluids Eng.* **1981**, *103*, 456–460. [CrossRef]
24. Wylie, E.B.; Streeter, V.L. *Fluid Transients in Systems*; Prentice-Hall: Englewood Cliffs, NJ, USA, 1993.
25. Zielke, W.; Perko, H.D.; Keller, A. Gas Release in Transient Pipe Flow. In *Pressure Surges, Proceedings of the 6th International Conference, Cambridge, UK, 4–6 October 1989*; BHRA: Cranfield, UK, 1990; pp. 3–13.
26. Santoro, V.C.; Crimì, A.; Pezzinga, G. Developments and limits of discrete vapor cavity models of transient cavitating pipe flow: 1D and 2D flow numerical analysis. *J. Hydraul. Eng.* **2018**, *144*, 04018047. [CrossRef]
27. Budny, D.D.; Wiggert, D.C.; Hatfield, F.J. The influence of structural damping on internal pressure during a transient pipe flow. *J. Fluids Eng.* **1991**, *113*, 424–429. [CrossRef]

Disclaimer/Publisher's Note: The statements, opinions and data contained in all publications are solely those of the individual author(s) and contributor(s) and not of MDPI and/or the editor(s). MDPI and/or the editor(s) disclaim responsibility for any injury to people or property resulting from any ideas, methods, instructions or products referred to in the content.

water

MDPI

Article

A Comparison of Different Methods for Modelling Water Hammer Valve Closure with CFD

Mehrdad Kalantar Neyestanaki [1,*], Georgiana Dunca [2], Pontus Jonsson [3] and Michel J. Cervantes [1]

1 Department of Engineering Sciences and Mathematics, Luleå University of Technology, 971 87 Luleå, Sweden; michel.cervantes@ltu.se
2 Department of Hydraulics, Hydraulic Machinery and Environmental Engineering, University Politehnica of Bucharest, 060042 București, Romania; georgianadunca@yahoo.co.uk
3 Vattenfall AB, 162 87 Stockholm, Sweden; pontus.jonsson@vattenfall.com
* Correspondence: mehrdad.kalantar.neyestanaki@ltu.se; Tel.: +46-700-721-475

Abstract: Water hammer is a transient phenomenon that occurs when a flowing fluid is rapidly decelerated, which can be harmful and damaging to a piping system. Three-dimensional computational fluid dynamics (CFD) with three-dimensional geometry is a common tool for studying water hammer, which is more accurate than numerical simulation with one-dimension approximation of the geometry. There are different methods with different accuracy and computational costs for valve closure modelling. This paper presents the result of water hammer 3D simulation with three main technics for modelling an axial valve closure: dynamic mesh, sliding mesh, and immersed solid methods. The variation of the differential pressure variation and the wall shear stress are compared with experimental results. Additionally, the 3D effects of the flow after the valve closure and the computational cost are addressed. The sliding mesh method presents the most physical results compared to the other two methods. The immersed solid method predicts a smaller pressure rise which may be the result of using a source term in the momentum equation instead of modelling the valve movement. The dynamic mesh method adds fluctuations to the primary phenomenon. Moreover, the sliding mesh is less expensive than the dynamic mesh method in terms of computational cost (approximately one-third), which was the primary method for axial valve closure modelling in the literature.

Keywords: CFD; water hammer; dynamic mesh; sliding mesh; immersed solid

Citation: Neyestanaki, M.K.; Dunca, G.; Jonsson, P.; Cervantes, M.J. A Comparison of Different Methods for Modelling Water Hammer Valve Closure with CFD. *Water* **2023**, *15*, 1510. https://doi.org/10.3390/w15081510

Academic Editors: Helena M. Ramos and Kamil Urbanowicz

Received: 10 March 2023
Revised: 4 April 2023
Accepted: 10 April 2023
Published: 12 April 2023

Copyright: © 2023 by the authors. Licensee MDPI, Basel, Switzerland. This article is an open access article distributed under the terms and conditions of the Creative Commons Attribution (CC BY) license (https://creativecommons.org/licenses/by/4.0/).

1. Introduction

Water hammer is a transient phenomenon caused by a sudden deceleration of the water in a closed system. Water hammer may cause a considerable pressure spike, just after the deceleration or acceleration of the fluid, followed by a pressure wave that travels periodically along the pipe. The wave is damped as it travels back and forth along the pipe. Water hammer may create strong vibrations, which put piping and equipment such as pumps and turbines in significant danger. Detailed information about pressure variation during this physical phenomenon can be used in designing pipe networks. Therefore, it is essential to accurately estimate the pressure rise during the water hammer.

An application which requires a decelerating flow, similarly to the water hammer, is the pressure-time method which is used for flow measurement in hydropower. This method takes advantage of the conversion of momentum into pressure during the deceleration of a liquid mass, caused by a valve or guide vane closure, to predict the flow rate [1]. Flow rate can be calculated by integrating the differential pressure and pressure losses due to friction during the water hammer in Equation (1).

$$Q = \frac{A}{\rho L}\int_0^{t_f}\left(\Delta p + \Delta p_f\right)dt + q, \tag{1}$$

where, Q, L, ρ, t_f, A, Δp, Δp_f, and q are the flow rate, length between the measuring cross-sections, density, final limit of integration, cross-section area, differential pressure, pressure losses due to friction, and leakage flow rate after valve closure. Transient viscous losses shall be estimated accurately during the water hammer to calculate the flow rate accurately. The existing evaluation method assumes one-dimensional flow limiting its applicability considerably. There is a need to extend the evaluation method to account for geometry variation as well as for second flows. Thus, 3-dimensional numerical simulations to evaluate the experimental data seem to be the next step for a better overall accuracy of the flow rate estimated.

Different boundary conditions were used in previous research to simulate transient flow regimes. Refs. [2,3] used a combination of 1D water hammer equations solved using the method of characteristics (1D-MOC) and CFD simulation to couple a pipe system to a more complex geometries such as a pump or turbine. Moreover, 1D-MOC was used to obtain the variation of the variables during valve closure. The mentioned variables were applied to the interface of the 1D and 3D domains for transient CFD simulation of the flow inside the pump and turbine. Saemi et al. [4] used 2D and 3D CFD simulations for modelling water hammer during a gate valve closure. They showed that a local recirculation zone appeared close to the gate for a downward valve closure, making the flow three-dimensional. However, 2D simulation can be used instead of 3D simulation for a distance larger than 2.33D from the gate as the recirculation zone created near the valve vanishes.

Refs. [5–8] changed the outlet boundary to the wall boundary condition for modelling the transient water hammer, simplifying the CFD simulation as the valve did not need to be modelled. However, this approach is not entirely accurate as the flow rate reduction is not instantaneous in reality. Ref. [9] used a velocity reduction at the outlet boundary instead of modelling the valve closure. They argued that a better agreement between CFD results and experimental data was obtained than with the MOC. However, the flow rate reduction curve may not always be available. The modelling of the valve closure may be necessary for more truthful simulation results.

There are several methods that can be used for modelling valve closure. The dynamic mesh method [4,10,11] has been used to model axial gate valve closure. In this method, the boundary moves, and the mesh deforms. Refs. [4,10,11] used total pressure at the inlet and static pressure at the outlet for the simulations. Remeshing with the dynamic mesh approach increases the simulation time and can cause divergence, especially at the end of the valve closure when re-meshing is performed in a smaller zone.

Refs. [12–15] used the sliding mesh method for modelling water hammer caused by the closure of a spherical valve, i.e., a circular movement of the valve. The results showed that the sliding mesh is an accurate tool for modelling the fast closure of ball valves rotational movement. In this method, separate zones move relative to each other. Despite the high capability of this method, no study using the sliding mesh method for a gate or sliding valve closure with vertical movement has been used yet.

Kalantar et al. [16] used the immersed solid method to model the valve closure. The immersed solid method defines a source term in the momentum equation to force velocity in the fluid domain to be the same as the immersed solid. Kalantar et al. [16] argued that only opening and total pressure at the inlet could predict oscillation [16]. This method is less time-consuming and more stable than the dynamic mesh method used for axial valve movement as the mesh deformation and re-meshing steps are removed.

The available literature presents results for the mentioned methods for modelling fast valve closure during a water hammer transient. They are applied to different cases with different types of valve closures. There is no study comparing the different methods for modelling valve closure during such a water hammer transient in terms of modelling accuracy and computational cost. Moreover, the sliding mesh has not been used for modelling the axial valve closure.

In this paper, the water hammer in a straight 3D pipe during an axial gate valve closure is modelled using CFD. Three methods: dynamic mesh, sliding mesh, and immersed solid methods are used for modelling the valve closure. The transient results are compared with experimental data from Ref. [17] that include the variation of the differential pressure between two cross-sections and the wall shear stress. Moreover, the three-dimensionality of the flow after the valve closure and the computational cost of mentioned methods are addressed.

2. Materials and Methods

2.1. Test Case

The analyzed test case of this study is based on an experimental investigation conducted by Sundstrom et al. [17]. The geometry is a straight pipe with a constant internal diameter of 300 mm. A schematic of the test apparatus used for the experiment is shown in Figure 1. The water flows by gravity from a head tank, situated $H = 9.75$ m above the measuring section. The maximum flow rate is $Q = 0.410$ m^3/s.

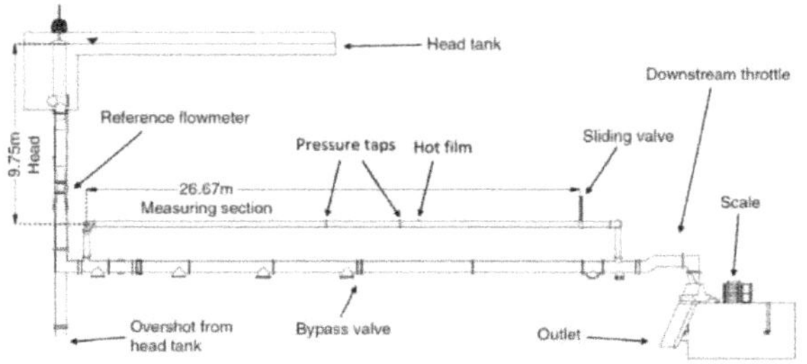

Figure 1. The water hammer test rig schematic. Figure courtesy of Sundstrom et al. [17].

A gate valve is used to decrease the flow rate which can be reduced to zero in 4.68 s. A differential pressure transducer with a range of 0–5 bar and an accuracy of 0.04% of full scale measures the pressure variation between two sections, 11 and 15 m upstream of the valve. In addition, the wall shear stress at the cross-section 10 m upstream of the valve was measured using a hot-film probe. The flow rate during the measurement was $Q = 0.169$ m^3/s, i.e., a Reynolds number $Re = 7 \times 10^5$.

The geometry is considered a straight pipe, and other parts such as elbows and fittings are eliminated in the CFD geometry for simplification. The length of the pipe is considered to be 36 m to match the water hammer period obtained in the experiments.

2.2. Mathematical Modelling

The continuity and momentum equations for a time-dependent isothermal compressible turbulent flow are given by:

$$\frac{\partial \rho}{\partial t} + \frac{\partial (\rho U_j)}{\partial x_j} = 0, \tag{2}$$

$$\frac{\partial (\rho U_i)}{\partial t} + \frac{\partial (\rho U_j U_i)}{\partial x_j} = -\frac{\partial P}{\partial x_i} + \frac{\partial}{\partial x_j}\left(\mu \frac{\partial U_i}{\partial x_j} - \rho \overline{u_i u_j}\right), \tag{3}$$

where P, ρ, U_i, and μ are the pressure, fluid density, mean velocity, and fluid dynamic viscosity, respectively. To model the Reynolds shear stress term $(-\rho \overline{u_i u_j})$ in the turbulent flow, the low Reynolds k-ω SST model [18] is used. Ref. [12] demonstrated that low-Re SST k-ω turbulence models predict more acceptable results for pressure variation during water hammer than high-Re turbulence models. The k-ε model with wall functions cannot capture the variation of the velocity profile close to the wall. This turbulence model was used in similar studies with satisfactory results [4,10–12]. To model the fluid compressibility, Hooke's law with Equation (4) describing the variation of the density with the pressure [19] is used. The effects of pipe elasticity [11] are accounted in modified bulk modulus K'_f,

Equation (5), where E, e, and D are the Young's modulus of elasticity, thickness, and pipe diameter, respectively.

$$dP/d\rho = K_f/\rho \tag{4}$$

$$K_f' = \frac{K_f}{1 + \frac{K_f D}{eE}} \tag{5}$$

To eliminate the impact of the outlet boundary condition, the pipe was extended for 6 m downstream the valve, i.e., 20 × D. The total pressure value at the inlet ($[P + \frac{1}{2}\rho U^2]_{inlet} = constant$) is adjusted in the steady-state simulation to match the experiment's flow rate, while at the outlet the atmospheric pressure ($P_{outlet} = p_{atm}$) is used. The boundary condition and the geometry used for the simulation is shown in Figure 2. A converged steady-state solution with a constant flow rate is employed for the initial condition of the transient simulation including the valve movement. A root-mean-square (RMS) residual level of 10-5 is considered for the convergence of the variables.

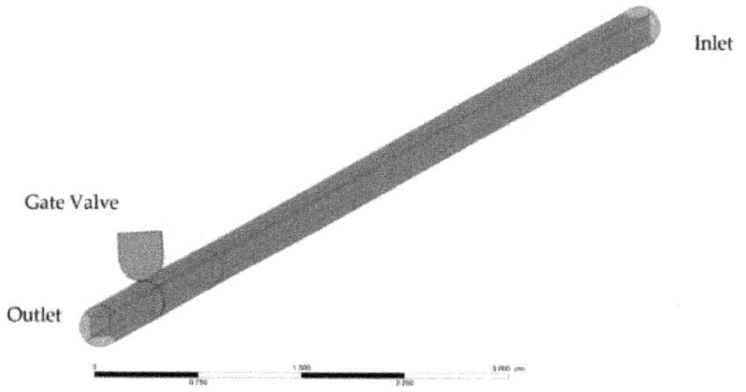

Figure 2. The boundary conditions and geometry used for the simulation.

2.3. *Valve Closure Modeling*

Three methods were used to model the valve closure: dynamic mesh, mesh motion, and immersed solid. Two codes were used: Ansys-Fluent and Ansys-CFX as immersed solid is unavailable in Ansys-Fluent and the sliding mesh is not available for domain translation in Ansys-CFX.

2.3.1. Immersed Solid Method

In the immersed solid method, the valve body is modelled by the domain named immersed solid domain. The immersed solid domain overlaps the fluid domain, represented by a pipe, as the valve enters the pipe, shown in Figure 3. In this method, there is no mesh deformation, re-meshing, or domain interface, making this method simple to implement and computationally effective. Instead, the region of overlap between the fluid domain and immersed solid domain is identified at each time step of the simulation. At the fluid cells overlapping with the immersed solid cells, a source term is defined in the momentum equation to match the fluid velocity to the solid velocity [20]. As the momentum equations enforce the fluid velocity in the fluid region to be the same as the velocity of the immersed solid, it will not precisely model the same physical phenomena. The part of the fluid domain that overlaps with the immersed solid has a downward velocity similar to the gate. However, there should be no water there. Moreover, the estimation of the source term by the solver could lead to some leakage through the immersed solid [16].

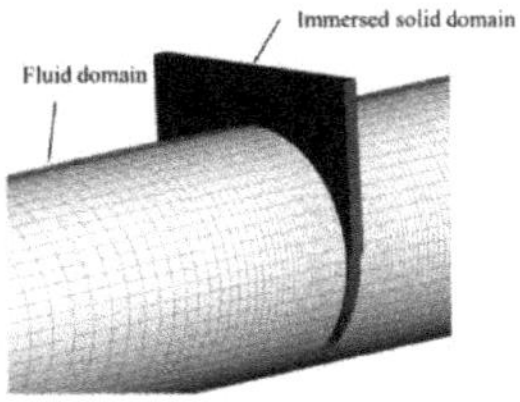

Figure 3. Overlapping of the fluid zone and immersed solid zone at time t =4 s (close to the end of the valve closure) in the immersed solid method for modelling valve movement.

2.3.2. Dynamic Mesh

In the dynamic mesh method, the complete geometry is modelled with 3 domains: the pipe upstream the valve (domain 1), the space to be occupied by the valve in the pipe (domain 2), and the pipe downstream the valve (domain 3), see Figure 4. The lower part of the valve is represented by the upper part of domain 2. As the valve moves inside the pipe, the upper part of domain 2 moves downward, decreasing the volume of domain 2. The movement shall be normal to the boundary and involves mesh deformation and remeshing. As the valve moves inside the pipe, a space modelling the valve, i.e., a solid, appears between domain 1 and domain 3. Interfaces connect the domains which are updated at each time step.

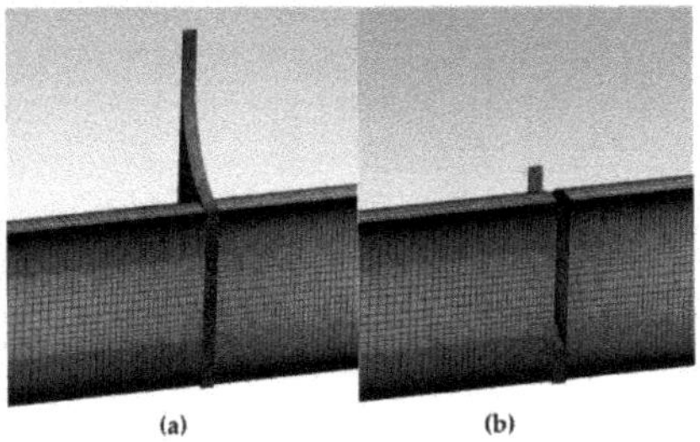

Figure 4. Dynamic mesh method grid cut in half: (**a**) before valve movement; (**b**) at t = 4 s close to the end of valve closure.

The dynamic mesh is available in Ansys-Fluent allowing smoothing and re-meshing at each time step. Re-meshing is not available at Ansys-CFX; therefore, Ansys-Fluent is used. Both smoothing and re-meshing are used in mesh deformation. The spring/Laplacian-based smoothing method [21] is employed. According to Hooke's law, a displacement at a moving boundary node will produce a force proportional to the displacement along the grid. The spring/Laplacian smoothing process moves each mesh vertex closer to its surrounding vertex's geometric center. Cell deformation with smoothing for considerable displacement becomes excessively skewed [21]. Therefore, re-meshing is used when the quality of the mesh decreases below thresholds; the maximum face skewness (0.5) and maximum cell skewness (0.7). Moreover, only triangular or tetrahedral mesh can be used for re-meshing in the dynamic mesh zone.

The re-meshing makes the simulation more time-consuming and expensive compared to the other two methods. At the end of the valve closure, as the space to re-mesh is smaller and smaller, there is a higher possibility of divergence as it is challenging to ensure a good quality. To solve the problem, a lower under-relaxation factor value and a higher number of iterations at each time step is considered, making the simulation even more expensive. Moreover, as the mesh is updated at each time step through deformation or a new mesh, the data from the previous time step will be interpolated, which could cause some errors.

2.3.3. Sliding Mesh

In the sliding mesh method, the computational domain is represented by three subdomains, like for the dynamic mesh. However, domain 2 is now sliding; thus, no deformation or mesh adjustment is performed. The other fluid zones are stationary. The mesh representing the volume to be occupied by the gate, domain 2, moves along the interface relative to the stationary mesh. After sliding the domain, the interface re-establishes the zone connectivity at each time step. Each zone (stationary or sliding) has at least one "interface zone" around it where it intersects with the neighboring cell zone. In the case of non-overlapping boundaries, wall boundaries are considered. Since the mesh does not deform, the downward movement of the valve zone extends at the bottom of the pipe, shown in Figure 5. Therefore, the bottom of the pipe at the position of domain 2 presents a cavity during the transient that does not present in reality. This could be why dynamic mesh is used for all previous research on gate valve closure [4,10,11]. However, as the valve body thickness is small compared to the pipe's diameter ($0.06 \times D$), it may have a neglectable effect on the results compared to the expensive method such as the dynamic mesh method. The axial and rotational sliding of mesh are available in Ansys-Fluent; however, the axial movement is not available at Ansys-CFX.

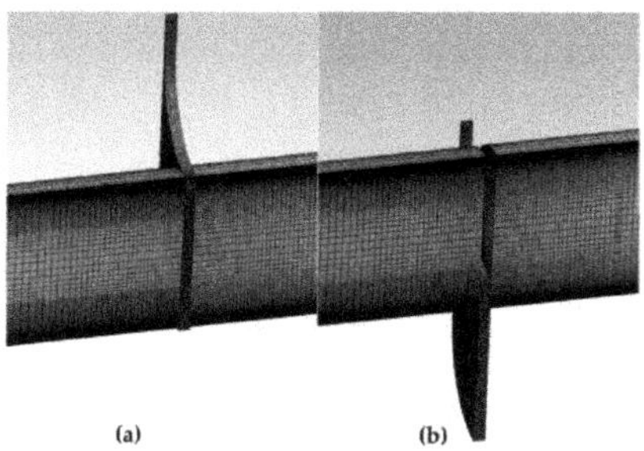

(a) (b)

Figure 5. Sliding mesh method grid cut in half: (**a**) at $t = 0$ s before valve movement; (**b**) at $t = 4$ s close to the end of valve closure.

2.4. Computational Setup

As mentioned, all methods are not available in both ANSYS-CFX and ANSYS-Fluent. Therefore, both solvers are employed based on the valve closure method. For the immersed solid method, ANSYS-CFX is used to solve the continuity, momentum, turbulence eddy frequency, and turbulence kinetic energy equations using the coupled finite volume method. The high-resolution scheme is used to solve the terms in the mentioned equations. The high-resolution scheme in CFX is equilibrium to upwind with the number of iterations to increase accuracy. Moreover, CFX uses a couple solver.

The pipe mesh is made of hexahedral elements with a finer mesh in the viscous sublayer to obtain $y^+ < 1$ at the wall. Moreover, a finer mesh close to the valve is also used to have better accuracy, see Figure 6.

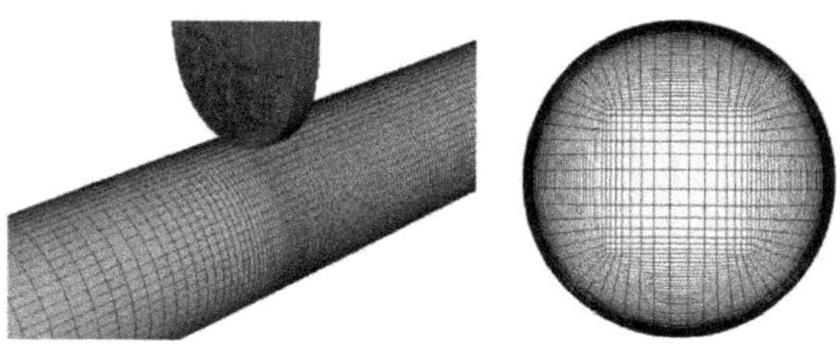

(a) (b)

Figure 6. Immersed solid method grid: (**a**) fine grid near the gate; (**b**) grid at the pipe cross-section.

Three different meshes, including both fluid and immersed solid, with grid nodes of 1×10^6, 2.4×10^6, and 4.5×10^6 are made using a time step size of 0.1 ms to study the effect of the mesh on the transient simulation solution. The average aspect ratio is around 1021, and the skewness is 0.13.

The average absolute pressure variation at the surface 11 m upstream of the valve is monitored for the three grids during the valve closure and is presented in Figure 7. This point is one of the points that were later used for validation in the differential pressure measurement. The coarse grids predict different results, frequency, and amplitude of the pressure oscillations. The denser and medium grids present similar results. Therefore, the mesh with 2.4×10^6 nodes is considered for the simulation.

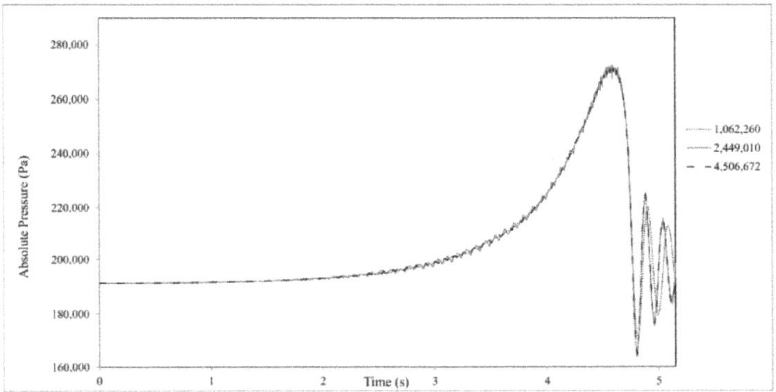

Figure 7. Average absolute pressure monitoring at section 11 m upstream of valve for three different grids.

Four time steps ranging from 1 ms to 0.1 ms were applied to the simulation to study the result's sensitivity to the time step size. The larger time step is not able to predict the pressure oscillations (Figure 8). By reducing the time step size, the oscillation's amplitude increases and becomes less sensitive to it. A time step size 0.1 ms is considered for independent simulation, which is used in a similar simulation by Kalantar et al. [16]. With a smaller time step, the simulation became too long and gave unphysical results.

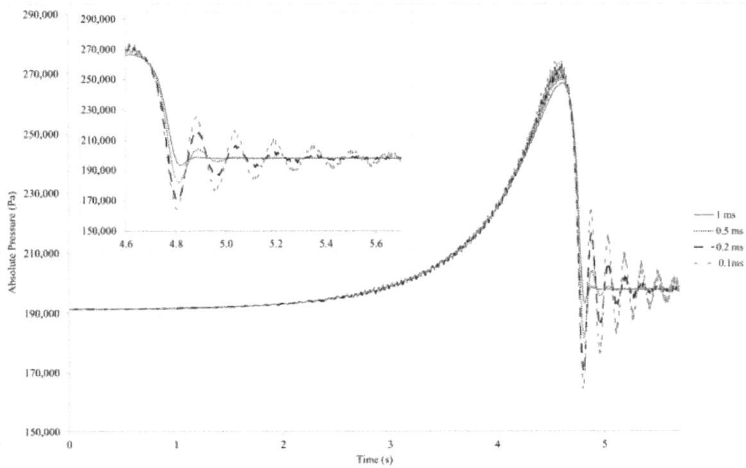

Figure 8. Average absolute pressure monitoring at section 11 m upstream of valve for four different time steps.

For modelling the valve movement with the dynamic mesh method, ANSYS-Fluent is employed. SIMPLE algorithm is used for solving the coupled equations of motion using the finite volume method. The third-order monotonic upwind method (MUSCL) is employed to discretize all transport equations' non-linear convective terms. A similar number of the elements (2.2 mil grid nodes) and time step (0.1 ms) to Refs. [4,10] are employed for the simulations. In the dynamic mesh method, the maximum face skewness of 0.5 and the maximum cell skewness of 0.7 are considered as the threshold for re-meshing.

For the sliding mesh, ANSYS-Fluent with a similar configuration is used. However, the domain slides and does not need any smoothing or re-meshing.

3. Results

3.1. Pressure Variation

The result of the simulations with different valve modelling has been validated with experimental data from Ref. [17]. The differential pressure variation at the pressure tap between two cross-sections, 11 m and 15 m upstream of the valve, is compared with experimental data in Figure 9. The numerical pressure is obtained at the position of the experimental pressure taps. The maximum peak is lower than the experimental one for the immersed solid method. After the supposed complete valve closure, there is a leakage flow rate, 0.06% of the initial flow rate. This leakage also happens during the valve movement, which leads to a delay in the flow deceleration and thus conversion of momentum into pressure which can be seen in the small delay of pressure the rise before t = 4 s in Figure 9. The underestimation of the maximum peak is certainly related to the leakage and delay in the flow reduction, which affects the maximum pressure rise. The pressure oscillations after the valve closure are also underestimated, which could be for the same reason. Moreover, possible source term estimation errors may add a fluctuation to the main phenomenon, as seen in Figure 9.

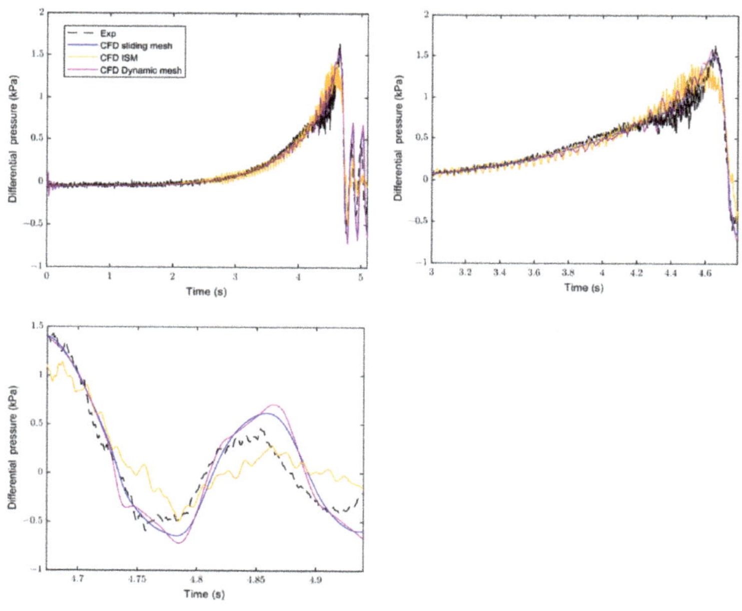

Figure 9. The variation of differential pressure with time during the valve closure between two cross-sections 11 m and 15 m upstream of the valve.

The maximum pressure peak in the immersed solid method happens earlier than the experimental one and other CFD results, about 0.1 s. It could be the effect of the downward movement of the water in the fluid zone overlapping the immersed solid. As mentioned, part of the fluid domain overlapping the immersed solid has a downward velocity similar to the gate. However, there should be no water there.

Fluctuations are observed at the start of the valve movement for the dynamic mesh. Changing the mesh at each time step, extrapolation of data to new mesh and skewed mesh during the smoothing leads to error in the simulation, which may be the reason for the instability. Smoothing and re-meshing lead to a higher deviation from the experiment compared to results from sliding mesh, with high-quality mesh.

The sliding mesh technic has a better agreement in predicting the maximum pressure peak and oscillation after the valve closure than the immersed solid and dynamic mesh methods. This method models the same phenomena that happen compared to the immersed solid method. Furthermore, the sliding mesh method uses a higher-quality grid and predicts pressure variation with less fluctuation than the dynamic mesh. Moreover, this method is less computationally intensive than the dynamic mesh.

Results from the sliding and dynamic mesh methods overestimate the pressure oscillations after the valve closure compared to the experimental ones. The reason could be the geometrical differences, simplification in the geometry, and the experimental results' sensitivity to the tubing in differential pressure measurement. The irregularities and the tank in the test rig may have a higher damping ratio compared to the geometry used in the simulation.

For a better comparison, the predicted differential pressure during the valve movement closure with the sliding mesh method, the closest model to the experiment minus the experimental value, is presented in Figure 10.

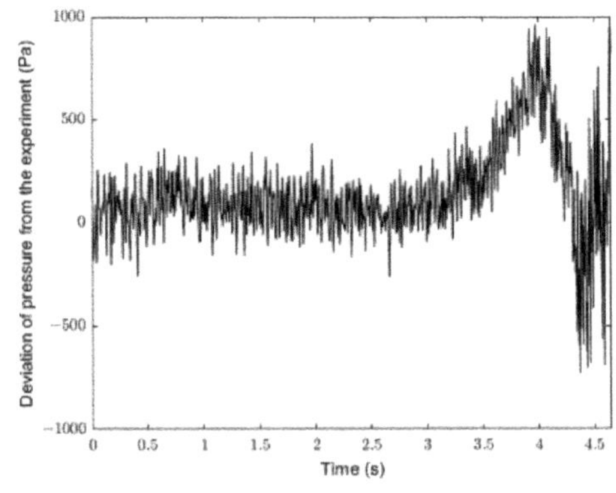

Figure 10. The deviation of estimated differential pressure by sliding mesh with the experiment between two cross-sections 11 m and 15 m upstream of the valve.

As it can be seen, the highest deviation to the experiment happens at the end of the valve closure, and the simulation underestimates the peak. At this time, the variation of the pressure during the water hammer is maximum and based on Equation (4), the highest fluid density variation occurs.

3.2. Wall Shear Stress Variation

The normalized magnitude of the wall shear stress is presented for the different methods used in Figure 11. The results from the sliding mesh agree better than the other methods with the experimental values. The immersed solid method overestimates the wall shear stress between 3 and 4.6 s. The reason could be the delay in flow rate reduction and possible leakage with the method. The dynamic mesh has a good agreement for predicting the wall shear stress during the valve closure. However, there is a slight overestimation of the wall shear stress from 3 to 4.6 s. The sliding mesh has the best agreement for predicting the wall shear stress during the valve closure compared to the other methods. All the methods overestimate the wall shear stress after the valve closure during the oscillations. The irregularities, such as elbow and contractions, and the tanks in the test rig may have a higher damping ratio compared to the simplified geometry used, a single pipe. Therefore, a lower flow rate will be expected during the oscillation compared to the geometry used in the simulation.

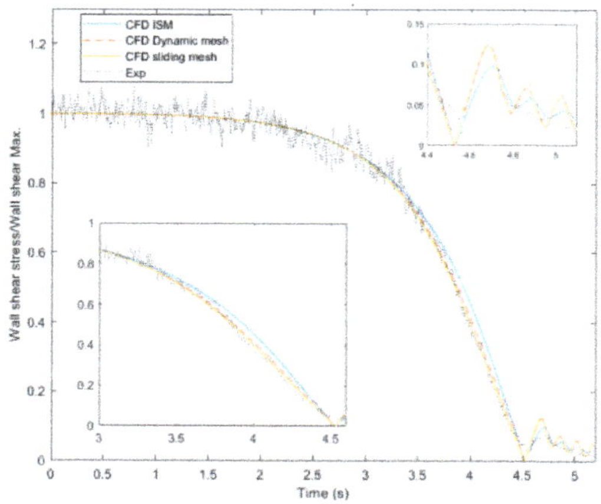

Figure 11. The time variation of the magnitude wall shear stress at cross-sections 10 m upstream of the valve.

3.3. 3D Effects

As mentioned, a non-symmetrical recirculation region appears near the gate valve with its movement. Therefore, the flow is 3D close to the valve; however, far from the valve, the flow is 2D. The 3D flow related to the valve closure during the water hammer is visualized by contours of the axial velocity close to the valve at the end of the closure. Streamlines and axial velocity contour of the flow inside the pipe are presented in Figure 12. This non-symmetrical recirculation region extends approximately two to three pipe diameters depending on the simulation method. In Figure 12, the streamlines are presented at time $t = 4.75$ s; however, the immersed solid method models may have a small time shift regarding the delay in the flow rate reduction. Immersed solid method predicts a shorter length of 3D effect because of possible leakage and weaker water hammer. Dynamic mesh indicates a negative axial velocity at the bottom of the pipe close to the valve, which other methods do not predict. The reason could be the skewed mesh at the end of valve closure, inducing an unphysical value for the velocity profile. The sliding mesh, the most physical method, predicts the most extended 3D effects.

The 3D effect can be observed more accurately with 3D flow streamlines close to the valve at the end of valve closure, shown in Figure 13. The leakage flow can be observed in the 3D streamlines obtained from the immersed solid method. The two-dimensional streamlines cannot show the recirculation zone along to the pipe wall. For the simulations performed using the sliding mesh and dynamic mesh techniques, the recirculation around the top and sides of the pipe wall is similar and recirculation zone at the bottom of the pipe is weaker. However, the recirculation zones close to the top of the pipe are weaker for simulation by the immersed solid technique than other results. It could be related to leakage through the immersed solid during the valve closure. The 3D streamlines predicted a similar 3D structure after the valve closure for simulation using the sliding mesh and dynamic mesh techniques.

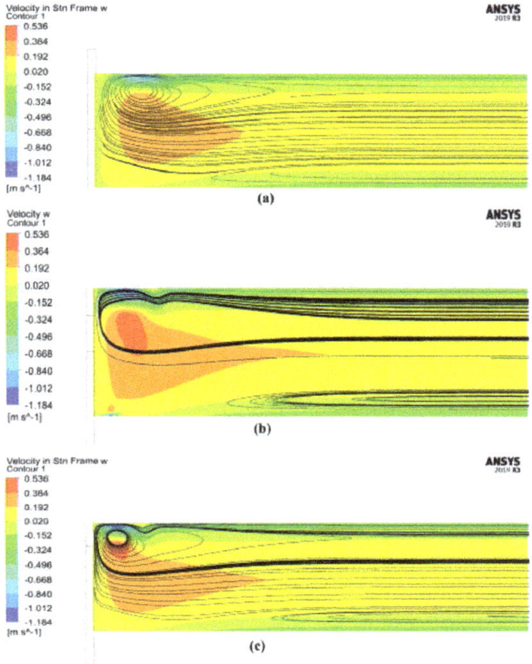

Figure 12. Streamlines and axial velocity contour of the flow inside the pipe at the end of the valve closure: (**a**) immersed solid method; (**b**) dynamic mesh; (**c**) sliding mesh.

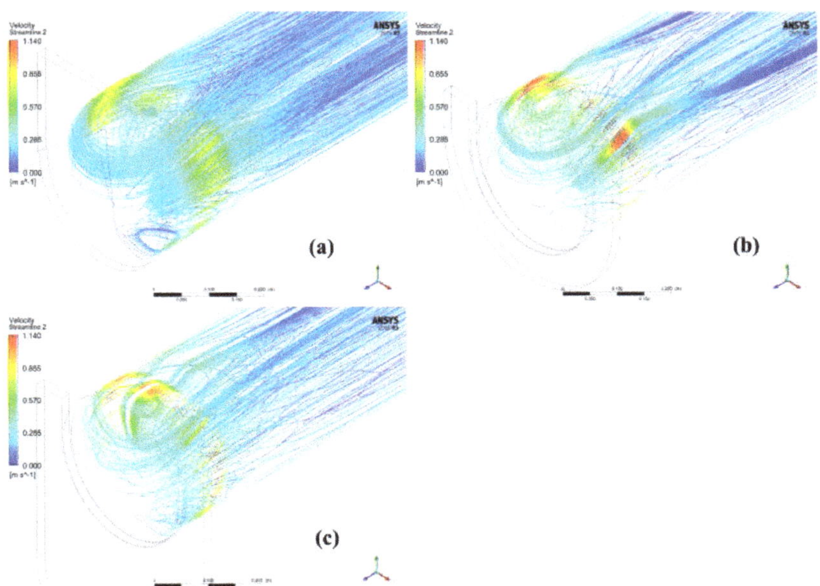

Figure 13. Streamlines of the flow inside the pipe at the end of the valve closure for the different valve modelling method: (**a**) immersed solid; (**b**) dynamic mesh; (**c**) sliding mesh.

The axial velocity profile at the vertical line with distances 25 cm, 50 cm, and 75 cm upstream of the valve at the end of the valve closure is presented in Figure 14. The velocity profiles are more asymmetrical closer to the valve. The flow field gradually becomes symmetrical away from the valve. By moving from the valve towards the inlet, the 3D effect decreases, and in line "c" (with a distance of 75 cm upstream of the valve), the 3D effect ends. Therefore, the pressure measurement for application, like the pressure-time method, shall be performed at a section before this area.

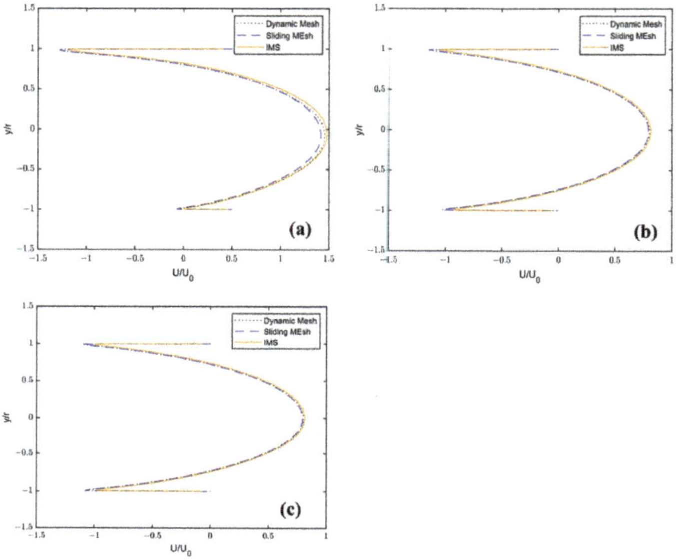

Figure 14. Axial velocity profile at the vertical line with distances of: (**a**) 25cm; (**b**) 50 cm; (**c**) 75 cm upstream of the valve at the end of the valve closure.

The 3D effect is more significant for the results from sliding mesh and the immersed solid method predicts the least 3D effects because of possible leakage and weaker water hamme

3.4. Computational Cost

Another essential aspect to consider is the computational cost of each method. Immersed solid method and sliding mesh have quite the same computational cost. For dynamic mesh, the process of re-meshing is added to the calculation. Moreover, the lower quality of regenerated mesh makes it more possible to diverge. Therefore, a lower under-relaxation factor and, consequently, a higher number of iterations per time step is applied in the simulation. The mentioned drawbacks made the computational cost of the dynamic mesh method around three times more than the sliding mesh and immersed solid method. The details of the computational resources and time used for the simulation with the different methods are summarized in Table 1.

Table 1. Computational resources and time allocated to the simulations.

Method	Number of CPU (2.60 GHz)	App. Time (h)
Immersed solid method	48	60
Dynamic mesh method	48	170
Sliding mesh method	48	65

4. Conclusions and Discussion

The current method presents a CFD simulation of the water hammer caused by the axial movement of a gate valve in a straight pipe. Three techniques for modelling valve closure are employed, including immersed solid (Ansys-CFX), dynamic, and sliding mesh methods (Ansys-Fluent). The results show that immersed solid method has a delay in flow rate reduction, which underestimates differential pressure rise and overestimates wall shear stress close to the end of valve closure. The dynamic mesh method models the same physical phenomenon. However, it is more time-consuming and three times more expensive in terms of computational cost than other methods. Furthermore, the dynamic mesh was unstable, with the possibility of divergence.

As an inexpensive and stable technique, the sliding mesh method predicts the closest result to the experimental value. It was proved that for a thin gate valve, the axial movement of the valve can be modelled by mesh movement without any mesh deformation. This method can predict more physical results for the conditions mentioned in the paper.

Author Contributions: Conceptualization, M.K.N. and M.J.C.; methodology, M.K.N. and M.J.C.; software, M.K.N.; validation, M.K.N.; formal analysis, M.K.N.; investigation, M.K.N., G.D. and M.J.C.; resources, M.J.C.; data curation, M.K.N.; writing—original draft preparation, M.K.N.; writing—review and editing, P.J., G.D. and M.J.C.; visualization, M.K.N.; supervision, P.J., G.D. and M.J.C.; project administration, M.J.C.; funding acquisition, M.J.C. All authors have read and agreed to the published version of the manuscript.

Funding: This research was funded by "Swedish Hydropower Centre-SVC" and "Luleå University of Technology-LTU". "Swedish Hydropower Centre-SVC" established by the Swedish Energy Agency, Elforsk and Svenska Kraftnät, together with the Luleå University of Technology, The Royal Institute of Technology, Chalmers University of Technology, and Uppsala University.

Data Availability Statement: Data available on request due to restrictions privacy.

Acknowledgments: The research presented was carried out as a part of "Swedish Hydropower Centre-SVC" established by the Swedish Energy Agency, Elforsk and Svenska Kraftnät, together with the Luleå University of Technology, The Royal Institute of Technology, Chalmers University of Technology, and Uppsala University.

Conflicts of Interest: The authors declare no conflict of interest.

References

1. *IEC 60041, I.*; Field Acceptance Tests to Determine the Hydraulic Performance of Hydraulic Turbines, Storage Pumps and Pump-Turbines. International Electrotechnical Commission: Geneva, Switzerland, 1991.
2. Zhang, X.; Cheng, Y.; Xia, L.; Yang, J. CFD Simulation of Reverse Water-Hammer Induced by Collapse of Draft-Tube Cavity in a Model Pump-Turbine during Runaway Process. In Proceedings of the IOP Conference Series: Earth and Environmental Science, Grenoble, France, 4–8 July 2016; Volume 49.
3. Wu, D.; Yang, S.; Wu, P.; Wang, L. MOC-CFD Coupled Approach for the Analysis of the Fluid Dynamic Interaction between Water Hammer and Pump. *J. Hydraul. Eng.* **2015**, *141*, 06015003. [CrossRef]
4. Saemi, S.; Raisee, M.; Cervantes, M.J.; Nourbakhsh, A. Numerical Investigation of the Pressure-Time Method Considering Pipe with Variable Cross Section. *J. Fluids Eng. Trans. ASME* **2018**, *140*, 101401. [CrossRef]
5. Saemi, S.D.; Raisee, M.; Cervantes, M.J.; Nourbakhsh, A. Computation of Laminar and Turbulent Water Hammer Flows. In Proceedings of the 11th World Congress on Computational Mechanics (WCCM XI) 5th European Conference on Computational Mechanics (ECCM V) 6th European Conference on Computational Fluid Dynamics (ECFD VI), Barcelona, Spain, 20–25 June 2014.
6. Martin, N.M.C.; Wahba, E.M. On the Hierarchy of Models for Pipe Transients: From Quasi-Two-Dimensional Water Hammer Models to Full Three-Dimensional Computational Fluid Dynamics Models. *J. Press. Vessel. Technol. Trans. ASME* **2022**, *144*, 021402. [CrossRef]
7. Nikpour, M.R.; Nazemi, A.H.; Dalir, A.H.; Shoja, F.; Varjavand, P. Experimental and Numerical Simulation of Water Hammer. *Arab. J. Sci. Eng.* **2014**, *39*, 2669–2675. [CrossRef]
8. Li, J.; Wu, P.; Yang, J. Cfd Numerical Simulation of Water Hammer in Pipeline Based on the Navier-Stokes Equation. In Proceedings of theV European Conference on Computational Fluid Dynamics, Lisbon, Portugal, 14–17 June 2010.
9. Mandair, S.; Magnan, R.; Morissette, J.-F.; Karney, B. Energy-Based Evaluation of 1D Unsteady Friction Models for Classic Laminar Water Hammer with Comparison to CFD. *J. Hydraul. Eng.* **2020**, *146*, 04019072. [CrossRef]

10. Saemi, S.; Cervantes, M.J.; Raisee, M.; Nourbakhsh, A. Numerical Investigation of the Pressure-Time Method. *Flow Meas. Instrum.* **2017**, *55*, 44–58. [CrossRef]
11. Saemi, S.; Sundström, L.R.J.; Cervantes, M.J.; Raisee, M. Evaluation of Transient Effects in the Pressure-Time Method. *Flow Meas. Instrum.* **2019**, *68*, 101581. [CrossRef]
12. SaemI, S.; Raisee, M.; Cervantes, M.J.; Nourbakhsh, A. Computation of Two- and Three-Dimensional Water Hammer Flows. *J. Hydraul. Res.* **2019**, *57*, 386–404. [CrossRef]
13. Wang, H.; Zhou, L.; Liu, D.; Karney, B.; Wang, P.; Xia, L.; Ma, J.; Xu, C. CFD Approach for Column Separation in Water Pipelines. *J. Hydraul. Eng.* **2016**, *142*, 04016036. [CrossRef]
14. Yang, S.; Wu, D.; Lai, Z.; Du, T. Three-Dimensional Computational Fluid Dynamics Simulation of Valve-Induced Water Hammer. *Proc. Inst. Mech. Eng. C J. Mech. Eng. Sci.* **2017**, *231*, 2263–2274. [CrossRef]
15. Han, Y.; Shi, W.; Xu, H.; Wang, J.; Zhou, L. Effects of Closing Times and Laws on Water Hammer in a Ball Valve Pipeline. *Water* **2022**, *14*, 1497. [CrossRef]
16. Kalantar, M.; Jonsson, P.; Dunca, G.; Cervantes, M.J. Numerical Investigation of the Pressure-Time Method. *IOP Conf. Ser. Earth Environ. Sci.* **2022**, *1079*, 012075. [CrossRef]
17. Sundstrom, L.R.J.; Cervantes, M.J. Transient Wall Shear Stress Measurements and Estimates at High Reynolds Numbers. *Flow Meas. Instrum.* **2017**, *58*, 112–119. [CrossRef]
18. Menter, F.R. Two-Equation Eddy-Viscosity Turbulence Models for Engineering Applications. *AIAA J.* **1994**, *32*, 1598–1605. [CrossRef]
19. Korteweg, D.J. Ueber Die Fortpflanzungsgeschwindigkeit Des Schalles in Elastischen Röhren. *Ann. Phys.* **1878**, *241*, 525–542. [CrossRef]
20. Yoon, Y.; Park, B.H.; Shim, J.; Han, Y.O.; Hong, B.J.; Yun, S.H. Numerical Simulation of Three-Dimensional External Gear Pump Using Immersed Solid Method. *Appl. Therm. Eng.* **2017**, *118*, 539–550. [CrossRef]
21. ANSYS, Inc. *ANSYS Fluent User's Guide, Release 17.2*; ANSYS, Inc.: Canonsburg, PA, USA, 2016.

Disclaimer/Publisher's Note: The statements, opinions and data contained in all publications are solely those of the individual author(s) and contributor(s) and not of MDPI and/or the editor(s). MDPI and/or the editor(s) disclaim responsibility for any injury to people or property resulting from any ideas, methods, instructions or products referred to in the content.

water

MDPI

Article

New Optimized Equal-Area Mesh Used in Axisymmetric Models for Laminar Transient Flows

Pedro Leite Ferreira [1,*] and Dídia Isabel Cameira Covas [2]

1 Department of Civil Engineering, Polytechnic of Porto, 4249-015 Porto, Portugal
2 CERIS, Instituto Superior Técnico, Universidade de Lisboa, 1049-001 Lisboa, Portugal
* Correspondence: plf@isep.ipp.pt

Abstract: The current paper aims at assessing the effect of the radial mesh on the description of the axial velocity in steady-state and transient conditions and at presenting the results of a new optimized equal-area mesh. For this purpose, a quasi-2D model is implemented and tested for different mesh configurations and sizes. A new two-region mesh geometry with 40 cylinders is proposed to optimize the description of the wall shear stress immediately after each pressure variation. This mesh is composed of two regions: one with a high-resolution near the pipe wall and the second with a coarser grid in the pipe core. Different configurations of this mesh are analysed for both steady and unsteady conditions. Results are compared with those obtained by a 1D model and with experimental data for laminar flows, discussed in terms of the computation effort and accuracy. The proposed two-region mesh has demonstrated: (i) a reduction in the simulation error by five times when compared with standard meshes for the same computational effort and for the instantaneous valve closure; (ii) an important improvement in accuracy for an experimental S-shape valve maneuver, particularly for meshes with few cylinders; and (iii) a correct description of the transient pressures collected in the experimental tests.

Keywords: hydraulic transient; unsteady friction; laminar flow; quasi-2D model; axisymmetric model; 1D model; radial mesh; experimental tests

Citation: Ferreira, P.L.; Covas, D.I.C. New Optimized Equal-Area Mesh Used in Axisymmetric Models for Laminar Transient Flows. *Water* **2023**, *15*, 1402. https://doi.org/10.3390/w15071402

Academic Editor: Giuseppe Pezzinga

Received: 1 February 2023
Revised: 27 March 2023
Accepted: 30 March 2023
Published: 4 April 2023

Copyright: © 2023 by the authors. Licensee MDPI, Basel, Switzerland. This article is an open access article distributed under the terms and conditions of the Creative Commons Attribution (CC BY) license (https://creativecommons.org/licenses/by/4.0/).

1. Introduction

A hydraulic transient corresponds to an intermediate state between two stationary flows, generated by valve maneuvers, pumps or turbines. This phenomenon can occur not only in water supply systems, but also in hydropower systems, the aircraft industry, railways tunnels, etc. [1–3]. Hydraulic transient analysis is particularly important in the design stage of pressurized systems in order to specify the pipe material and wall thickness, and, whenever necessary, to design surge protection devices.

Classic transient analysis, considering steady-state friction formulas (amongst other assumptions), is commonly used for design purposes, as it describes the maximum and minimum pressure variations reasonably well [2–4]. However, these models cannot accurately represent the complete transient phenomenon due to the wave propagation and complex diffusion mechanisms that significantly affect the pressure wave dissipation, dispersion and shape. This is particularly relevant in fast and high-frequency transient events, in which unsteady friction has a dominant effect [5,6]. Several efforts have been made to develop models with more accurate, simpler and lower computational demands, and the description of energy dissipation in pressurized transient flows has been a major challenge over the past decades; these models can be one-dimensional (1D), two-dimensional (2D) and three dimensional (3D) models, as briefly described in the following paragraphs.

In 1D models, several formulations have been proposed to compute unsteady friction for both laminar [5,7–9] and turbulent flows [10–15]. These formulations can be classified as instantaneous acceleration-based [16–18] or convolution-based [5,9,14,19–22].

Acceleration-based approaches introduce an additional term relating the local and the convective acceleration to the momentum equation. This formulation is quick to compute, but has a low accuracy; some formulations (e.g., [16–18]) must be calibrated based on experimental results. Convolution-based formulations consider the complete history of the local acceleration. This approach is consistent and theoretically more robust, but is very time-consuming, which is why several approximate solutions have been developed (e.g., [12,20,22]).

Axisymmetric models, also referred as quasi-two-dimensional models (Q2D), have been traditionally considered a good compromise between accuracy and computational time [6,23]. Q2D models have demonstrated consistency, regarding the physics of the phenomenon, and numerical robustness, ensuring the calculation of energy dissipation with high accuracy. More efficient versions than that initially proposed by Vardy and Hwang [6] have also been developed [24–27]. A second promising feature is the enhanced definition of the radial grid, which can ensure higher accuracy simulations with fewer cylinders and reduced computation time.

Finally, tri-dimensional computational fluid dynamics (3D-CFD) models, integrating turbulent models, are the most powerful and comprehensive hydraulic analysis models in pressurized pipes, though extremely computationally expensive in terms of time and data storage for regular use in engineering practice [28–32].

In any numerical model (1D, Q2D or 3D-CFD), the mesh size and configuration strongly affects the computational effort, as well as the accuracy of the results [6,33]. Not so intuitive, but equally important, is that the mesh must adapt to geometrical boundaries and to the flow physics to ensure that the velocity variation is of the same order along the numerical mesh. In practice, a non-uniform grid should be generated, with a higher resolution near the pipe wall, due to the high gradients observed. Calibrating the grid according to the velocity gradient history can reduce the number of mesh points at the same accuracy level [34]. Q2D models have a high potential for future improvements, particularly as concerns the mesh definition and optimization to be used in engineering practice.

This paper aims at the assessment of radial mesh influence on the computation of unsteady energy dissipation in pressurized pipes using a Q2D model. An extensive numerical analysis of the effect of the numerical schemes and of the radial mesh on the computation of unsteady energy is carried out. Several radial meshes are defined, as a compromise between flow dynamics and total mesh points. A comprehensive analysis and comparison of the Q2D results obtained with experimental data with results from 1D model are conducted for laminar flows and for two valve closure maneuvers (i.e., an instantaneous and a S-shape closure).

The key innovative features of the current paper are: (i) the analysis of a new optimized equal-area mesh (divided into high and low-resolution regions) applying transient laminar flows created by an instantaneous valve closure; (ii) the application of the new mesh to transient tests observed in an experimental facility, created by a S-shaped valve closure; and (iii) the assessment of the advantages of neglecting the lateral velocity transfer for both instantaneous and calibrated valve maneuvers.

2. Numerical Models

2.1. The Vardy and Hwang (1991) Model: Basic Equations, Assumptions and Numerical Scheme

The Q2D model makes the same traditional 1D model assumptions concerning the fluid and the pipe, namely: (i) the fluid is one-phase, homogenous, almost incompressible and isothermal (i.e., fluid properties—density and viscosity—are constant); and (ii) the pipe is uniform, is completely constrained from any axial or lateral movement and has linear-elastic rheological behavior. The axisymmetric flow assumptions allow neglect of the radial velocity and the viscous terms not perpendicular to the flow, simplifying the momentum equation in the conduit direction—Equation (1)—and eliminating the radial momentum equation. Consequently, the pressure is constant in the radial direction (the reason why axisymmetric models are also referred to as quasi-2D models). The governing equations constitute a system of two partial differential equations with three unknown parameters (p, u, v) with the following form [6,24]:

$$\frac{\partial H}{\partial t}+\frac{c^2}{g}\frac{\partial u}{\partial x}=-\frac{c^2}{gr}\frac{\partial (rv)}{\partial r} \tag{1}$$

$$\frac{1}{g}\frac{\partial u}{\partial t}+\frac{\partial H}{\partial x}=\frac{1}{r\rho}\frac{\partial r\tau}{\partial r} \tag{2}$$

$$\tau=\rho v\frac{\partial u}{\partial r} \tag{3}$$

where x is the distance along the pipe; r is the distance from the axis in the radial direction; t is time; H is the piezometric head; $u(x, r, t)$ is the local axial velocity; $v(x, r, t)$ is the local radial velocity; g is the gravitational acceleration; c is the wave speed; ρ is the liquid density; τ is the shear stress; and ν is the liquid kinematic viscosity.

The Q2D model divides the pipe radial section into N_C concentric and hollow cylinders, extended along the whole pipe length (Figure 1). The pipe length, L, is divided into N_X equal reaches, such that $\Delta x = L/N_X$, and each hollow cylinder has Δx length. Each cylinder is defined by the respective index, j, increasing from the centerline ($j = 0$) to the pipe wall ($j = N_C - 1$). The wall thickness of the jth cylinder is denoted by Δr_j ($= r_j - r_{j-1}$), the external radius by r_j ($= \sum_0^j \Delta r_j$) and the cylinder center radius is $\overline{r_j} = \left(r_{j-1} + r_j\right)/2$. The time step is determined according to the Courant condition, $\Delta t = \Delta x/c$. Each conduit section, referred to as x_i, is located at a distance $x = \Delta x.i$ from the upstream end of the pipe.

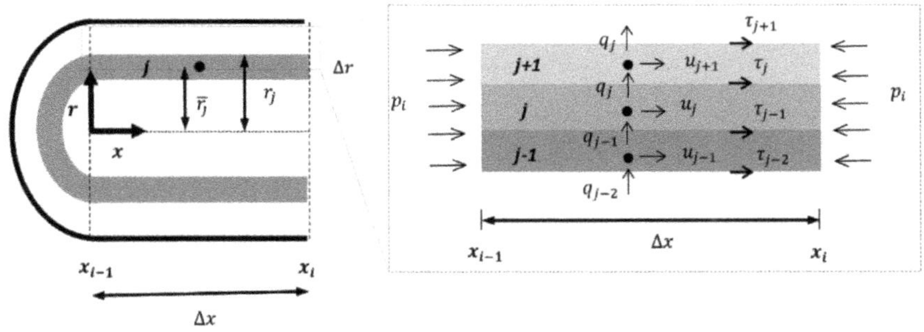

Figure 1. Radial grid system for numerical solution with Q2D model.

The Q2D model uses the Method of Characteristics to numerically solve the axial propagation of the pressure wave in each cylinder, defined by the terms on the left-hand side of Equations (4) and (5). The terms on the right-hand side of these equations are approximated by a second-order central difference formula. Accordingly, the Q2D model equations at each cylinder are:

$$\begin{aligned}&H_i^{t+1}-D_j\theta_2 q_{i,j-1}^{t+1}+D_j\theta_2 q_{i,j}^{t+1}-A_j\theta_1 u_{i,j-1}^{t+1}+\left(\tfrac{a}{g}+C_j\theta_1\right)\times u_{i,j}^{t+1}-\theta_1 B_j u_{i,j+1}^{t+1}\\&\quad=H_{i-1}^{t}+D_j\times(1-\theta_2)\times q_{i-1,j-1}^{t}-D_j\times(1-\theta_2)\times q_{i-1,j}^{t}+A_j\times(1-\theta_1)\times u_{i-1,j-1}^{t}\\&\quad+\left(\tfrac{a}{g}-(1-\theta_1)\times C_j\right)\times u_{i-1,j}^{t}+(1-\theta_1)\times B_j\times u_{i-1,j+1}^{t}\end{aligned} \tag{4}$$

$$\begin{aligned}&H_i^{t+1}-D_j\theta_2 q_{i,j-1}^{t+1}+D_j\theta_2 q_{i,j}^{t+1}+A_j\theta_1 u_{i,j-1}^{t+1}-\left(\tfrac{a}{g}+C_j\theta_1\right)\times u_{i,j}^{t+1}+\theta_1 B_j u_{i,j+1}^{t+1}\\&\quad=H_{i+1}^{t}+D_j\times(1-\theta_2)\times q_{i+1,j-1}^{t}-D_j\times(1-\theta_2)\times q_{i+1,j}^{t}-A_j\times(1-\theta_1)\times u_{i+1,j-1}^{t}\\&\quad-\left(\tfrac{a}{g}-(1-\theta_1)\times C_j\right)\times u_{i+1,j}^{t}-(1-\theta_1)\times B_j\times u_{i+1,j+1}^{t}\end{aligned} \tag{5}$$

$$A_j = \begin{cases} 0\,; j = 0 \\ \frac{c\Delta t v_{j-1}}{\overline{r_j}\,\Delta r_j g} \times \frac{r_{j-1}}{\left(\overline{r_j} - \overline{r_{j-1}}\right)}; j > 0 \end{cases} \tag{6}$$

$$B_j = \begin{cases} \frac{c\Delta t v_j}{\overline{r_j}\,\Delta r_{j+1} g} \times \frac{r_j}{\overline{r_{j+1}} - \overline{r_j}};\ j < N_C - 1 \\ \frac{c\Delta t v_j}{\overline{r_j}\,\Delta r_{j+1} g} \times \frac{r_j}{\overline{r_{j+1}} - \overline{r_j}}; j = N_C - 1 \end{cases} \tag{7}$$

$$C_j = A_j + B_j \tag{8}$$

$$D_j = \frac{c^2 \Delta t}{g} \times \frac{1}{\overline{r_j}\,\Delta r_j} \tag{9}$$

where q is the lateral mass transfer ($q = rv$); θ_1 and θ_2 are the weighting coefficients (θ_1 and θ_2 vary between 0 and 1, while $\theta_1 + \theta_2 = 1$); subscripts i and j are the conduit grid point and radial grid point, respectively; and t refers to the time.

The general central difference formula truncation error is given by:

$$\left(\frac{\partial \omega}{\partial r}\right)_j = \frac{\omega_{j+1} - \omega_{j-1}}{r_{j+1} - r_{j-1}} \underbrace{- \frac{\Delta r_{j+1} - \Delta r_j}{2} \times \left(\frac{\partial^2 \omega}{\partial r^2}\right)_j - \frac{\Delta r_{j+1}{}^2 + \Delta r_{j+1}.\Delta r_j + \Delta r_j{}^2}{6} \times \left(\frac{\partial^3 \omega}{\partial r^3}\right)_j}_{\textit{Truncation error}} \tag{10}$$

This formula is second-order accuracy, $O\left(\Delta r^2\right)$, if the variation of the grid size is very small ($\Delta r_{j+1} \approx \Delta r_j$) and, consequently, the first term of the truncation error tends to zero. In a general non-uniform scheme, the second-order central difference scheme loses one order of accuracy and the truncation error is $O(\Delta x)$. The main conclusion is that, generally, different formulas can lose at least one order of accuracy in non-uniform grids.

2.2. The Complete Zhao and Ghidaoui (2003) Model

The original Vardy and Hwang (1991) [6] model is numerically inefficient, since it requires the inversion of a ($2N_C \times 2N_C$) sparse matrix. Zhao and Ghidaoui (2003) [24] proposed an improved higher efficiency numerical model, developed by algebraic manipulation of Equations (3) and (4) into two smaller systems with tridiagonal matrices. The use of tridiagonal matrices allows the use of a faster form of Gaussian elimination, named the Thomas algorithm [35]. This algorithm obtains the solution via N_C operations instead of ${N_C}^3$ operations needed by the original Vardy and Hwang (1991) model, where N_C is the matrix dimension.

A tridiagonal system is defined and allows elimination of the piezometric-head and mass transfer from the coefficient matrix, $\mathbf{A_u}$, and unknown vector, $\mathbf{z_u}$, at time t + 1 by subtracting Equations (4) to (5) for each cylinder.

$$\mathbf{A_u} \times \mathbf{z_u} = \mathbf{b_u} \tag{11}$$

The axial velocities matrix, $\mathbf{A_u}$, for each cylinder at time t + 1 is described as follows:

$$\mathbf{A_u} = \begin{pmatrix} 2 \times \left(\frac{a}{g} + \theta_1 C_{j=0}\right) & -2\theta_1 B_{j=0} & & \frac{a}{g} & \\ & \ddots & & \frac{a}{g}\ \frac{a}{g}\ \frac{a}{g} & \\ \frac{a}{g} & \frac{a}{g} - 2A_j\theta_1 & 2 \times \left(\frac{a}{g} + \theta_1 C_j\right) & -2\theta_1 B_j \frac{a}{g} & \frac{a}{g} \\ \frac{a}{g} & \frac{a}{g} & \frac{a}{g} & \frac{a}{g}\ \ddots & \frac{a}{g} \\ \frac{a}{g} & \frac{a}{g} & \frac{a}{g} & -2A_{j=Nc}\theta_1 \frac{a}{g} & 2 \times \left(\frac{a}{g} + \theta_1 C_{j=Nc}\right) \end{pmatrix} \tag{12}$$

The $\mathbf{z}_u = (u_{i,j=0}^{t+1}, \ldots, u_{i,j}^{t+1}, \ldots, \ldots, u_{i,j=NC}^{t+1})$ is the axial velocities' unknown vector. The result vector, $\mathbf{b_u}$, at time t depends on the piezometric head, axial velocities and lateral mass transfer and is defined as follows:

$$\mathbf{b_u} = \begin{Bmatrix} H^t_{i-1} - H^t_{i+1} + \left(\frac{a}{g} - (1-\theta_1) \times C_j\right) \times u^t_{i-1,j} + \left(\frac{a}{g} - (1-\theta_1) \times C_j\right) \times u^t_{i+1,j} + (1-\theta_1) \times B_j \times u^t_{i-1,j+1} + (1-\theta_1) \times B_j \times u^t_{i+1,j+1} \\ -D_j \times (1-\theta_2) \times q^t_{i-1,j} + D_j \times (1-\theta_2) \times q^t_{i+1,j} \\ \vdots \\ H^t_{i-1} - H^t_{i+1} + A_j \times (1-\theta_1) \times u^t_{i-1,j-1} + A_j \times (1-\theta_1) \times u^t_{i+1,j-1} + \left(\frac{a}{g} - (1-\theta_1) \times C_j\right) \times u^t_{i-1,j} + \left(\frac{a}{g} - (1-\theta_1) \times C_j\right) \times u^t_{i+1,j} \\ +(1-\theta_1) \times B_j \times u^t_{i-1,j+1} + (1-\theta_1) \times B_j \times u^t_{i+1,j+1} + D_j \times (1-\theta_2) \times q^t_{i-1,j-1} - D_j \times (1-\theta_2) \times q^t_{i+1,j-1} - D_j \times (1-\theta_2) \times q^t_{i-1,j} + D_j \times (1-\theta_2) \times q^t_{i+1,j} \\ \vdots \\ H^t_{i-1} - H^t_{i+1} + A_j \times (1-\theta_1) \times u^t_{i-1,j-1} + A_j \times (1-\theta_1) \times u^t_{i+1,j-1} + \left(\frac{a}{g} - (1-\theta_1) \times C_j\right) \times u^t_{i-1,j} + \left(\frac{a}{g} - (1-\theta_1) \times C_j\right) \times u^t_{i+1,j} \\ +D_j \times (1-\theta_2) \times q^t_{i-1,j-1} - D_j \times (1-\theta_2) \times q^t_{i+1,j-1} \end{Bmatrix} \quad (13)$$

The lateral velocities and the piezometric-head are calculated by adding Equations (4) and (5) for each cylinder, and the equation obtained for the first cylinder is subtracted from the second equation and so on, obtaining the coefficient matrix, $\mathbf{A_q}$, and unknown vector, $\mathbf{z_q}$, at time t + 1.

$$\mathbf{A_q} \times \mathbf{z_q} = \mathbf{b_q} \quad (14)$$

The mass transfer and piezometric-head coefficient matrix, $\mathbf{A_q}$, for each cylinder at time t + 1 is described as follows:

$$\mathbf{A_q} = \begin{pmatrix} \frac{a}{g} & 2 & 2\theta_2 D_{j=0} & & & -2.\theta_2.(D_{j=n-1} - D_{j=n-2}) & \\ \frac{a}{g} & \vdots & \ddots & & \frac{a}{g}\ \frac{a}{g} & -2.\theta_2.(D_{j=n-1} - D_{j=n-2}) & \\ \frac{a}{g} & 0 & \frac{a}{g}.\theta_2.D_{j=0} & \frac{a}{g}2\theta_2 D_{j-1} & \frac{a}{g} - 2\theta_2 \times (D_{j-1} + D_j) & +\frac{a}{g}2\theta_2 D_j + & -2.\frac{a}{g}.\left(\frac{a}{g}D_{j=n-1} - D_{j=n-2}\right) \\ \frac{a}{g} & \vdots & \frac{a}{g}.\theta_2.D_{j=0} & \frac{a}{g}.\theta_2.D_{j=0} & -2.D_{j=Nc-1}\theta_2\frac{a}{g} & +2.\theta_2.\frac{a}{g_j} \ddots + & -2.\frac{a}{g}.\left(\frac{a}{g}D_{j=n-1} - D_{j=n-2}\right) \\ \frac{a}{g} & 0 & \frac{a}{g}.\theta_2.D_{j=0} & \frac{a}{g}.\theta_2.D_{j=0} & -2.D_{j=Nc-1}\theta_2\frac{a}{g} & \frac{a}{g}2\theta_2 D_{j=Nc-2} & -2\theta_2 \times (D_{j=Nc-1} - D_{j=Nc-2})\frac{a}{g} \end{pmatrix} \quad (15)$$

The $\mathbf{z_q} = (H_i^{t+1}, q_{i,0}^{t+1}, \ldots, q_{i,j}^{t+1}, \ldots, q_{i,j=NC-1}^{t+1})$ is the unknown vector. The result vector, $\mathbf{b_q}$, depends on the piezometric-head and on the velocities (axial and lateral) at the time t and is defined by:

$$\mathbf{b_q} = \begin{Bmatrix} H^t_{i-1} - H^t_{i+1} + \left(\frac{a}{g} - (1-\theta_1) \times C_{j=0}\right) \times \left(u^t_{i-1,j=0} - u^t_{i+1,j=0}\right) + (1-\theta_1) \times B_{j=0} \times \left(u^t_{i-1,j=1} - u^t_{i+1,j=1}\right) - D_{j=0} \times (1-\theta_2) \times \left(q^t_{i-1,j=0} + q^t_{i+1,j=0}\right) \\ \vdots \\ A_j \times (1-\theta_1) \times \left(u^t_{i-1,j-1} - u^t_{i+1,j-1}\right) - A_{j-1} \times (1-\theta_1) \times \left(u^t_{i-1,j-2} - u^t_{i+1,j-2}\right) + \left(\frac{a}{g} - (1-\theta_1) \times C_j\right) \times \left(u^t_{i-1,j} - u^t_{i+1,j}\right) \\ -\left(\frac{a}{g} - (1-\theta_1) \times C_{j-1}\right) \times \left(u^t_{i-1,j-1} - u^t_{i+1,j-1}\right) + (1-\theta_1) \times B_j \times \left(u^t_{i-1,j+1} - u^t_{i+1,j+1}\right) - (1-\theta_1) \times B_j \times \left(u^t_{i-1,j} - u^t_{i+1,j}\right) \\ +D_j.(1-\theta_2) \times \left(q^t_{i-1,j-1} + q^t_{i+1,j-1} - q^t_{i-1,j} - q^t_{i+1,j}\right) - D_{j-1} \times (1-\theta_2) \times \left(q^t_{i-1,j-2} + q^t_{i+1,j-2} - q^t_{i-1,j-1} - q^t_{i+1,j-1}\right) \times q^t_{i+1,j} \\ \vdots \\ A_{j=n-1} \times (1-\theta_1) \times \left(u^t_{i-1,j=n-2} - u^t_{i+1,j=n-2}\right) - A_{j=n-2} \times (1-\theta_1) \times \left(u^t_{i-1,j=n-3} - u^t_{i+1,j=n-3}\right) \\ +\left(\frac{a}{g} - (1-\theta_1) \times C_{j=n-1}\right) \times \left(u^t_{i-1,j=n-1} - u^t_{i+1,j=n-1}\right) - \left(\frac{a}{g} - (1-\theta_1) \times C_{j=n-2}\right) \times \left(u^t_{i-1,j=n-2} - u^t_{i+1,j=n-2}\right) \\ -(1-\theta_1) \times B_{j=n-1} \times \left(u^t_{i-1,j=n-1} - u^t_{i+1,j=n-1}\right) + D_{j=n-1} \times (1-\theta_2) \times \left(q^t_{i-1,j=n-2} + q^t_{i+1,j=n-2} - q^t_{i-1,j=n-1} - q^t_{i+1,j=n-1}\right) \end{Bmatrix} \quad (16)$$

2.3. *The Simplified Zhao and Ghidaoui (2003) Model*

Usually, the radial velocity is of significantly lower magnitude than the axial velocity, thus justifying the respective elimination from Equation (1), with further computation simplifications [26]. Using the scheme of Zhao and Ghidaoui (2003) [24], the axial velocity for each cylinder is calculated by Equations (12) and (13) and the piezometric head uses the tradition 1D model discretization form of Equation (17), obtained by summing the MOC characteristics equations [2,26]. The wall shear stress is defined by the axial velocity in the wall cylinder using Equation (18) and the mean velocity is calculated by the radial velocity integration, as presented in Equation (19).

$$H_i^{t+1} = \frac{1}{2}\left(H^t_{i-1} + H^t_{i+1} + \frac{c}{g} \times (U^t_{i-1} + U^t_{i+1})\right) - (1-\theta^*) \times \frac{c\Delta t}{\rho g R} \times \left(\tau_{w^t_{i-1}} - \tau_{w^t_{i+1}}\right) \quad (17)$$

$$\tau_w = \mu_{j=Nc} \times \frac{u_{j=Nc}}{R - \overline{r}_{j=Nc}} \tag{18}$$

$$U = \frac{1}{S} \sum_{j=0}^{j=N_C+1} u_j \Delta s_j \tag{19}$$

where U is the mean velocity in the complete pipe cross section; R is the pipe radius; θ^* is a weighting coefficient considered equal to 0.5, $\Delta s_j = 2\pi \overline{r}_j \Delta r_j$ is the cross section of each hollow cylinder, and S is the complete pipe cross section.

Additionally, neglecting the radial velocity allows the use of $\theta_1 = 1$ in Equations (12) and (13), which further simplifies the calculation of vector $\mathbf{z_q}$.

2.4. Radial Mesh Discretization

The radial mesh (or grid) generation is crucial for the accurate simulation of the flow, both in steady and unsteady conditions. A computer code was implemented to allow the definition of the three traditional radial meshes (Figure 2): (i) geometric sequence cylinders (GS); (ii) equal area cylinders (EAC); and (iii) equal thickness cylinders (ETC).

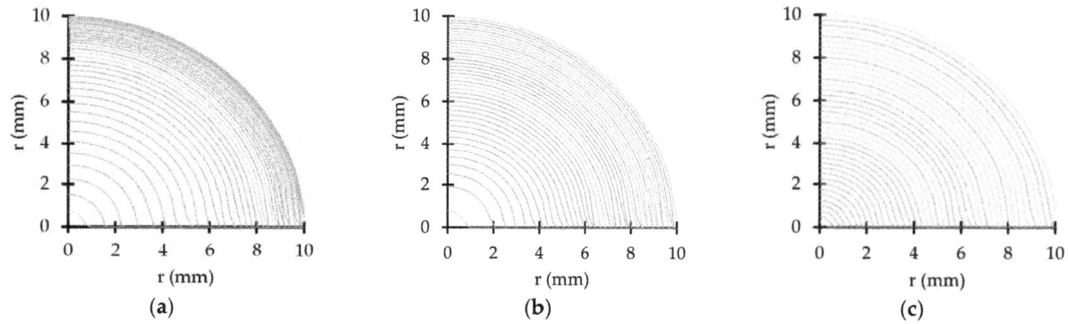

Figure 2. Radial meshes with different geometries for N_C = 40: (**a**) GS5%40; (**b**) EAC40; and (**c**) ETC40.

The geometric sequence (GS) mesh is defined by fixing the first cylinder thickness near the pipe wall, $\Delta r_{j=N_C-1}$, and the geometric sequence common ratio, C_R. The thickness of cylinder j is calculated as follows:

$$\Delta r_j = \Delta r_{j=N_C-1} \times (1 + C_R)^{N_C - j} \tag{20}$$

A particular GS mesh is referred to as "GS $\mathbf{C_R}$ N_C", in which N_C represents the total number of cylinders. This radial geometry has been widely used by several researchers [24,25,27].

The equal area cylinder (EAC) mesh assumes cylinders with a uniform area along the pipe radius, as follows:

$$\Delta r_{j+1}{}^2 - \Delta r_j{}^2 = const \tag{21}$$

Bratland [23] and Vardy and Hwang [6] used a mesh that is only defined by $\boldsymbol{N_C}$, that is "EAC N_C".

The equal thickness cylinder (ETC) mesh, also implemented by Vardy and Hwang [6], ensures a constant thickness of the radial grid and is also only described by the total number of cylinders, N_C, as "ETC N_C", as follows:

$$\Delta r_{j+1} - \Delta r_j = const \tag{22}$$

3. Experimental Facility Description

Transient tests for laminar flow conditions were carried out in the experimental facility assembled at the Laboratory of Hydraulics and Water Resources of the Instituto Superior Técnico (Figure 3). These tests were conducted for a discharge of 0.016 L/s, corresponding to a Reynolds number ($Re = UD/\nu$) equal to 1000. Water was at 20 °C with a density $\rho = 1000$ kg/m^3 and a kinematic viscosity $\nu = 1.01 \times 10^{-6}$ m^2/s.

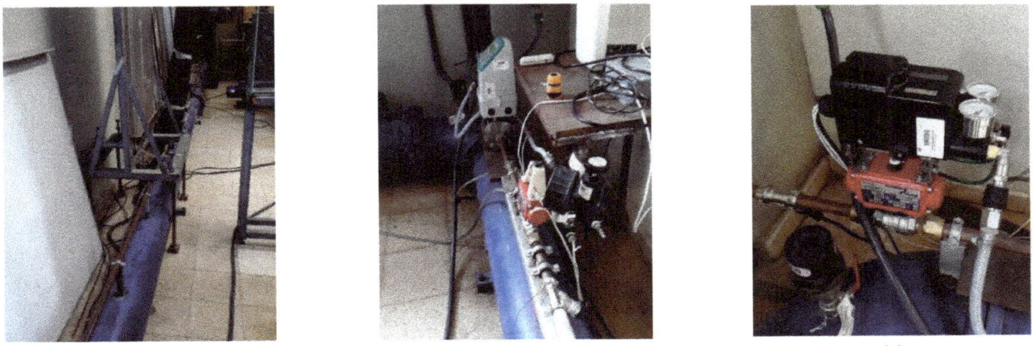

(a) (b) (c)

Figure 3. Experimental test facility: (**a**) upstream anchored copper pipe; (**b**) downstream pipe end; and (**c**) pneumatically actuated ball valve.

The system has a reservoir-pipe-valve configuration, with a rigidly supported straight copper pipe with an inner diameter, D, of 20 mm, wall thickness, e, of 0.001 m and a total length, L, of 15.2 m. A centrifugal pump with a hydro-pneumatic vessel is installed at the upstream end of the pipe to simulate a constant level reservoir. Two in-line ball valves are installed at the downstream end of the pipe: one valve, manually actuated, is used to control the flow rate and the other valve, pneumatically actuated, is used to generate the transient event. Different closure times are attained by varying the operating pressure of the pneumatic valve [36].

The data acquisition system (DAS) is composed of a computer, a Picoscope 3424 oscilloscope, a trigger-synchronizer, an electromagnetic flowmeter (with a 0.4% accuracy) and two strain-gauge type pressure transducers (WIKA S-10, 25 bars nominal pressure and 0.25% full-scale span accuracy). The first transducer is at the upstream end of the ball valve and the second is at the pipe mid-length.

The experimental wave speed, c, estimated based on the travelling time of the pressure wave between the two transducers, is 1250 m/s, which is consistent with values of previous studies carried out in the same experimental facility [36–41].

The valve closure law of the ball valve was thoroughly studied by Ferreira et al. [36]. The authors concluded that, despite the valve total closure time being higher than the system characteristic time, $2L/c$, the effective closure of the ball valve corresponds to 1/10 to 1/8 of the total closure time, and a rapid maneuver is always attained. The maneuver can be described by the sigmoidal-type closure law, as follows:

$$\frac{Q}{Q_0} = \left(\frac{1}{1 + e^{\xi(\tau - 95)}}\right)^{\eta} \tag{23}$$

where τ is the valve closure percentage, and ξ and η are the coefficients that best fit the numerical results with collected pressure-head data.

In the current study, the considered manoeuvre has total and effective closure times of 0.045 and 0.005 s, respectively. Accordingly, the experimental and Q2D models' calibrated results are presented in Figure 4. The calibrated flow rate variation is shown in Figure 4a.

The piezometric head obtained in the Q2D model fits almost perfectly with the collected data (Figure 4b).

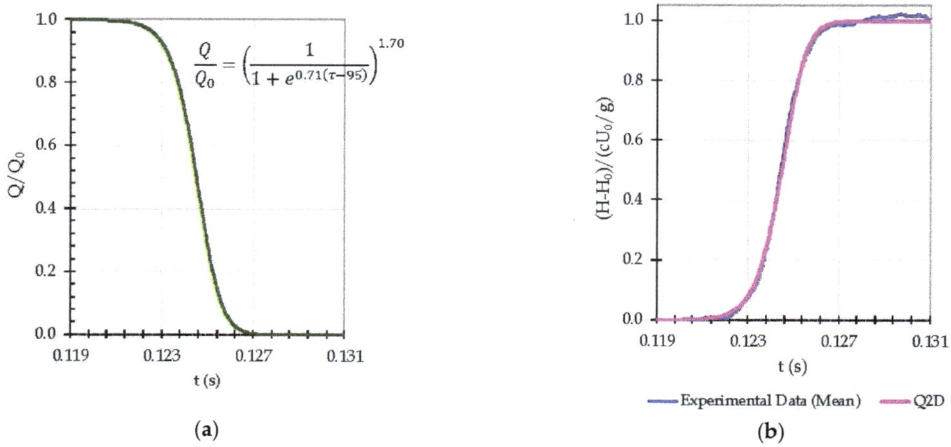

Figure 4. Calibrated valve maneuver for laminar flow: (**a**) Q2D model flow rate variation; (**b**) experimental and Q2D model piezometric head variation.

4. Numerical Results for the Instantaneous Valve Closure

4.1. Introduction

This section presents the results obtained by the Q2D model proposed by Zhao and Ghidaoui (2003) applied to an experimental pipe system for simulation of an instantaneous maneuver in a laminar flow. First, the effect of the mesh size near the pipe wall is analyzed for several standard equal area meshes, and results are compared with the reference Zielke's solution implemented in 1D models. Secondly, results from the Q2D model with an optimized equal-area cylinder, and with 40 cylinders, are compared with those from standard equal area meshes.

4.2. Standard Equal-Area Cylinder Mesh: Effect of the Mesh Discretization

The numerical Q2D results for the wall shear stress and the piezometric-head with the standard equal-area cylinder mesh (EAC) and a variable number of cylinders (Nc = 20, 40, 60 and 150) are depicted in Figure 5. Zielke's 1D model is also plotted in this figure and used as the benchmark solution for comparison.

The effect of the number of cylinders, N_C, is observed immediately after the first pressure wave arrival. The 1D model shows a high wall shear stress peak after each pressure surge, followed by an exponential decay (Figure 5a). Low N_C meshes show a lower peak and a linear decay with time. Increasing the N_C value brings the Q2D model simulation closer to Zielke's 1D solution. The N_C = 20 mesh reduces the Zielke's peak more than 17 times; the N_C = 150 mesh increases the previous value almost eight times but is still approximately half the 1D model peak.

As expected, a more precise wall shear stress calculation leads to a better piezometric description. For the N_C = 20 mesh, the Q2D model piezometric-head results (Figure 5b–d) do not depict the correct wave shape (the round-shape), showing a square shape geometry, and present a poor calculation of the wave damping (e.g., for t/T = 0.25, the difference to the 1D model is noticeable). Increasing the number of cylinders allows progressive approximation of Zielke's results.

The Q2D model computes the wall shear stress and the piezometric-head at a given time based on the axial velocity calculation. Figure 6 shows the axial velocity profile for the EAC150 mesh simulation at five-time steps during the first pressure surge ($0 \leq t/T \leq 0.25$). The time steps are equally spaced: t_0 is the velocity profile immediately after the pressure wave arrival, and each new time adds $t/16T$ to the previous one ($t_0 = 0$, $t_1 = t/16T$, $t_2 = t/8T$, $t_3 = 3/16T$ and $t_4 = 1/4T$).

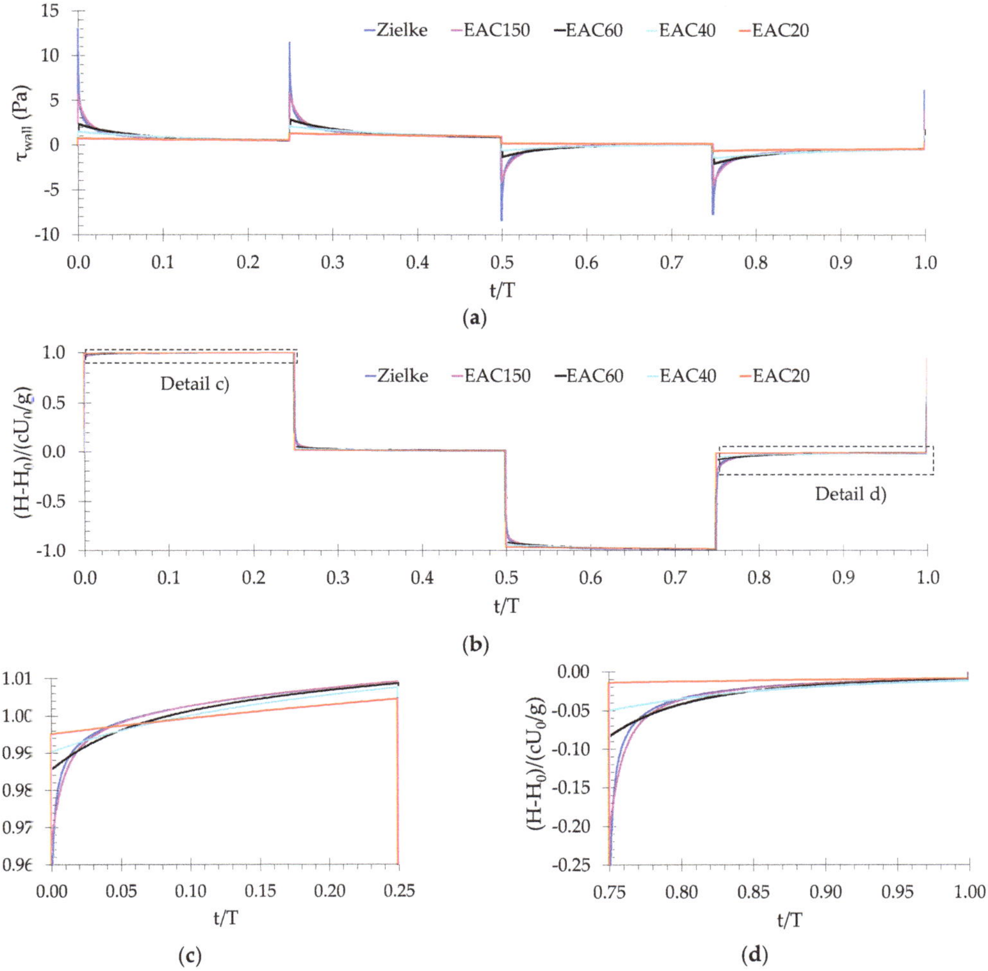

Figure 5. Q2D simulations for EAC meshes with increasing N_C using Zielke's 1D model as benchmark: (**a**) wall shear stress; (**b**) piezometric-head; (**c**,**d**) details of the piezometric-head.

The pressure surge arrival shifts the steady-state axial velocity profile and an inflection point with a high axial velocity gradient appears near the pipe wall because of the no-slip condition. From t_0 to t_5, the velocity profile gradient near the wall reduces as the inflection point moves into the pipe core. The transient flow changes occur close to the pipe wall. At t_4, just before the next pressure surge, the inflection point maximum displacement from the pipe wall is less than 4% of the pipe radius. Therefore, a high percentage of the pipe profile is unaffected and the Q2D model's cylinders maintain a steady-state profile axial velocity.

The Q2D model numerical scheme accuracy is estimated by the finite difference truncation error, Equation (10). This means that high axial velocity variations must be followed by considerably lower grid intervals, Δr, to achieve the same accuracy. The formation of the inflection point leads to a high numerical error in the pipe wall area and a low error in the pipe core, because it maintains the steady-state axial velocity profile with a parabolic variation ($\partial^2 u/\partial^2 r \approx constant$). The mesh refinement should focus on the wall area where the high axial velocity variations occur.

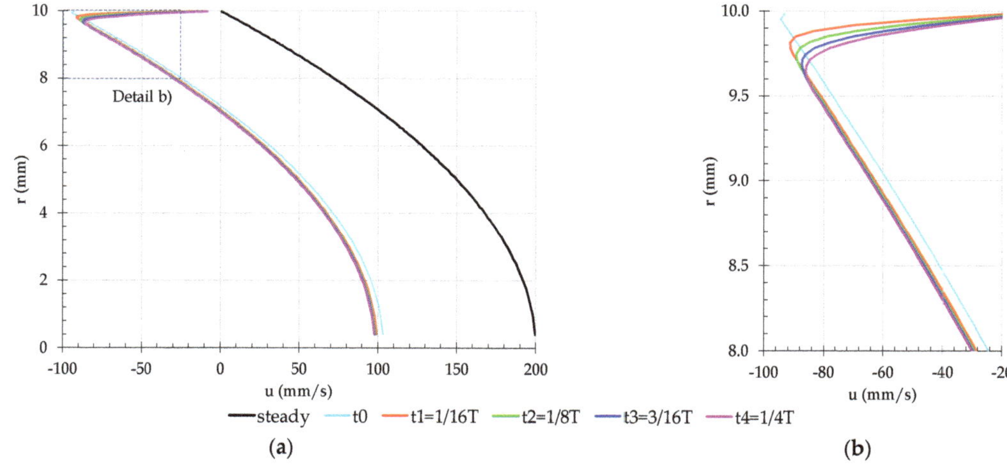

Figure 6. Axial velocity for four instances equally spaced in time during the first pressure wave: (**a**) complete pipe radius, $0 \leq r$ (mm) ≤ 10; (**b**) close to the pipe wall, $8 \leq r$ (mm) ≤ 10.

The EAC mesh geometry uniformly reduces the grid space in the pipe section, implying a considerable computational effort to approximate the 1D model results, especially immediately after each pressure surge. The use of EAC150 mesh corresponds to almost eight times the EAC20 computational effort, as it increases linearly with the number of cylinders for the implemented numerical scheme.

4.3. Optimized Equal-Area Cylinder Mesh

4.3.1. Mesh Description

The optimized equal-area cylinder (OEAC) mesh, innovatively proposed in this research, has a high-resolution grid close to the pipe wall and a low-resolution grid in the pipe core, as presented in Figure 7. Two equal-area cylinder meshes are implemented, with the lowest cylinder area near the wall. The underlying principle is to use a finer mesh in areas with high-velocity gradients and a coarse mesh in the low-velocity gradient region, thus, reducing the computational effort.

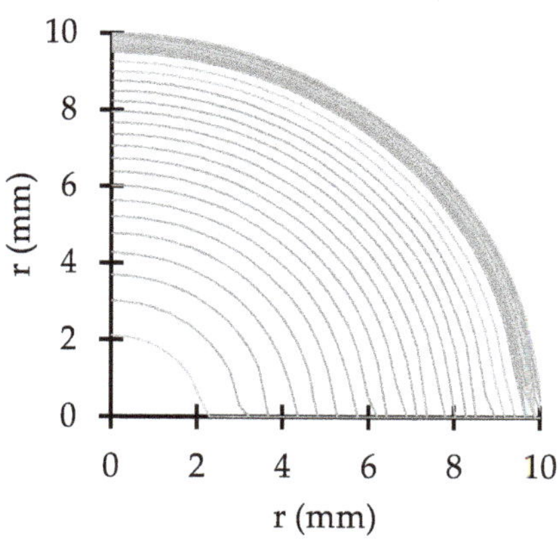

Figure 7. Radial mesh for the OEAC40 $N_{5\%} = 20$.

The new mesh is characterized by three parameters: (i) the total number of cylinders, N_C; (ii) the limit of the high-resolution grid, r, defined as a percentage of the pipe radius and measured from the pipe wall; and (iii) the number of cylinders in the high-resolution region, N_{HR}. Thus, the optimized equal area cylinder (OEAC) meshes are expressed as follows, OEACN_C Nr = N_{HR}. For instances, OEAC40 $N_{5\%}$ = 20 stands for a mesh that has a total of 40 cylinders (N_C = 40); the limit of the high-resolution mesh occurs at r = 9.5 mm (r/R = 5%) and has 20 cylinders in this high-resolution region.

This mesh is referred to as optimized since it provides better results than those for the standard EAC meshes for the same number of cylinders, as demonstrated in the following sections. This analysis will be carried out for N_C = 40, because the EAC mesh with 40 cylinders does not lead to an accurate simulation of the piezometric head (Figure 5b), representing a good comparison with optimized mesh.

4.3.2. Shear Stress Results

The wall shear stress obtained by the Q2D model for the optimized equal-area cylinder (OEAC) meshes with N_{HR} between 10 to 30, r = 5% and N_C = 40 are presented in Figure 8. The r = 5% corresponds to the inflection point maximum displacement determined in the previous analysis (Figure 6). All simulations' accuracy has significantly improved compared to the EAC results (Figure 4), for the same total number of cylinders (N_C = 40). The new mesh with more than 20 cylinders in the wall region (OEAC40 $N_{5\%}$ = 20) also presents better results than the 150 equal area cylinder meshes (EAC150). These results are compared with those obtained by the Zielke formulation used in the 1D model.

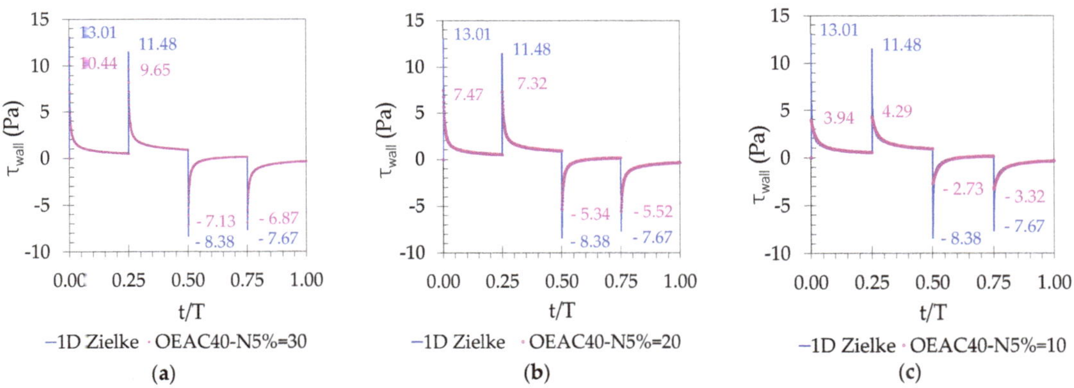

Figure 8. Wall shear stress obtained for OEAC meshes and for the Zielke's 1D model: (**a**) OEAC40 $N_{5\%}$ = 30; (**b**) OEAC40 $N_{5\%}$ = 20; (**c**) OEAC40 $N_{5\%}$ = 10.

The OEAC40 $N_{5\%}$ = 30 mesh allows an almost perfect adjustment to Zielke's results (Figure 8a), whereas the OEAC40 $N_{5\%}$ = 10 mesh leads to a maximum wall shear stress four times lower than the Zielke's value. This shows that the higher the number of cylinders in the high-resolution area, the more accurate the prediction of the maximum wall stress and of the subsequent decay. A second aspect concerns the relation between the wall shear stress peak values in the deceleration (t/T = 0 and t/T = 0.25) and the acceleration (t/T = 0.50 and t/T = 0.75) phases. For the meshes with higher resolution near the wall ($N_{HR} \geq 20$), the peaks decay with time, as in the Zielke model; on the contrary, for the low resolution mesh (Figure 8c), the peaks increase compared with the previous surge for t/T = 0.25 and t/T = 0.75, as observed for the standard meshes (Figure 8a).

The mean absolute error (MAE) is used to measure the Q2D accuracy, using paired observation from the 1D Zielke model, and is given by:

$$\text{MAE} = \frac{\sum_{i=1}^{n} |y_i - x_i|}{n} \tag{24}$$

in which n is the total number of time step at the pipe middle section, y is the Zielke model value (assumed as exact) and x is the Q2D predicted value, both for the same instance i.

The MAE calculation for the wall shear stress simulation during the first pressure wave period ($0 < t/T \leq 1$) confirms the previous assessment (Figure 9). MAE shows an almost exponential increase along with the N_{HR} reduction. For the same number of cylinders, the EAC40 mesh results have an error almost 10 times higher than that of the best OEAC mesh (OEAC40 $N_{5\%} = 30$). The EAC150 results present the same MAE as the OEAC40 $N_{5\%} = 15$ and a higher error than when using a higher N_{HR} value. This analysis highlights that a higher accuracy can be achieved with one-quarter of the number of standard mesh cylinders (40/150). OEAC mesh's results are clearly better than those obtained for the EAC grid, considering the same (EAC40), or even a higher (EAC150), value of the total number of cylinders.

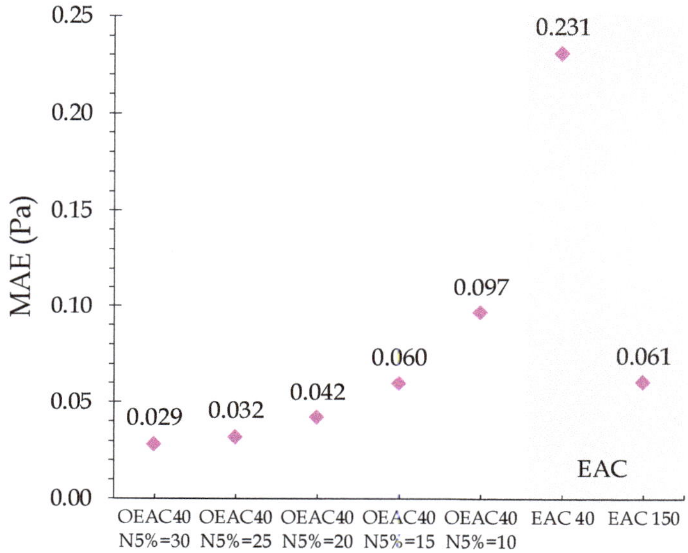

Figure 9. Mean absolute error (MAE) for the wall shear stress for different meshes.

4.3.3. Mean Velocity and Piezometric Head Results

The Q2D model simulation accuracy for the mean velocity and piezometric head depends not only on the unsteady wall shear stress calculation, but also on the steady-state velocity profile calculation. Figure 10 shows the relative steady-state mean velocity error for EAC and OEAC meshes, described by $(U_{2D} - U_{1D})/U_{1D}$, in which U_{1D} is the steady-state mean velocity obtained by Hagen–Poiseuille solution and U_{2D} is the Q2D model steady-state flow approximation for each mesh. Firstly, OEAC meshes have a worse steady-state velocity representation than the EAC meshes (i.e., the EAC40 mesh leads to a lower error). Secondly, this error increases exponentially with the N_{HR} increase. The OEAC40 $N_{5\%} = 10$ mesh has a 0.08% error, which is close to EAC40 (0.06%), but the OEAC40 $N_{5\%} = 30$ error rises to 0.64% (i.e., 8 to 10 times higher).

All meshes show similar wall shear stress values close to the pipe wall (see $r = 10$ mm in Figure 11b). Nevertheless, the highest resolution OEAC meshes tend to have higher shear stress values in the pipe core (see $r < 2$ mm in Figure 11a). Increasing the distance between grid points, Δr, towards the pipe axis implies a higher numerical error in the pipe core. Additionally, the OEAC meshes have a discontinuity between the high and the low-resolution regions, this discontinuity being more pronounced for higher values of N_{HR} (Figure 11b; see black and green curves for $r = 9.5$ mm). This discontinuity occurs in the transition between the two regions of the OEAC meshes (i.e., at *r*= 9.5 mm), being more noticeable for the OEAC40 $N_{5\%}$ = 30 mesh.

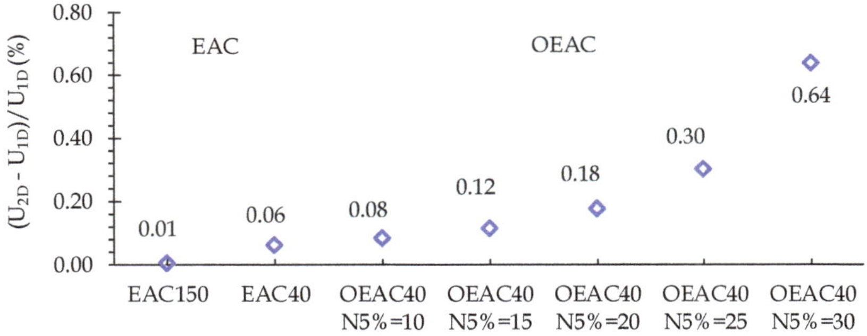

Figure 10. Steady-state velocity profile error for standard and optimized meshes.

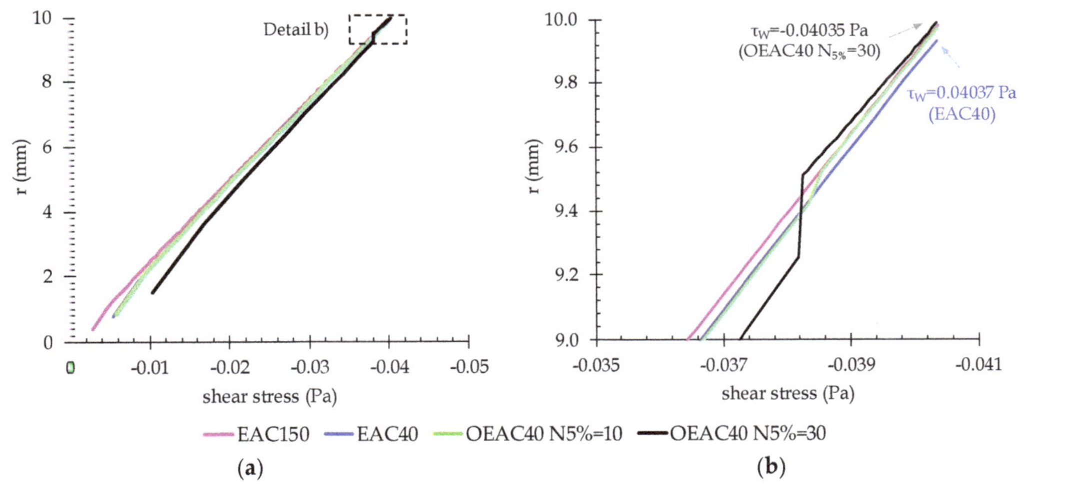

Figure 11. Shear stress profile for EAC40, EAC150, OEAC40 $N_{5\%}$ = 10 and OEAC40 $N_{5\%}$ = 30: (**a**) $0 \leq r \leq 10$ mm; and (**b**) $9.0 \leq r \leq 10$ mm.

The OEAC meshes with high N_{HR} values allow a considerable improvement in the wall shear stress simulation results without increasing N_C; however, a balance with higher steady-state error must be accomplished. The MAE of the mean velocity simulation is calculated in the first wave period ($t/T < 1$) using Zielke's model as benchmark, as depicted in Figure 12a. The EAC meshes (EAC40 and EAC150) and OEAC meshes with $N_C = 40$, and varying the N_{HR} values from 30 to 10, are considered. OEAC40 $N_{5\%}$ = 20 has the lowest error (MAE = 0.15), close to that of the standard EAC150 (MAE = 0.16) and five times lower than that for the standard EAC40 mesh (with the same number of cylinders). Increasing the number of cylinders in the wall region (N_{HR} from 20 to 30) increases the MAE value because a higher wall shear stress accuracy does not compensate for a lower steady velocity profile accuracy; decreasing the number of cylinders in the wall region also increases the error for the opposite reason. The piezometric-head results lead to the same conclusions (Figure 12b).

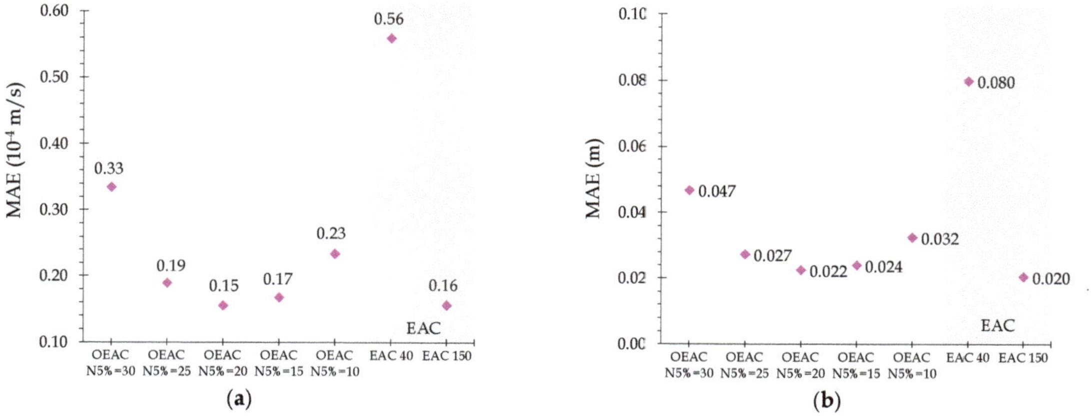

Figure 12. MAE for EAC and varying the N_{HR} value in OEAC meshes: (**a**) wall shear stress; (**b**) piezometric head.

5. Numerical Results for the Calibrated Valve Closure

5.1. Introduction

This section presents the analysis of the Q2D model results for a non-instantaneous fast valve maneuver. This maneuver has a S-shape and resulted from the calibration of the 1D and Q2D models with the collected data (see Figure 4a). In this analysis, the $N_{HR}/N_C = 0.5$ ratio is considered, as it presented the best results for an instantaneous valve closure. First, results obtained with traditional radial meshes (i.e., EAC, ETC and GS) are compared and discussed for both steady and unsteady flow conditions. Second, the Q2D results with an OEAC20 mesh are compared with the experimental data and with results from 1D models.

5.2. Comparison with Traditional Meshes' Results

5.2.1. Steady-State Mesh Assessment

The steady-state mean velocity relative error is assessed for the four analyzed mesh geometries (ETC, EAC, OEAC and GS), according to the total number of cylinders (N_C), and presented in Figure 13. The relative error is given by $(U_{2D} - U_{1D})/U_{1D}$ in which U_{1D} is the input mean velocity and U_{2D} is the Q2D model steady-state mean velocity (i.e., after the convergence process).

These results show that the error decreases exponentially with the N_C increase. The equal thickness cylinders mesh (ETC) has the best performance. The ETC mesh error drops 25 times with the Δr reduction to $1/5$ (increasing N_C from 20 to 100), as expected by the finite difference scheme truncation error, $O(\Delta r^2)$; see Equation (10). The cylinders have a uniform thickness along the pipe radius, and there is no error associated with the grid non-uniformity.

The non-uniform grids (GS, EAC and OEAC) show a lower error decrease with the N_C increase, because the cylinder thickness reduces faster in the wall area than in the pipe core. The OEAC meshes give overall worse results. Only with $N_C = 60$ does this geometry present a higher accuracy than the ETC20 (i.e., with three times the N_C value and computer effort). The equal area cylinder mesh (EAC) presents better results than those of the OEAC meshes. The GS mesh does not show the same accuracy improvement with N_C increase; for $N_C = 20$, this mesh is close to the best results (ETC20); on the contrary, for $N_C = 100$, it tends to the mesh with the worst results (OEAC100). With the N_C increase, the GS geometry focuses its refinement on the wall area, where the axial velocity has lower values. The mesh refinement in the wall area explains the GS slower error reduction with N_C and the lower overall performance of OEAC.

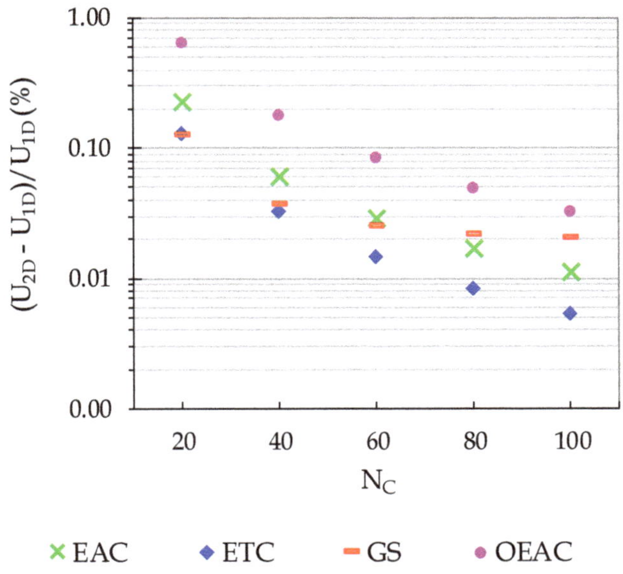

Figure 13. Steady-state mean velocity error according to the total number of cylinders (log scale).

5.2.2. Unsteady-State Mesh Assessment

The simulations are carried out for the analyzed four mesh geometries (ETC, EAC, OEAC and GS) and for a 10-wave period. A high-resolution mesh with N_C = 300 is considered for benchmark comparison. The number of cylinders for each geometry varies from 20 to 100 (N_C = 20, 40, 60, 80 and 100). Figure 14 shows the piezometric head results for the different meshes, plotting only results with significant differences from those of the benchmark mesh.

The OEAC mesh results approach those of the reference mesh, with only 20 cylinders, OEAC20 (Figure 14a). The GS mesh doubles the previous N_C value (GS40) to ensure the same global accuracy (Figure 14b). The EAC mesh requires 80 cylinders (i.e., four times the OEAC effort) to achieve the same accuracy (Figure 14c). Only the ETC grid with the maximum number of 100 cylinders (ETC100) and five times the OEAC20 mesh computational effort can ensure the same simulation accuracy (Figure 14d). The wave damping is correctly calculated by most meshes, except for non-OEAC meshes with the lowest N_C.

The wall shear stress and piezometric head mean that absolute error, MAE, is calculated at the pipe midsection for the previous four meshes (Figure 15) and for the complete simulation ($0 \leq t/T \leq 10$). The quasi-steady and Trikha [9] 1D model results are also presented, as these represent low computation time solutions. The OEAC grids show an exceptional energy dissipation accuracy for low N_C values ($N_C \leq 60$). Compared with the second-best results (EAC20, GS40 and GS60), this mesh allows an overall error reduction of 86%, 88% and 66% for N_C = 20, 40 and 60, respectively. Traditional meshes with a low number of cylinders are not capable of correctly describing the τ_{wall} peak and its subsequent exponential decay. For higher Nc values ($N_C \geq 80$), the GS meshes have higher accuracy than the OEAC meshes. EAC and ETC meshes have the worst results: for instance, the OEAC20 has a similar MAE value to the EAC100 and is three times smaller than ETC100, both with five times the new mesh computer effort. The quasi-steady 1D model results are incompatible with an accurate energy dissipation and with double the worst Q2D mesh error (i.e., ETC20). Trikha's 1D results are comparable with those of low resolution meshes but do not match the best Q2D model's results.

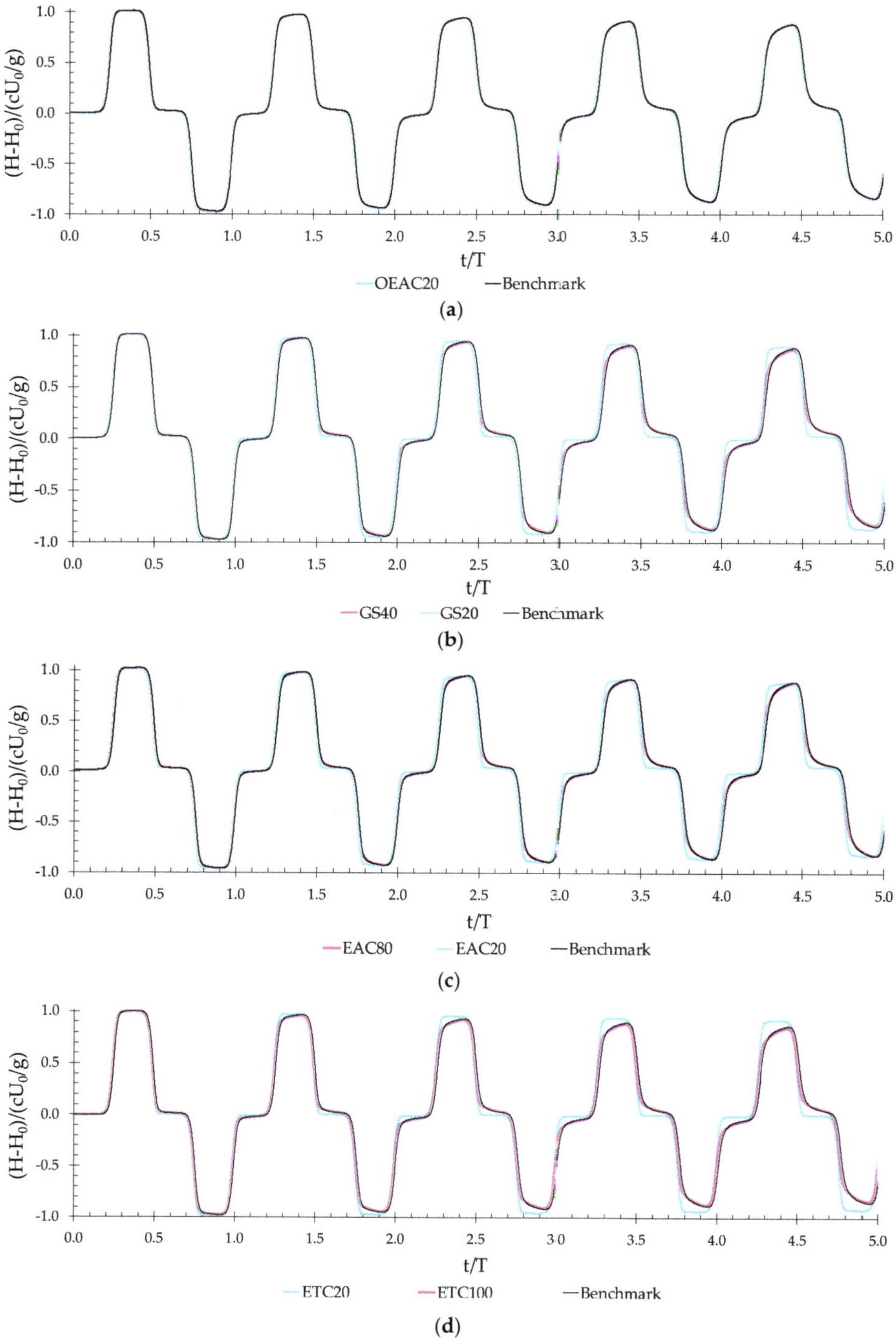

Figure 14. Piezometric head time history for the S-shape valve closure: (**a**) OEAC, (**b**) GS, (**c**) EAC, and (**d**) ETC.

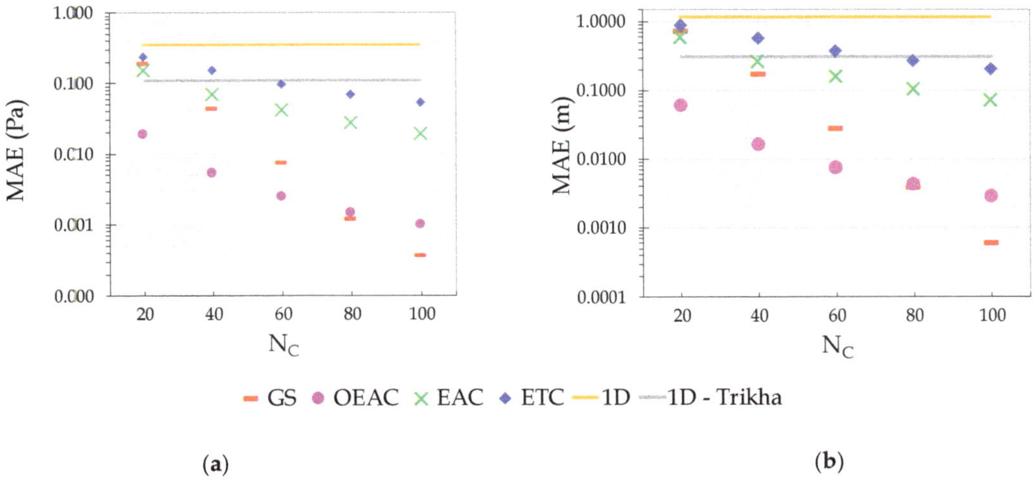

Figure 15. MAE according to *Nc* and radial geometry. (**a**) Wall shear stress, (**b**) piezometric head.

The OEAC mesh has demonstrated itself to be the best radial geometry in order to calculate unsteady energy dissipation with fewer cylinders; however, these meshes have the highest steady-state error due to the low mesh resolution in the pipe core (Figure 13). The piezometric head MAE calculation (Figure 15b) accounts for both the steady and unsteady simulation accuracy. The piezometric head and wall shear stress MAE results shows a similar behavior. This similarity is due to the minor importance of the steady velocity profile error concerning overall unsteady energy dissipation and OEAC mesh improvement.

5.3. Comparison with Experimental Data

Experimental data are compared with the results from the Q2D model with the OEAC20 mesh (OEAC20 $N_{5\%} = 10$). The quasi-steady and Trihka's 1D models are also included, for comparison purposes. The analysis excludes Zielke's 1D model, as it produces similar results to those obtained by the Q2D model, requiring higher computation time. All model simulations consider the calibrated valve closure presented in Figure 4a.

Figure 16a shows the results for the first wave period ($0 \leq t/T \leq 1$). All models present comparable results. This similarity reinforces the idea that the simplest models (quasi-steady and Trikha's 1D models) are applicable for a first pressure surge simulation or maximum pressure calculation because of the low energy dissipation observed during this short period.

The results are considerably different for a 10-wave periods analysis (Figure 16a,c). The quasi-steady 1D model is not capable of describing the pressure wave attenuation, nor the wave round shape over time, and this model underestimates the pressure damping by 75% at $t/T = 10$. The Trikha's model ensures the correct wave damping but not the correct wave shape, with a rectangular shaped pressure front wave, due to this model's inability to correctly compute the high energy dissipation period immediately after each pressure surge (Figure 16c,d). The previous conclusion, regarding the practical adequacy of 1D models for the estimation of transient pressures in the first pressure wave period, is not valid for longer time periods.

The Q2D model simulation using the optimized mesh and the lowest N_C value (OEAC20) ensures both the wave attenuation (compare the Q2D model and the test data for the middle point of each wave peak) and the front wave shape (see Figure 16c,d). However, the Q2D model cannot correctly describe the symmetric wave shape observed in the experimental data.

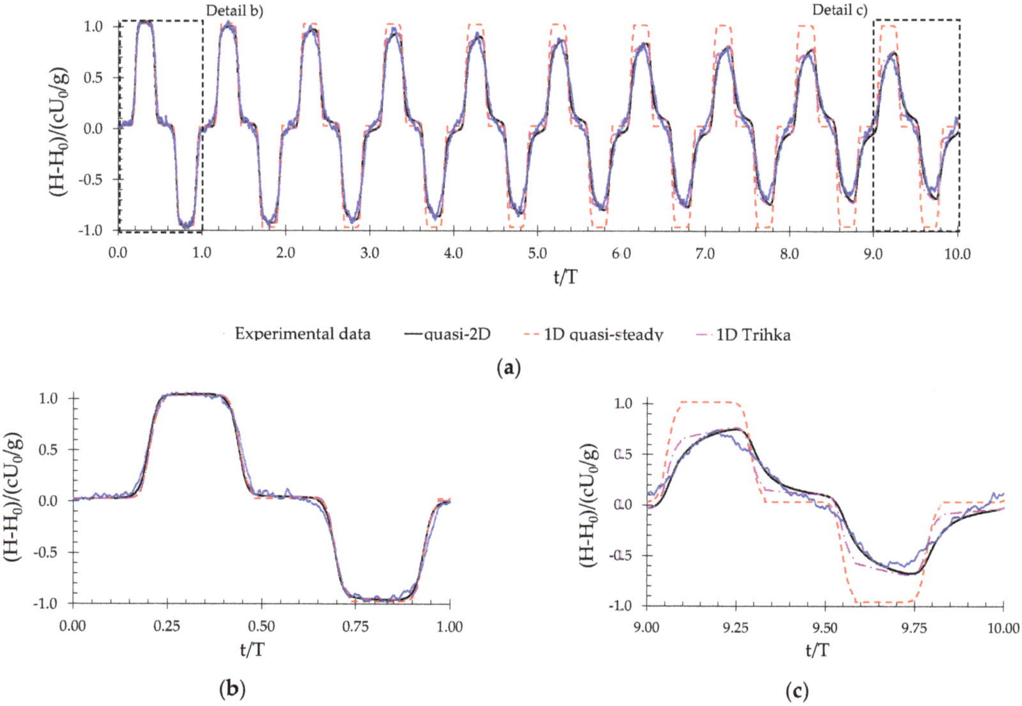

Figure 16. Piezometric head time history at conduit midsection, comparing experimental data, quasi-2D and 1D models: (**a**) $0.0 \leq t/T \leq 10.0$; (**b**) $0.0 \leq t/T \leq 1.0$; (**c**) $9.0 \leq t/T \leq 10.0$.

6. Complete Versus Simplified Q2D Models

Results presented in previous sections refer to the complete Q2D model from Vardy and Hwang (1981) [6], with Zhao and Ghidaoui's implementation [24]. The simplified Q2D model neglects the lateral mass transfer and allows a computation effort reduction of 30% and 40% with θ_1 = 0.5 and θ_1 = 1.0, respectively. The accuracy of the three models—complete Q2D, simplified Q2D (θ_1 = 0.5) and simplified Q2D (θ_1 = 1)—is assessed herein for the instantaneous and for the calibrated valve maneuvers.

6.1. Instantaneous Valve Manoeuvre

The wall shear stress results are analysed for N_C values from 20 to 100 and compared with the Zielke's analytical solution for Δx = 0.01m. For N_C = 100, a slight difference is registered between the three models during the high velocity-gradient immediately after the pressure surge and fades shortly after (Figure 17). The complete model has the best accuracy, and the simplified model with θ_1 = 1.0 the worst. Nevertheless, for the lowest N_C meshes (N_C = 60 and N_C = 80), the differences between the three models are hardly observed because the period immediately after the pressure surge is not described so accurately.

Figure 18 shows the product between the accuracy (described by MAE) and the CPU time for the different N_C values. The rationale is to combine these two aspects in a single parameter for best selection of the numerical model. The MAE value exponentially decreases with N_C. Increasing the N_C value ensures a better simulation accuracy independently of the numerical model selected. Increasing the model complexity is not justifiable for meshes with low N_C, and the simplified model with θ_1 = 1.0 is the best option. For higher N_C values the simplified model with θ_1 = 0.5 compensates for the additional computation time. The complete model is only recommended for a detailed evaluation in the period immediately after the pressure surge.

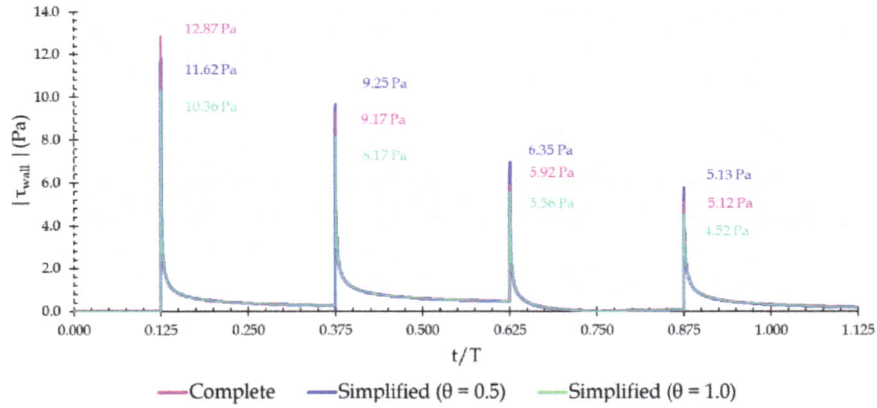

Figure 17. Wall shear stress simulation at pipe midsection for N_C = 100. $0.0 \leq t/T \leq 1.25$.

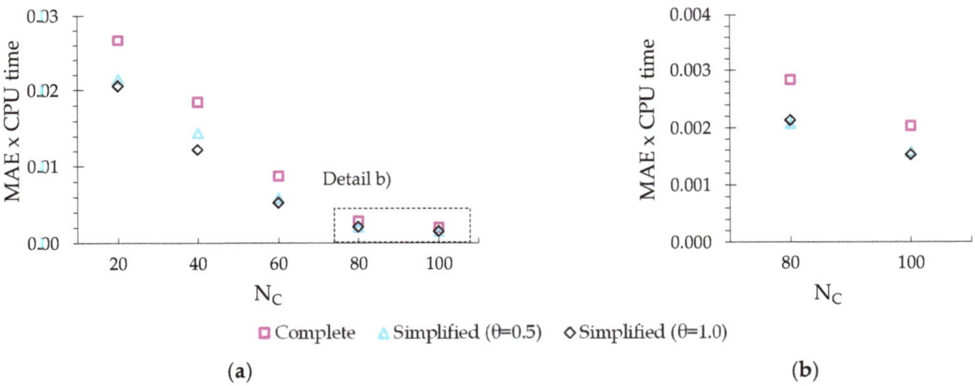

Figure 18. MAE and CPU time product calculation for the three numerical schemes using 1D Zielke model as benchmark: (**a**) for $20 \leq N_C \leq 100$; and (**b**) detail for N_C = 80 and 100.

6.2. Calibrated Valve Manoeuvre

The calibrated valve closure (Figure 4) is used to analyze the accuracy of the three numerical models results for the traditional mesh geometries (GS, EAC and ETC) and the new optimized mesh (OEAC). For the sake of simplicity, only the N_C = 100 meshes are considered, since the results for lower N_C values were negligible for instantaneous valve closure.

A slower valve closure maneuver reduces the wall shear stress peak and, consequently, approximates the results for the three traditional mesh geometry models, compared to the instantaneous valve closure. Additionally, the EAC and ETC meshes show a low accuracy and the simulation results are independent of the selected Q2D model, thus, the fastest numerical scheme should be considered (simplified model with θ_1 = 0.5). On the contrary, the results of the high-resolution grids (OEAC and GS) are influenced by the model, and the conclusions are similar to those obtained for the instantaneous valve maneuver.

7. Conclusions

An optimized equal area cylinder (OEAC) radial mesh is proposed to ensure the Q2D model accuracy without increasing the computation effort. The new mesh, with 40 cylinders, is defined by a high-resolution grid near the pipe wall and a lower-resolution grid in the pipe core. For an instantaneous valve closure, this mesh reduced the calculation time by four times with similar simulation error, compared to the standard equal area cylinder (EAC) mesh.

The results of the new mesh are compared with those obtained for the traditional mesh geometries (GS, ETC and EAC) for a calibrated S-shape valve closure. Compared to the geometric sequence (GS) mesh, the new geometry achieved a significant improvement in accuracy for meshes with fewer cylinders ($N_C \leq 60$). The other two radial meshes (ETC and EAC) do not provide the same accuracy. The new mesh correctly describes the experimental data with only 20 cylinders, which is not achieved by the fastest 1D models.

The simplified Q2D model (i.e., neglecting the lateral velocity) reduces the computational effort without significant loss of accuracy for low resolution meshes in the wall region. If a higher accuracy simulation is required, it is recommended to increase the number of cylinders instead of using the complete Q2D model.

The new two-region mesh has been applied herein to laminar flows, for which there is an exact analytical solution for unsteady friction (Zielke's formulation) that is used as a benchmark solution for comparison. In future works, the same approach can be applied to turbulent flows by including an adequate turbulent model in the fundamental equations, and the results obtained can be analyzed and compared with experimental data. Additionally, future research could explore the possibility of using a dynamic-mesh, reconfigured during the transient event according to the axial velocity gradient, in order to reduce the computation effort, but maintaining a high accuracy; this will imply the development of a new specific numerical scheme, adequate for non-uniform meshes, to avoid the loss of accuracy due to high grid space variations.

Author Contributions: Conceptualization, P.L.F. and D.I.C.C.; methodology, P.L.F. and D.I.C.C.; software, P.L.F.; validation, P.L.F.; formal analysis, P.L.F.; investigation, P.L.F. and D.I.C.C.; resources, P.L.F. and D.I.C.C.; data curation, D.I.C.C.; writing—original draft preparation, P.L.F.; writing—review and editing, P.L.F. and D.I.C.C.; visualization, P.L.F.; supervision, D.I.C.C.; project administration, D.I.C.C.; funding acquisition, D.I.C.C. All authors have read and agreed to the published version of the manuscript.

Funding: The authors would like to thank the Fundação para a Ciência e Tecnologia (FCT) through funding UIDB/04625/2020 from the research unit CERIS.

Data Availability Statement: The data presented in this study are available upon request from the corresponding author.

Acknowledgments: The authors would like to acknowledge João Paulo Ferreira for assisting the data collection in the experimental facility and for providing additional data.

Conflicts of Interest: The authors declare no conflict of interest.

Nomenclature

A_j, B_j, C_j, D_j	quasi-2D model coefficients (-)
C_R	geometric sequence common ratio (-)
c	pressure wave speed (m/s)
D	pipe inner diameter (m)
g	acceleration due to gravity (m/s^2)
H	piezometric head (m)
L	pipe length (m)
N_C	total number of cylinders of the radial mesh (-)
N_{HR}	total number of cylinders in the high-resolution area (OEAC meshes)
N_X	number of axial mesh points in the pipe direction (-)
Q	flow-rate or discharge (m^3/s)
R	pipe inner radius (m)
Re	Reynolds number, $Re = UD/\nu$ (-)
r	distance from the axis in the radial direction (m)

r_j	external radius of the jth cylinder measured from the pipe axis (m)
$\bar{r}_j$	center radius of the jth cylinder measured from the pipe axis (m)
T	pressure wave period, $T = 4L/c$ (s)
t	time (s)
U	mean velocity of the fluid in the pipe cross-section (m/s)
u	axial velocity (m/s)
v	radial velocity (m/s)
x	distance along the pipe (m)
Δt	numerical time step (s)
Δx	numerical spatial step (m)
Δr_j	wall thickness of the jth cylinder $\left(\Delta r_j = r_j - r_{j-1}\right)$
ε	pipe wall roughness (m)
θ_1, θ_1	weighting coefficient $(\theta_1 + \theta_2 = 1)$
ν	kinematic viscosity of liquid (m^2/s)
ρ	liquid density (kg/m^3)
τ	wall shear stress (Pa)

Abbreviations

CFD	Computation fluid dynamics
CFL	Courant-Friedrich-Lewy
EAC	Equal area cylinder mesh
ETC	Equal thickness cylinder mesh
GS	Geometric sequence cylinder mesh
MAE	Mean absolute error
MOC	Method of Characteristics
Q2D	Quasi-2D model

References

1. Leishear, R.A. *Fluid Mechanics, Water Hammer, Dynamic Stresses, and Piping Design*; ASME: New York, NY, USA, 2012.
2. Chaudhry, M.H. *Applied Hydraulic Transients*, 3rd ed.; Springer: Berlin/Heidelberg, Germany, 2014.
3. Wylie, E.B.; Streeter, V.L. *Fluid Transients in Systems*; Prentice Hall: Hoboken, NJ, USA, 1993.
4. Almeida, A.B.; Koelle, E. *Fluid Transients in Pipe Networks*; Elsevier Applied Science: New York, NY, USA, 1992.
5. Zielke, W. Frequency-dependent friction in transient pipe flow. *J. Basic Eng.* **1968**, *90*, 109–115. [CrossRef]
6. Vardy, A.E.; Hwang, K.-L. A characteristic model of transient friction in pipes. *J. Hydraul. Res.* **1991**, *29*, 669–684. [CrossRef]
7. Shuy, E.B. Approximate wall shear equation for unsteady laminar pipe flows. *J. Hydraul. Res.* **2010**, *33*, 457–469. [CrossRef]
8. Brunone, B.; Ferrante, M.; Cacciamani, M. Decay of Pressure and Energy Dissipation in Laminar Transient Flow. *J. Fluids Eng.* **2004**, *126*, 928–934. [CrossRef]
9. Trikha, A.K. An efficient method for simulating frequency dependent friction in transient liquid flow. *J. Fluids Eng.* **1975**, *97*, 97–105. [CrossRef]
10. He, S.; Ariyaratne, C.; Vardy, A.E. Wall shear stress in accelerating turbulent pipe flow. *J. Fluid Mech.* **2011**, *685*, 440–460. [CrossRef]
11. Pothof, I. A turbulent approach to unsteady friction. *J. Hydraul. Res.* **2010**, *46*, 679–690. [CrossRef]
12. Vardy, A.E.; Brown, J.M.B. Approximation of Turbulent Wall Shear Stresses in Highly Transient Pipe Flows. *J. Hydraul. Eng.* **2007**, *133*, 1219–1228. [CrossRef]
13. Vardy, A.E.; Brown, J.M.B. Transient, turbulent, smooth pipe friction. *J. Hydraul. Res.* **1995**, *33*, 435–456. [CrossRef]
14. Vardy, A.E.; Brown, J.M.B. Transient Turbulent Friction in Smooth Pipe Flows. *J. Sound Vib.* **2003**, *259*, 1011–1036. [CrossRef]
15. Vardy, A.E.; Hwang, K.-L. A weighting function model of transient turbulent pipe friction. *J. Hydraul. Res.* **1993**, *31*, 533–548. [CrossRef]
16. Vitkovsky, J.P.; Lambert, M.F.; Simpson, A.R. Advances in unsteady friction modelling in transient pipe flow. In Proceedings of the 8th International Conference on Pressure Surges, The Hague, The Netherlands, 12–14 April 2000; Pub. BHR Group Ltd.: The Hague, The Netherlands, 2000; pp. 471–498.
17. Ramos, H.M.; Covas, D.; Loureiro, D.; Borga, A. Surge damping analysis in pipe systems: Modelling and experiments. *J. Hydraul. Res.* **2004**, *42*, 413–425. [CrossRef]
18. Brunone, B.; Karney, B.W.; Mecarelli, M.; Ferrante, M. Velocity Profiles and Unsteady Pipe Friction in Transient Flow. *J. Water Resour. Plan. Manag.* **2000**, *126*, 236–244. [CrossRef]
19. Vardy, A.E.; Brown, J.M.B. On turbulent, unsteady, smooth-pipe friction. In Proceedings of the 7th International Conference on Pressure Surges and Fluid Transients in Pipelines and Open Channels, Harrogate, UK, 16–18 April 1996; Pub. BHR Group Ltd.: Harrogate, UK, 1996; pp. 289–311.

20. Vardy, A.; Brown, J. Efficient Approximation of Unsteady Friction Weighting Functions. *J. Hydraul. Eng.* **2004**, *130*, 1097–1107. [CrossRef]
21. Urbanowicz, K.; Zarzycki, Z. Improved Lumping Friction Model for Liquid Pipe Flow. *J. Theor. Appl. Mech.* **2015**, *53*, 295–305. [CrossRef]
22. Urbanowicz, K. Analytical expressions for effective weighting functions used during simulations of water hammer. *J. Theor. Appl. Mech.* **2017**, *55*, 1029–1040. [CrossRef]
23. Bratland, O. Frequency-dependent friction and radial kinetic energy variation in transient pipe flow. In Proceedings of the 5th International Conference on Pressure Surges, Hannover, Germany, 22–24 September 1986; BHRA, The Fluid Engineering Centre: Hannover, Germany; pp. 95–101.
24. Zhao, M.; Ghidaoui, M. Efficient quasi-two-dimensional model for water hammer problems. *J. Hydraul. Eng.* **2003**, *129*, 1007–1013. [CrossRef]
25. Pezzinga, G. Quasi-2D model for unsteady flow in pipe networks. *J. Hydraul. Eng.* **1999**, *125*, 676–685. [CrossRef]
26. Korbar, R.V. Efficient solution method for quasi two-dimensional model of water hammer. *J. Hydraul. Res.* **2014**, *52*, 575–579. [CrossRef]
27. Korbar, R.V. Truncated method of characteristics for quasi-two dimensional water hammer model. *J. Hydraul. Eng.* **2014**, *140*, 04014013. [CrossRef]
28. Alizadeh Fard, M.; Baruah, A.; Barkdoll, B.D. CFD modeling of stagnation reduction in drinking water storage tanks through internal piping. *Urban Water J.* **2021**, *18*, 608–616. [CrossRef]
29. Martins, N.M.C.; Delgado, J.N.; Ramos, H.M.; Covas, D.I.C. Maximum transient pressures in a rapidly filling pipeline with entrapped air using a CFD model. *J. Hydraul. Res.* **2017**, *55*, 506–519. [CrossRef]
30. Martins, N.M.C.; Soares, A.K.; Ramos, H.M.; Covas, D.I.C. CFD modeling of transient flow in pressurized pipes. *Comput. Fluids* **2016**, *126*, 129–140. [CrossRef]
31. Simao, M.; Mora-Rodriguez, J.; Ramos, H.M. Computational dynamic models and experiments in the fluid-structure interaction of pipe systems. *Can. J. Civil. Eng.* **2016**, *43*, 60–72. [CrossRef]
32. Gidaspow, D.; Li, F.; Huang, J. A CFD simulator for multiphase flow in reservoirs and pipes. *Powder Technol.* **2013**, *242*, 2–12. [CrossRef]
33. Martins, N.M.C.; Carriço, N.J.G.; Ramos, H.M.; Covas, D.I.C. Velocity-distribution in pressurized pipe flow using CFD: Accuracy and mesh analysis. *Comput. Fluids* **2014**, *105*, 218–230. [CrossRef]
34. Hirsch, C. *Numerical Computation of Internal & External Flows*, 2nd ed.; Elsevier: Oxford, UK, 2007.
35. Ferziger, J.H.; Perić, M. *Computational Methods for Fluid Dynamics*; Springer: Berlin/Heidelberg, Germany, 1996.
36. Ferreira, J.P.; Martins, N.M.C.; Covas, D.I.C. Ball Valve Behavior under Steady and Unsteady Conditions. *J. Hydraul. Eng.* **2018**, *144*, 04018005. [CrossRef]
37. Ferreira, J.P.; Buttarazzi, N.; Ferras, D.; Covas, D.I.C. Effect of an entrapped air pocket on hydraulic transients in pressurized pipes. *J. Hydraul. Res.* **2021**, *59*, 1018–1030. [CrossRef]
38. Ferras, D.; Manso, P.A.; Schleiss, A.J.; Covas, D.I.C. Fluid-structure interaction in straight pipelines with different anchoring conditions. *J. Sound Vib.* **2017**, *394*, 348–365. [CrossRef]
39. Ferràs, D.; Manso, P.A.; Schleiss, A.J.; Covas, D.I.C. Experimental distinction of damping mechanisms during hydraulic transients in pipe flow. *J. Fluids Struct.* **2016**, *66*, 424–446. [CrossRef]
40. Martins, N.M.C.; Brunone, B.; Meniconi, S.; Ramos, H.M.; Covas, D.I.C. Efficient Computational Fluid Dynamics Model for Transient Laminar Flow Modeling: Pressure Wave Propagation and Velocity Profile Changes. *J. Fluids Eng.* **2018**, *140*, 011102. [CrossRef]
41. Martins, N.M.C.; Brunone, B.; Meniconi, S.; Ramos, H.M.; Covas, D.I.C. CFD and 1D Approaches for the Unsteady Friction Analysis of Low Reynolds Number Turbulent Flows. *J. Hydraul. Eng.* **2017**, *143*, 04017050. [CrossRef]

Disclaimer/Publisher's Note: The statements, opinions and data contained in all publications are solely those of the individual author(s) and contributor(s) and not of MDPI and/or the editor(s). MDPI and/or the editor(s) disclaim responsibility for any injury to people or property resulting from any ideas, methods, instructions or products referred to in the content.

water

MDPI

Article

Water Hammer Simulation Using Simplified Convolution-Based Unsteady Friction Model

Kamil Urbanowicz [1,*], Anton Bergant [2,3], Michał Stosiak [4], Adam Deptuła [5], Mykola Karpenko [6], Michał Kubrak [7] and Apoloniusz Kodura [7]

1 Faculty of Mechanical Engineering and Mechatronics, West Pomeranian University of Technology in Szczecin, 70-310 Szczecin, Poland
2 Litostroj Power d.o.o., 1000 Ljubljana, Slovenia
3 Faculty of Mechanical Engineering, University of Ljubljana, 1000 Ljubljana, Slovenia
4 Faculty of Mechanical Engineering, Wrocław University of Science and Technology, 50-370 Wrocław, Poland
5 Faculty of Production Engineering and Logistics, Opole University of Technology, 45-758 Opole, Poland
6 Faculty of Transport Engineering, Vilnius Gediminas Technical University, LT-10223 Vilnius, Lithuania
7 Faculty of Building Services, Hydro and Environmental Engineering, Warsaw University of Technology, 00-653 Warsaw, Poland
* Correspondence: kamil.urbanowicz@zut.edu.pl

Abstract: Omission of frequency-dependent hydraulic resistance (skin friction) during modelling of the water hammer phenomenon is unacceptable. This resistance plays a major role when the transient liquid flow occurs in rigid-walled pipes (steel, copper, etc.). In the literature, there are at least two different modelling approaches to skin friction. The first group consists of models based on instantaneous changes in local and convective velocity derivatives, and the second group are models based on the convolution integral and full history of the flow. To date, more popular models are those from the first group, but their use requires empirical coefficients. The second group is still undervalued, even if based on good theoretical foundations and does not require any empirical coefficients. This is undoubtedly related to the calculation complexity of the convolution integral. In this work, a new improved effective solution of this integral is further validated, which is characterised with the use of a simplified weighting function consisting of just two exponential terms. This approach speeds the numerical calculations of the basic flow parameters (pressure and velocity) significantly. Presented comparisons of calculations using the new procedure with experimental pressure runs show the usefulness of the proposed solution and prove that it maintains sufficient accuracy.

Keywords: water hammer; hydraulic transients; unsteady friction; convolution-based model; numerical simulation

Citation: Urbanowicz, K.; Bergant, A.; Stosiak, M.; Deptuła, A.; Karpenko, M.; Kubrak, M.; Kodura, A. Water Hammer Simulation Using Simplified Convolution-Based Unsteady Friction Model. *Water* **2022**, *14*, 3151. https://doi.org/10.3390/w14193151

Academic Editor: Armando Carravetta

Received: 14 September 2022
Accepted: 3 October 2022
Published: 6 October 2022

Publisher's Note: MDPI stays neutral with regard to jurisdictional claims in published maps and institutional affiliations.

Copyright: © 2022 by the authors. Licensee MDPI, Basel, Switzerland. This article is an open access article distributed under the terms and conditions of the Creative Commons Attribution (CC BY) license (https://creativecommons.org/licenses/by/4.0/).

1. Introduction

In water supply networks, power hydraulics systems, transmission and heating lines, etc., unsteady flows are common. Sudden changes in flow velocity are the source of pressure waves which propagate in these systems. Conditions caused by breakdowns or those related to incorrectly set operating conditions of the components (valves, pumps, motors, distributors, pipelines, etc.) are particularly dangerous in the event of a power failure. Large pressures may occur in the case of liquid column separation and unwanted wave interference. Their values may even exceed the Joukowsky pressure rise Δp:

$$\Delta p = \rho c \Delta v, \tag{1}$$

where: ρ—liquid density; c—pressure wave speed; Δv—velocity change at the valve after its closure.

An interesting practical example can be drawn using the dependency graph for pressure wave speeds in water flows presented in Pothof and Karney's Chapter 1 of *Guidelines for Transient Analysis in Water Transmission and Distribution Systems* [1]. It is shown that a typical pressure wave speed in a steel pipe with elasticity modulus $E = 2{\cdot}10^{11}$ N/m^2 and inner diameter to wall thickness ratio (D/e) equal to $2{\cdot}10^2$ is about 1300 m/s, while for the same ratio of D/e in a PVC pipe ($E = 3.5{\cdot}10^{10}$ N/m^2) c is about 400 m/s and in a HDPE pipe ($E = 8{\cdot}10^9$ N/m^2) c is just about 200 m/s. These values show that the pressure wave speed in metal pipes is more than three times larger than in PVC pipes and more than six times than in HDPE ones. Therefore, the initial pressure rise resulting from the Equation (1) is significantly larger in metal pipes than in plastic pipes, and that is why metal pipes are the subject of this research. During water hammer events, several accompanying phenomena may occur, including: cavitation [2–4] (when the pressure drops to the vapour pressure of the liquid), unsteady friction [5–7] (resistance of the liquid during unsteady flow against the pipe wall), and fluid–structure interaction [8–10] (interactions of movable or deformable pipe structure with an internal or surrounding fluid flow). Assuming the adequate restraint of the pipe elements and pressure above the liquid vapour pressure, then the modelling of the unsteady friction remains the greatest challenge. To date, most of the hydraulic resistance models can be classified into one of two groups: (a) instantaneous acceleration-based (IAB) models or (b) convolution-based models (CBM).

IAB-type models were introduced by Daily et al. [11], Carstens and Roller [12], and Safwat and van der Polder [13]. Chronologically, this model approach was refined by Brunone et al. [14], Vítkovský et al. [15], Ramos et al. [16], Reddy et al. [17], and Cao et al. [18]. Currently, it is widely used [19–23], despite a serious drawback which is the necessity to experimentally calibrate the dissipation coefficient k.

The CBM-type models are derived theoretically. A pioneering work has been done by Zielke [24]. The model is based on the convolutional integral. The solution of the convolutional integral requires a continuous return to the historical values of the local fluid accelerations, which are multiplied by analytical weighting factors. Such a procedure in its original form requires a large number of calculations, which translates into a large load for computer processors in the analysis of long transient runs ($t > 4$ s). Trikha [25] developed a method that simplifies these calculations significantly. It requires an approximation form of the weighting function. Trikha's method was improved by Kagawa et al. [26], Schohl [27], and recently by Urbanowicz [28]. In this work, the procedure simplifying the CBM model is verified by referring to the experimental studies of water hammer carried out at the Institute of Fluid-Flow Machinery of the Polish Academy of Sciences by Adamkowski and Lewandowski [29]. The simplification of CBM consists in filtering the weighting function to just two exponential terms. The CBM solution requires simplifications, as the review of commercial programs [22] for modelling transients in pressurized conduits has shown that the quasi-steady model and IAB are widely used, and the CBM model has still not been implemented. However, the CBM model is characterised by high model consistency in a wide range of Reynolds numbers (transient laminar and turbulent pipe flows—the weighting function for laminar flow was developed by Zielke [24] and for turbulent flows by Vardy and Brown [30]). The objective of this paper is aimed to further test a computationally effective and accurate CBM model developed by Urbanowicz [31]. In an earlier work, this approach was verified only for the case of unsteady flows with cavitation [31]; therefore, in this paper, we validate the model against the experimental results without cavitation [29]. The second objective is verification of the effectiveness of Johnston's lumped friction model [32], according to which the unsteady friction can be concentrated only at the boundary nodes of the numerical grid.

2. Basic Equations

The basic continuity (2) and momentum (3) equations describing the unsteady pipe flow in horizontal pipes [33] follow:

$$\frac{\partial p}{\partial t} + \rho c^2 \frac{\partial v}{\partial x} = 0, \tag{2}$$

$$\frac{\partial p}{\partial x} + \rho \frac{\partial v}{\partial t} + \frac{2}{R}\tau = 0, \tag{3}$$

where: p—pressure; t—time; v—average liquid velocity; R—inner pipe radius; τ—wall shear stress.

The system of Equations (2) and (3) above contains three unknowns: v, p, and τ. In order to close the system, an additional relationship should be established, which is most often the relationship between the wall shear stress τ and the average flow velocity $\tau = f(v)$. Numerical details of modelling the wall stress on the pipe wall are the subject of the next section in this work.

Using the commonly known method of characteristics [33], Equations (2) and (3) can be led to the form:

$$C^+ : \begin{cases} \frac{dx}{dt} = +c \\ \frac{1}{c\rho}\frac{\mathrm{dp}}{\mathrm{dt}} + \frac{dv}{dt} + \frac{2}{\rho R}\tau = 0 \end{cases} \qquad C^- : \begin{cases} \frac{dx}{dt} = -c \\ -\frac{1}{c\rho}\frac{\mathrm{dp}}{\mathrm{dt}} + \frac{dv}{dt} + \frac{2}{\rho R}\tau = 0 \end{cases}. \tag{4}$$

At any internal point D of the characteristics grid (Figure 1), through which two characteristics C^+ and C^- pass, between points D and A as well as D and B, the integration can be performed using the finite linear differences. As a result, the following equations are obtained:

$$\left.\begin{array}{l} \frac{1}{c\rho}(p_D - p_A) + (v_D - v_A) + \frac{2\Delta t}{\rho R}\tau_A = 0 \\ -\frac{1}{c\rho}(p_D - p_B) + (v_D - v_B) + \frac{2\Delta t}{\rho R}\tau_B = 0 \end{array}\right\}. \tag{5}$$

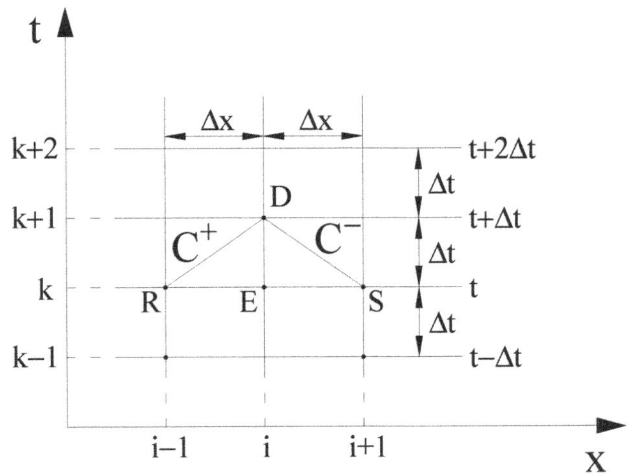

Figure 1. Method of characteristics grid.

Solving the system of Equation (5), one can find the final formulas for the calculated values of pressure and velocity at the inner node D of the characteristics grid in the following form:

$$p_D = \frac{1}{2}\left[(p_A + p_B) + c\rho(v_A - v_B) + \frac{2c\Delta t}{R}(\tau_B - \tau_A)\right], \tag{6}$$

$$v_D = \frac{1}{2}\left[(v_A + v_B) + \frac{1}{c\rho}(p_A - p_B) - \frac{2\Delta t}{\rho R}(\tau_A + \tau_B)\right]. \quad (7)$$

In order to develop a complete solution of the presented task, it is necessary to know the boundary conditions (Figure 2).

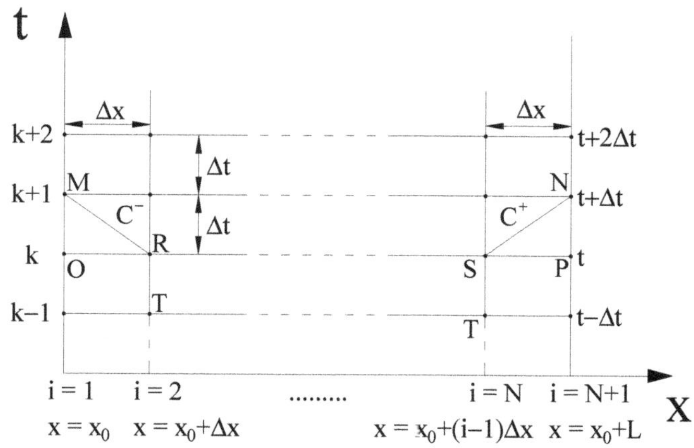

Figure 2. Boundary conditions.

When at the $i = 1$ node (cross-section) of the characteristics grid, the flow velocity v is determined (quickly closing valve) for time $t > 0$, and at the $i = N + 1$ node the pressure p is known (reservoir pressure), then:

$$p_M = p_R + c\rho(v_M - v_R) + \frac{2c\Delta t}{R}\tau_R, \quad (8)$$

$$v_N = v_S - \frac{1}{c\rho}(p_N - p_S) - \frac{2\Delta t}{\rho R}\tau_S. \quad (9)$$

Conversely, if the pressure p was determined as the boundary condition at the $i = 1$ node of the characteristics grid (reservoir section), and the value of the flow velocity v at the $i = N + 1$ node (valve section), then:

$$v_M = v_R + \frac{1}{c\rho}(p_M - p_R) - \frac{2\Delta t}{\rho R}\tau_R, \quad (10)$$

$$p_N = p_S - c\rho(v_N - v_S) - \frac{2c\Delta t}{R}\tau_S. \quad (11)$$

3. Modelling Wall Shear Stress

Commonly used quasi-steady, one-dimensional model of friction losses based on the Darcy–Weisbach formula can be used in the case of slow changes in liquid velocity at the pipe cross-section. However, it fails in the case of simulation for fast-changing flow, i.e., in the case of water hammer, the calculated results significantly differ from the results of measurements [29,34,35].

Models of unsteady friction losses, as mentioned in the introduction, can be divided into two groups. The first group consists of models based on the instantaneous values of velocity and acceleration (in literature often named Instantaneous Accelerated Based (IAB) model). The forerunner in this group was the model proposed by Daily et al. [11]. The term associated with the unsteady shear stress at pipe wall is proportional to the acceleration of liquid. This model was later improved by other researchers [12,13]. In this group falls the

Brunone et al. model [14], in which the wall shear stress is proportional not only to local derivative of flow velocity but also to its convective derivative:

$$\tau = \frac{f_q \rho v|v|}{8} + \frac{k\rho D}{4}\left(\frac{\partial v}{\partial t} - c\frac{\partial v}{\partial x}\right), \tag{12}$$

where: f_q—Darcy–Weisbach friction factor; k—empirical unsteady friction coefficient of the IAB model; D—inner pipe diameter.

This model underwent further modifications. Vítkovský et al. [15] rightly pointed out that the acoustic convection term $c(\partial v/\partial x)$ should be added or subtracted depending on the type of the flow:

$$\tau = \frac{f_q \rho v|v|}{8} + \frac{k\rho D}{4}\left(\frac{\partial v}{\partial t} + c\frac{|v|}{v}\left|\frac{\partial v}{\partial x}\right|\right). \tag{13}$$

The next major change was the introduction of separate unsteady friction coefficients for the local derivative k_t and convective derivative k_x by Ramos et al. [16]:

$$\tau = \frac{f_q \rho v|v|}{8} + \frac{\rho D}{4}\left(k_t\frac{\partial v}{\partial t} + k_x c\frac{|v|}{v}\left|\frac{\partial v}{\partial x}\right|\right). \tag{14}$$

Ramos et al. [16] proved numerically that the expression $k_t(\partial v/\partial t)$ affects the phase shift of pressure waves and that $k_x(\partial v/\partial x)$ affects the rate of attenuation of these waves. The coefficients k_t and k_x can be calculated on the basis of known experimental results using the method presented by Reddy et al. [17]. The main disadvantage of this approach is the need to determine k_t and k_x empirically, and that the shape of simulated pressures differs significantly from the shape observed in experiments. Owing to its simplicity, the expression above is often cited and used in practise. It should be noted, however, that the details of the implementation of Equation (14) in the method of characteristics have been described in a comprehensive and clear manner only in one conference article, namely, in reference [15] written by Vítkovský et al. In all other papers the procedure to determine the spatial derivative $\frac{\partial v}{\partial x}$, in particular at the boundary, is unclear. The most recent improvement of this model has been presented by Cao et al. [18]:

$$\tau = \frac{f_q \rho v|v|}{8} + \frac{k\rho D}{4}\left(\frac{\partial v}{\partial t} + c\frac{|v|}{v}\left|\frac{\partial v}{\partial x}\right|\right) - \frac{k_d \rho D}{4}\left|\frac{\partial^2 v}{\partial x^2}\right|, \tag{15}$$

where: $k_d = \frac{\mu'}{\rho} \approx 716.1\cdot ln(0.135\cdot ln(Re))$; μ'—is the second viscosity coefficient.

This model is a further modification of Vítkovský et al. model Equation (13). It takes into account an additional energy dissipation term describing a compression–expansion effect of the fluid. Although the Cao et al. model is an interesting alternative, but this model has a problem with guaranteeing the appropriate dispersion (delay, phase shift) of the pressure wave for low Reynolds numbers [18].

The second group consists of models based on the history of the flow (CBM—convolution-based models). The wall shear stress (and hence the instantaneous coefficient of friction losses) depends here on the frequency of changes in flow and pressure. These models reflect relatively well not only the degree of dissipation of pressure waves but also dispersion. They treat the pressure histories in detail. The forerunner in this group of models has been proposed by Zielke [24], who developed the wall shear stress for transient laminar pipe flow in the form of the sum of quasi-steady shear stress and unsteady contribution, which is an integral convolution of the mean local acceleration of the liquid and a weighting function $w(t)$:

$$\tau(t) = \tau_q + \tau_u = \frac{4\mu}{R}v + \frac{2\mu}{R}\int_0^t w(t-u)\frac{\partial v(u)}{\partial t}du, \tag{16}$$

where: μ—dynamic viscosity; u—time, used in convolution integral; $w(t)$—weighting function.

The wall shear stress time domain solution given above is an inverse Laplace transform of the WSS function written in the frequency domain. For laminar flow, this function has a form based on multiplication of a certain frequency-dependent function $\hat{F}(s)$ with a partial time derivative of velocity transform. This form was firstly derived and presented by Zielke in his doctoral thesis [36]:

$$\hat{\tau}(s) = \hat{F}(s)\frac{\partial \hat{v}(s)}{\partial t} = \frac{\rho R}{\frac{j\sqrt{s\frac{R^2}{\nu}}J_0\left(j\sqrt{s\frac{R^2}{\nu}}\right)}{J_1\left(j\sqrt{s\frac{R^2}{\nu}}\right)} - 2}\frac{\partial \hat{v}(s)}{\partial t}, \tag{17}$$

where: s—Laplace parameter; ν—kinematic viscosity of liquid; j—imaginary unit; J_0 and J_1—Bessel functions of the first kind (order 0 and 1). Zielke calculated the inverse Laplace transform of $F(s)$, which gives the following time domain function:

$$F(t) = \frac{4\mu}{R} + \frac{2\mu}{R}\sum_{n=1}^{\infty} e^{-\kappa_n^2 \hat{t}}. \tag{18}$$

A time-domain solution of multiplication of two frequency-dependent functions is a convolutional integral, Equation (16). According to Equation (18), the weighting function in Equation (16) is an infinite series of exponential terms that has the following form for the laminar flow [36]:

$$w_{lam}\left(\hat{t}\right) = \sum_{n=1}^{\infty} e^{-\kappa_n^2 \hat{t}}. \tag{19}$$

where κ_n in the power of exponent are nth zeros of the Bessel function of type J_2. Zielke approximated this function [24,36] in the following way:

$$w_{lam,classic}\left(\hat{t}\right) = \sum_{i=1}^{\epsilon} m_i \hat{t}^{(i-2)/2}, \text{ for } \hat{t} \le 0.02, \tag{19a}$$

$$w_{lam,classic}\left(\hat{t}\right) = \sum_{i=1}^{5} e^{-m_i \hat{t}}, \text{ for } \hat{t} > 0.02, \tag{19b}$$

where: $m_1 = 0.282095$; $m_2 = -1.25$; $m_3 = 1.057855$; $m_4 = 0.9375$; $m_5 = 0.396696$; $m_6 = -0.351563$; $n_1 = 26.3744$; $n_2 = 70.8493$; $n_3 = 135.0198$; $n_4 = 218.9216$; and $n_5 = 322.5544$.

For turbulent flow, much more complicated formulas for impedance have been derived by Vardy and Brown [30] and Zarzycki [37]. Both Zarzycki and Vardy and Brown concluded that in time domain the solution of Equation (16) can be used for turbulent flow, the only difference is that in this flow the weighting function shape depends not only on dimensionless time but also on the initial Reynolds number and characteristic roughness size. In this work, the Vardy and Brown weighting function is used for transient turbulent pipe flow:

$$w_{turb,classic}\left(\hat{t}, Re\right) \approx \frac{A^* e^{-B^* \hat{t}}}{\sqrt{\hat{t}}}, \tag{20}$$

where: $A^* = \sqrt{1/4\pi}$, $B^* = Re^{\kappa}/12.86$, $\kappa = \log_{10}(15.29/Re^{0.0567})$—for smooth pipes [30] and $A^* = 0.0103\left(\frac{\varepsilon}{D}\right)^{0.39}\sqrt{Re}$, $B^* = 0.352Re\left(\frac{\varepsilon}{D}\right)^{0.41}$ for rough pipes [38]; the ratio ε/D is a relative roughness.

In the method of characteristics based on a rectangular grid, the classical numerical solution of the convolution integral Equation (16) can be expressed as:

$$\tau_u = \frac{2\mu}{R}\sum_{j=1}^{n-1}\left(v_{i,j+1} - v_{i,j}\right) \cdot w\left((n-j)\Delta\hat{t} - \frac{\Delta\hat{t}}{2}\right) = \frac{2\mu}{R}\sum_{j=1}^{n-1}\left(v_{i,n-j+1} - v_{i,n-j}\right) \cdot w\left(j\Delta\hat{t} - \frac{\Delta\hat{t}}{2}\right). \tag{21}$$

In the above equation, $\Delta\hat{t}$ is a dimensionless time step, which is:

$$\Delta\hat{t} = \Delta t \cdot \frac{\nu}{R^2} = \frac{\Delta x}{c} \cdot \frac{\nu}{R^2} = \frac{L}{N} \cdot \frac{\nu}{c \cdot R^2} = \frac{Wh}{N}, \tag{22}$$

where: Δt—numerical time step; Δx—reach length between the nodes; L—pipe length; N—number of reaches (number of analysed pipe cross-sections—spatial nodes); $Wh = \frac{\nu L}{cR^2}$—water hammer number [39].

One can see in Equation (21) that the number of iterations required to determine the shear stress increases with the time of simulation of the transient event. In the last forty years, a number of authors showed at least three distinct effective solutions. A simplified recursive solution was first presented by Trikha in 1975 [25]. Its drawback is due to an excessive number of simplifications; thus, it is not suitable for the calculation in a wide range of dimensionless times. Improved forms of recursive formulas have been presented by Kagawa et al. [26] and Schohl [27], respectively:

$$\tau_u(t+\Delta t) \approx \frac{2\mu}{R}\sum_{i=1}^{j}\left[\underbrace{y_i(t)\cdot e^{-n_i\cdot\Delta\hat{t}} + m_i \cdot e^{-n_i\cdot\left(\frac{\Delta\hat{t}}{2}\right)}\cdot\left[v_{(t+\Delta t)} - v_t\right]}_{y_i(t+\Delta t)}\right], \tag{23}$$

$$\tau_u(t+\Delta t) \approx \frac{2\mu}{R}\sum_{i=1}^{j}\left[\underbrace{y_i(t)\cdot e^{-n_i\cdot\Delta\hat{t}} + \frac{m_i}{\Delta\hat{t}\cdot n_i}\cdot\left[1-e^{-n_i\cdot\Delta\hat{t}}\right]\cdot\left[v_{(t+\Delta t)} - v_t\right]}_{y_i(t+\Delta t)}\right]. \tag{24}$$

Kagawa et al. [26] assumed that the integral of the weighting function can be approximated in the following form:

$$\int_t^{t+\Delta t} e^{n_i\cdot\frac{\nu}{R^2}\cdot u}du \approx e^{n_i\cdot\frac{\nu}{R^2}\cdot\left(t+\frac{\Delta t}{2}\right)}\cdot\int_t^{t+\Delta t} du. \tag{25}$$

Schohl [27] calculated the same integral symbolically:

$$\int_t^{t+\Delta t} e^{n_i\cdot\frac{\nu}{R^2}\cdot u}du = \frac{R^2}{n_i\cdot\nu}\cdot\left[e^{n_i\cdot\frac{\nu}{R^2}\cdot(t+\Delta t)} - e^{n_i\cdot\frac{\nu}{R^2}\cdot t}\right]. \tag{26}$$

It is worth noting that in all efficient solutions the weighting function needs to be written as a finite sum of exponential terms:

$$w_{eff.} = m_i e^{n_i\cdot\hat{t}}. \tag{27}$$

Recently, Vardy-Brown [40] pointed out an overlooked error in a classical computationally inefficient methodology of Equation (21) and suggested calculating the wall shear stress by using the following equation:

$$\tau_u = \frac{2\mu}{R}\sum_{j=1}^{n-1}\left[\left(v_{i,n-j+1} - v_{i,n-j}\right)\cdot\int_{(j-1)\Delta\hat{t}}^{j\Delta\hat{t}} w(\hat{t})d\hat{t}\right]. \tag{28}$$

That is why in this work a corrected solution of CBM is used, which is an effective counterpart of the above-corrected Equation (28):

$$\tau_u = \frac{2\mu}{R}\sum_{i=1}^{j}\underbrace{\left[y_{i(t)}\cdot A_i + \eta\cdot B_i\cdot\left[v_{(t+\Delta t)} - v_{(t)}\right] + [1-\eta]\cdot C_i\cdot\left[v_{(t)} - v_{(t-\Delta t)}\right]\right]}_{y_{i(t+\Delta t)}}, \tag{29}$$

where: η—correction factor. The details of the derivation of Equation (29) can be found in [28]. The constants in the formula above are calculated as follows:

$$A_i = e^{-n_i \cdot \Delta\hat{t}}; \quad B_i = \frac{m_i}{\Delta\hat{t} \cdot n_i} \cdot [1 - A_i]; \quad C_i = A_i \cdot B_i, \tag{30}$$

where: n_i and m_i—coefficients describing the effective weighting functions. The algorithm for determining the values of these coefficients is presented in Appendix A. In this efficient formula, Equation (29), the effective weighting function, does not need to have an extended range of applicability in dimensionless time to correctly model transient flows. For the dimensionless time range from 0 to $\Delta\hat{t}$, the integral for the effective weighting function is replaced with either the integral from the classical laminar-flow weighting function according to the Zielke Equation (19) or the turbulent-flow weighting function according to Vardy-Brown Equation (20) (depending on the type of flow that takes place: laminar or turbulent) as presented in Figure 3. In addition to the standard model in which the friction term is calculated in the same way at each node of the numerical grid of characteristics, this work also investigates a model lumping the unsteady friction factor only at the boundary nodes of the pipe. The author of this approach is Johnston, who described its basics in [32]. The lumping of τ_u at the sections $i = 1$ and $i = N + 1$ significantly shortens the numerical computational time, because in all other nodes calculations are based on the quasi-steady solution (τ_q). However, this approach requires modification of the velocity values $v_{M,c}$ and $v_{N,c}$ at the boundary nodes, as follows:

$$v_{M,c} = \frac{1}{2}\left(v_M + \frac{p_M}{\rho c}\right); \; v_{N,c} = \frac{1}{2}\left(v_N + \frac{p_N}{\rho c}\right), \tag{31}$$

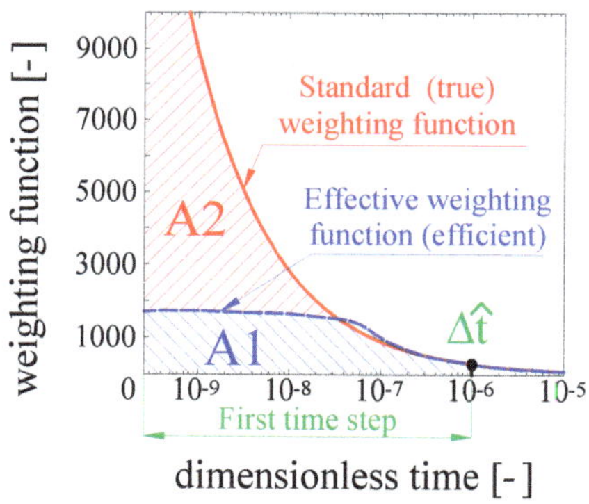

Figure 3. Areas under classic and efficient weighting function for low dimensionless times.

Equation (31) is used to determine the lumped values of the wall shear stress at the boundary nodes. This model has been recently investigated by Xu et al. [41] with an objective to develop an ultrafast numerical solution based on a gridless scheme.

In some recent works [31,42], the impact of (i) the number of terms describing the effective weighting function, (ii) the scope of applicability in dimensionless time, and (iii) the lumped friction model, were analysed. The main conclusion from these studies was that the time range of applicability of the effective weighting function in order to model unsteady pressure events with sufficient accuracy should be from $\Delta\hat{t}$ to $\Delta\hat{t} \cdot 10^3$. This indicates that the effective weighting functions do not need to be composed of many exponential terms, as only two are sufficient and it is less than in the well-known effective weighting function presented by Trikha [25]. In addition, Bergant et al. [43] found that CBM cannot produce a small-frequency shift in pressure history observed in experimental results.

This deficiency can be eliminated either by inclusion of the momentum correction factor in the inertia term of Equation (3) or by using the measured pressure wave propagation speed.

4. Analysis of the Results

The experimental tests of Adamkowski and Lewandowski [29], in which a simple water hammer event in a reservoir–pipeline–valve system occurred due to rapid closure of the valve, were selected for our comparison analysis. A test stand was located at the Institute of Fluid-Flow Machinery in Gdańsk, Poland, the main element of which was a long metal copper pipe. The pipe was 98 m long and a large part of it was wound on a steel cylinder (with a diameter of about 1.6 m; please note that pipe was rigidly mounted to the cylinder coating in order to minimise its vibrations), as can be seen in Figure 4. Horizontal parts of the pipeline (not coiled) were constrained with the help of steel clamps, spaced at about every 0.4 m to the concrete base of the laboratory. The upstream end tank is a pressure reservoir with a capacity of 1.6 m^3. Its main role was to maintain constant pressure during steady-flow conditions and near-constant pressure under transient operation. The test rig was equipped with absolute semiconductor pressure transducers (measuring range from 0 to 4 MPa; transmitted frequency band from 0 to 2 kHz, and precision class equal to 0.2%), turbine flowmeter (range of 1.5 m^3/h and precision class of 1%), ball valve (installed between the quick-closing valve and flowmeter), and feed pump (with adjustable rotational speed). The two elements mentioned (ball valve and feed pump) were used to adjust required initial conditions in the system. A water hammer event was generated by a quick-closing valve in which the closing time was minimised using a specially designed spring driving mechanism.

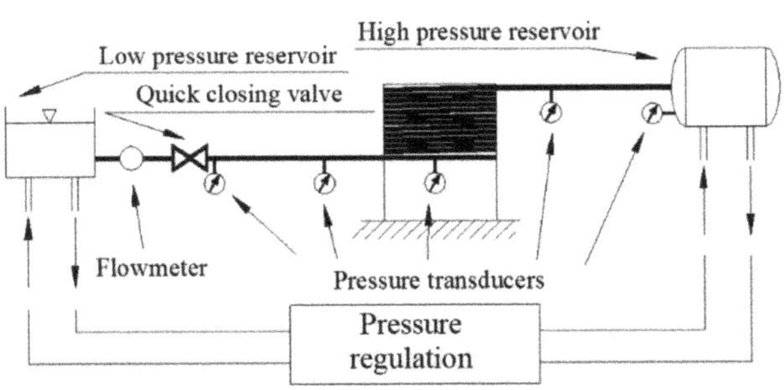

Figure 4. Gdańsk Institute of Fluid-Flow Machinery test rig.

Detailed values describing the basic parameters of the experimental apparatus are presented in Table 1, where: T_{vc}—valve full closing time; T—temperature; e—pipe wall thickness. These parameters were used as input parameters in a proprietary computer program written in the MATLAB environment.

Table 1. Test rig details.

L [m]	D [m]	e [m]	T_{vc} [s]	T [°C]	ν [m^2/s]	ρ [kg/m^3]
98.11	0.016	0.001	0.003	22.6	$9.493 \cdot 10^{-7}$	997.65

Comparative analysis was performed for nine test cases. Additional details on the boundary and initial conditions necessary to model these cases are summarized in Table 2.

Table 2. Analysed flow cases.

Case	v_0 [m/s]	Re_0 [−]	p_R [Pa]	c [m/s]
01	0.066	1100	$1.265 \cdot 10^6$	1300
02	0.162	2750	$1.264 \cdot 10^6$	1300
03	0.340	5750	$1.265 \cdot 10^6$	1300
04	0.467	7900	$1.253 \cdot 10^6$	1305
05	0.559	9400	$1.264 \cdot 10^6$	1300
06	0.631	10,650	$1.264 \cdot 10^6$	1303
07	0.705	11,900	$1.263 \cdot 10^6$	1300
08	0.806	13,600	$1.263 \cdot 10^6$	1300
09	0.940	15,850	$1.264 \cdot 10^6$	1300

v_0 and Re_0—initial velocity and Reynolds number, respectively; p_R—reservoir pressure.

Water hammer simulation, especially with the use of the classical full-convolutional integral and its computationally ineffective solution, takes a long time. Therefore, the comparative studies were limited to computational time of $t = 5.5$ s. This time covers eighteen water hammer periods ($t/(4L/c)$), more than enough for an adequate comparison study.

The influence of the mesh refinement of the method of characteristics on the obtained results was also examined. The results obtained for the simplified CBM model (SM—Equation (29)) and the lumped friction model (LFM—Equation (31)) were analysed for meshes with the following densities: coarse mesh $N = 32$ (nodes $\approx$ 77,000); $N = 52$ (nodes $\approx$ 201,000), $N = 102$ (nodes $\approx$ 766,000), and very fine mesh $N = 202$ (nodes $\approx$ 2,989,000). The N parameter influences not only the mesh refinement along its length, but also the time step Δt, which determines the mesh refinement in time (due to Courant–Friedrichs–Lewy CFL stability condition):

$$\Delta x = L/N \quad \text{and} \quad \Delta t = \Delta x/c. \tag{32}$$

All the results using the classical computationally ineffective solution of the convolutional integral (FULL CONV.) were realised only for $N = 32$. In this case, to perform 5.5 s simulation required about an hour; thus, in order to save the time, it was decided not to repeat these tests for fine meshes.

When analysing the results of experimental studies by Adamkowski and Lewandowski, one can notice an atypical pressure peak at the first amplitude of all the runs (Figure 5) to a value much higher than the predicted value, which can be calculated from the Joukowsky Equation (1). These short-duration peaks at the first pressure amplitude plateau are most probably the result of undesired mechanical vibrations produced by the valve closing drive [44]. They are quickly damped out for all types of supports and are present only at the first pressure pulse and do not influence further water hammer pressure oscillations. The other reason for these peaks (initial disturbances) can be probably linked to the system response due to the excitation from the step-load induced by the fast-closing valve [45]. Another source of such peaks can be explained to be the result of the type of valve used [46,47]. The use of the globe valve instead of the ball valve allows elimination of their presence in experiments. These peaks, however, with the correct restraint of the valve and pressure measurement sections, should not occur; therefore, the maximum pressure values from these peaks are not taken into account in the quantitative analysis. The maximum bulk pressure pulse is taken into consideration, as illustrated in Figure 5. At subsequent amplitudes, the observed maximum pressure values and the times in which these maximums appeared were taken into account (see Figure 6).

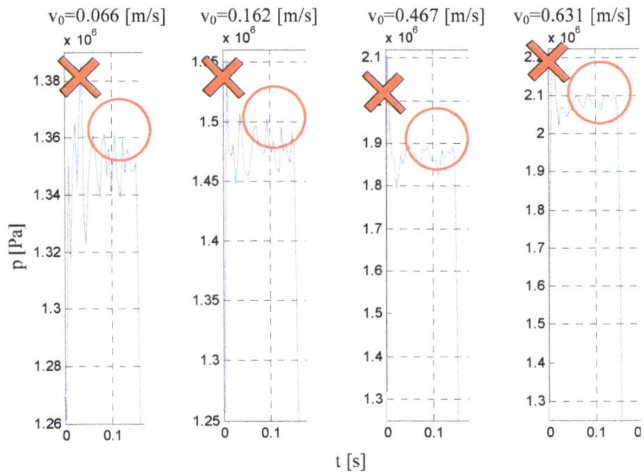

Figure 5. Pressure overshoot at the first bulk pressure amplitude.

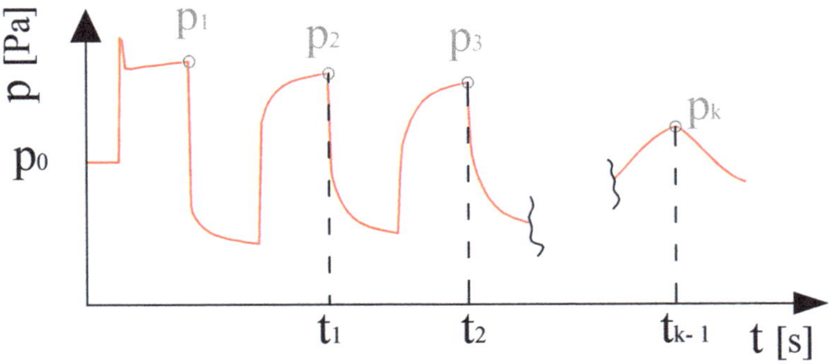

Figure 6. Analysed bulk pressure peaks in the quantitative comparison.

As an example, the simulation results (N = 102) for Case 02 (Re = 2750) are shown in Figure 7. On the other hand, Figure 8 shows the enlargement of the three initial pressure amplitude crests (Figure 8a) and valleys (Figure 8b). It can be seen that the LFM slightly underestimates the pressure in the initial period of the water hammer event and delicately distorts the valleys of these amplitudes. However, from the fourth amplitude to the eighteenth amplitude, there is a reasonable match. The analysed quantitative parameters were calculated from the following formulas:

$$Ep = \frac{\sum_{i=1}^{18} \left| \frac{p_{is} - p_{ie}}{p_{ie}} \right| \cdot 100\%}{18}; \quad Et = \frac{\sum_{i=1}^{17} \left| \frac{t_{is} - t_{ie}}{t_{ie}} \right| \cdot 100\%}{17}. \tag{33}$$

Note: In the time analysis, while calculating the Et parameters, the focus was on the times of the peaks at successive amplitudes starting from the second (excluding first). It is related to the registered fact of "overpressures" and their influence on this parameter on the first amplitude; if they were taken into account, the error Et value would be distorted.

The final results of the Ep errors from all simulation tests are summarised graphically (Figure 9), while the results of the Et errors are summarised in Table 3.

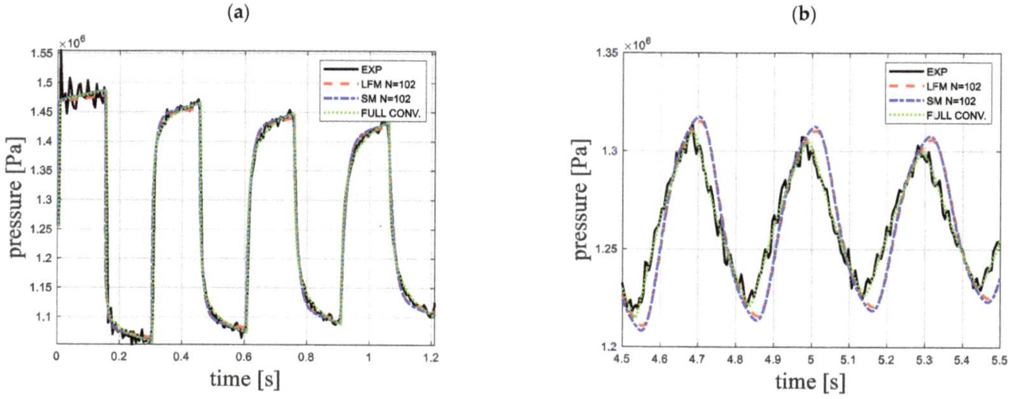

Figure 7. Selected results of pressures histories for Case 02 (Re = 2750): (**a**) initial phase; (**b**) later phase.

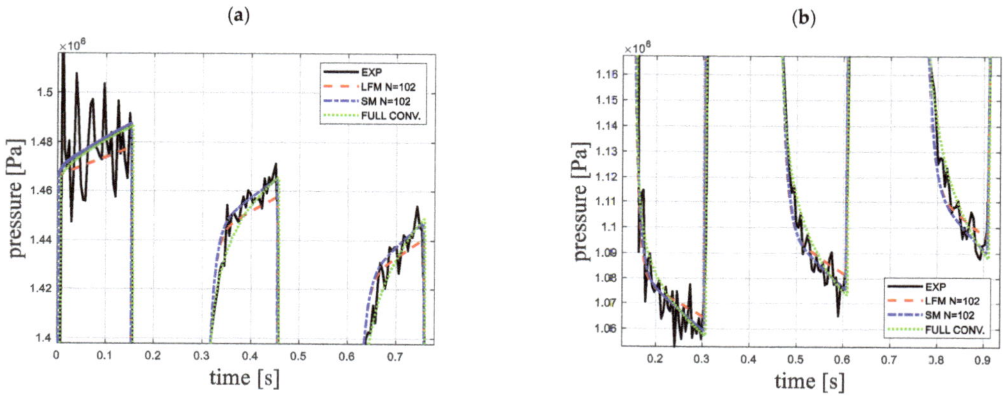

Figure 8. Enlargement of pressure histories of the first three amplitudes for Case 02 (Re = 2750): (**a**) crest, (**b**) valley.

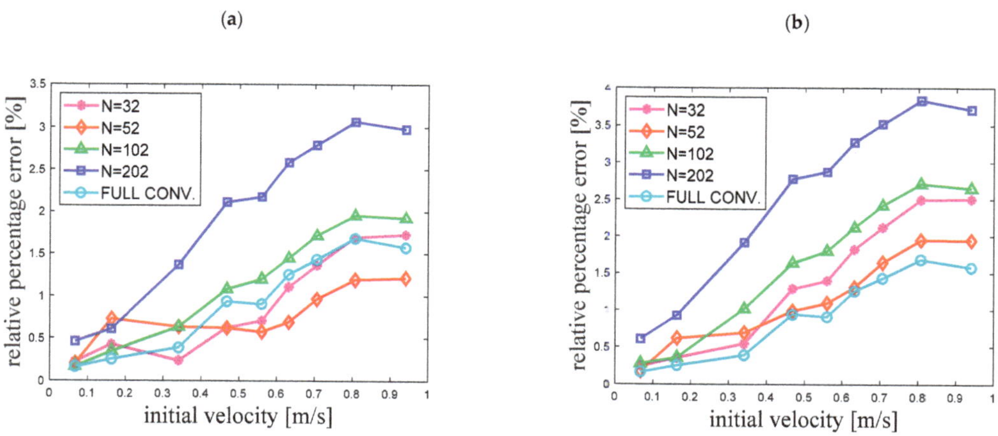

Figure 9. Variation in Ep error coefficient for: (**a**) standard method (SM); (**b**) lumped friction method (LFM).

Table 3. Quantitative results of the *Et* coefficients.

Case	Velocity [m/s]	SM—Standard Method				LFM—Lumped Friction Method				Full Conv.
		$N = 32$	$N = 52$	$N = 102$	$N = 202$	$N = 32$	$N = 52$	$N = 102$	$N = 202$	
01	0.066	1.72	1.59	1.49	1.49	1.82	1.66	1.54	1.53	1.48
02	0.162	0.96	0.82	0.70	0.70	1.02	0.87	0.75	0.75	0.66
03	0.340	0.92	0.78	0.67	0.66	0.98	0.84	0.73	0.72	0.63
04	0.467	1.10	0.97	0.86	0.85	1.16	1.02	0.91	0.90	0.86
05	0.559	1.16	1.03	0.93	0.91	1.22	1.09	0.98	0.97	0.86
06	0.631	0.94	0.81	0.71	0.69	1.00	0.87	0.77	0.75	0.69
07	0.705	0.72	0.61	0.50	0.48	0.78	0.65	0.56	0.54	0.48
08	0.806	1.32	1.21	1.11	1.09	1.39	1.26	1.16	1.14	1.01
09	0.940	1.03	0.92	0.82	0.80	1.10	0.98	0.88	0.85	0.86

Table 3 shows that the time consistency *Et* of the transient pressure waveforms simulated in the way proposed in this work was worse than the waveforms simulated with the full-convolutional integral. However, it was noticed during the implementation of these simulations that this disadvantage representing the simplified simulations can be easily minimised. Namely, during the simulation for $N = 32$, assuming only one parameter other than in the case of the waveform simulated with the full convolution (ineffective), this parameter is a speed of pressure wave propagation c. Assuming the value of $c_e = 1.01 * c_{fc}$ (one percent higher) during effective simulations, a significant improvement in the temporal consistency of the simulated waveforms is obtained (compare exemplary results presented in Figures 10 and 11 for Case 09—Re = 15,850), while maintaining very good agreement of the modelling of the maximum pressures (Figure 11). This necessity to modify the speed of pressure wave propagation can be explained by the use of a simplified weighting function in the calculations (made up of only two exponential terms). The quantitative results obtained from the additional simulations performed, presented in Figure 12, also indicate the improvement of the compliance fit. This improvement confirms similar findings by Bergant et al. [44].

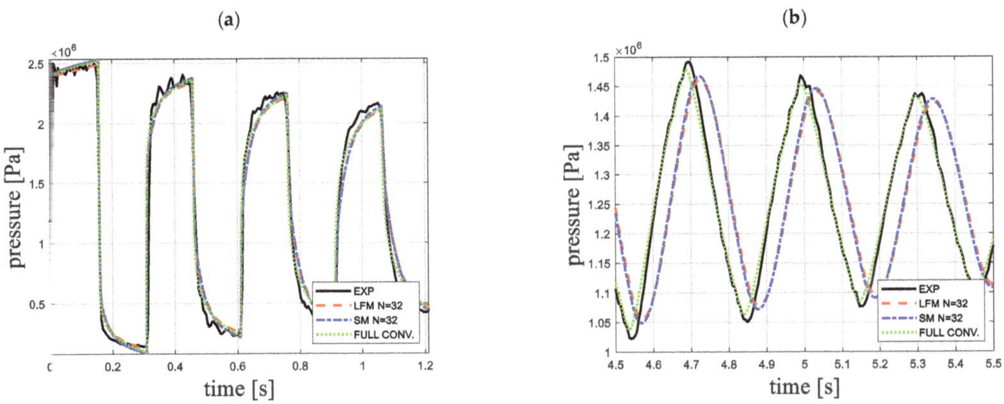

Figure 10. Results before correction of pressure wave (Case 09, Re = 15,850): (**a**) initial phase; (**b**) later phase.

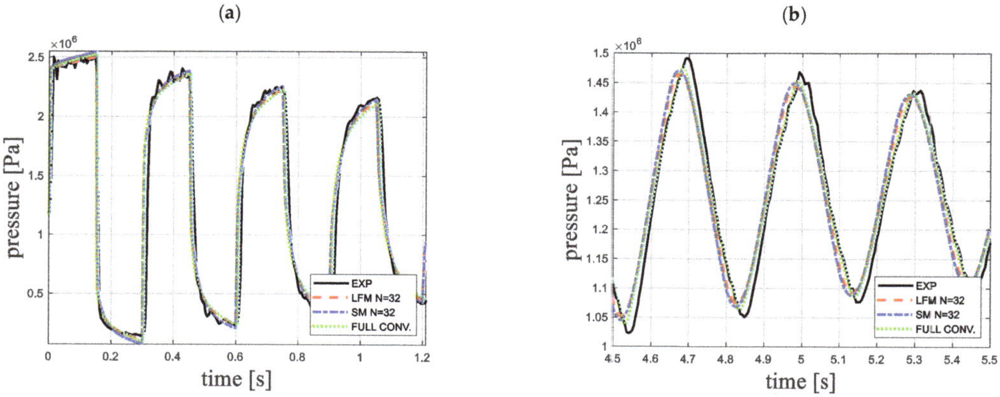

Figure 11. Simulation results after pressure wave correction (Case 09, Re = 15,850): (**a**) initial phase; (**b**) later phase.

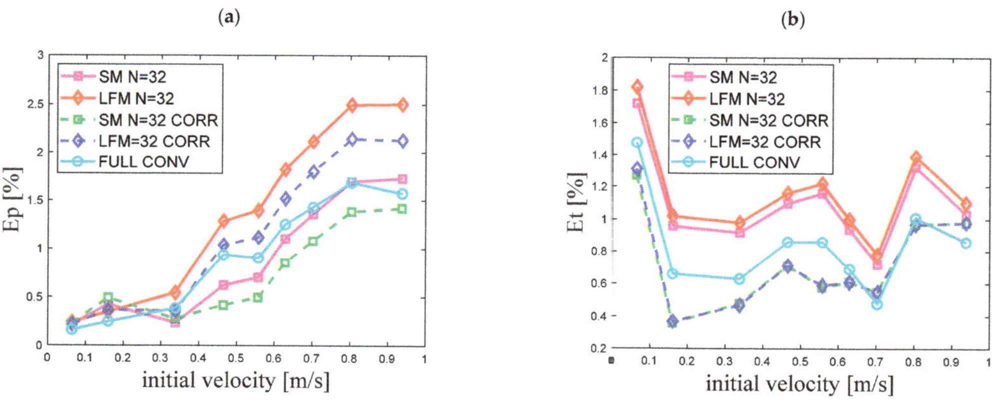

Figure 12. Variation in error coefficients: (**a**) Ep; (**b**) Et.

Completed extensive simulations have shown that modelling of hydraulic resistance during water hammer using the SM does not have to be a very complicated issue. The main conclusions of the research carried out are as follows:

- Use of simplified weighting functions, as shown in this paper, built from only two exponential terms, guarantees the results of a high agreement with the experimental results;
- Division of the pipeline along its length into 52 computational reaches guarantees the results with the lowest Ep errors;
- The smallest errors of parameter Et representing the time compliance of the simulated amplitudes were obtained using the largest division, i.e., 202 elements. It should be noted, however, that the application of a simple correction in the form of a slight increase (decrease) in the value of the pressure wave speed c significantly reduces this error.

Apart from the advantages, there are also disadvantages of the above-examined procedure:

- Necessity to use a constant time step (in a way, it is also a disadvantage of the characteristics method);
- Necessity of one-time analytical calculation of appropriate values of the weighting function coefficients (from the formulas presented in the Appendix A);
- Owing to the filtering of the upper range of the weighting function (from $10^3 \cdot \hat{t}$ to ∞), this method can only be used for modelling water hammer. Thus, preliminary analyses

showed that it is not suitable for modelling typically unidirectional flows (accelerated and delayed).

Further work should be aimed at an attempt to completely replace the weighting function built from a sum of exponential expressions with another simple function.

5. Conclusions

This paper investigates the performance of the computationally effective and accurate convolutional-based unsteady skin friction model (CBM) developed recently by Urbanowicz [31]. The weighting function is constructed from just two exponential terms, although then the coefficients m_i and n_i need to be calculated from the formulas given in the Appendix A. These coefficients are a function of the assumed dimensionless time step $\Delta\hat{t}$ in the numerical method. The simplification of the weighting function in conjunction with the corrected effective method for solving the convolution integral enables the determination of resistances from the final formulas of mathematical complexity similar to the IAB model. Contrary to the IAB models, in the analysed CBM approach, there is no need to calibrate the parameters describing the wall shear stress. A further possibility to simplify the modelling of unsteady resistance may be to use a model that lumps unsteady friction at the boundary nodes. The simulations carried out with the use of Johnston's model showed that the analysed transient waveforms were simulated with sufficient compliance with this model, which also used the two-term weighting function. Thus, we do hope that the validated simplifications of the CBM model implemented in this paper will find wider practical application, for example, in commercial programs for modelling transient flows in hydraulic pipe networks.

Author Contributions: Conceptualisation, K.U., A.B. and M.S.; methodology, K.U., A.B. and M.S.; software, K.U., A.D. and M.K. (Mykola Karpenko); validation, M.K. (Michał Kubrak) and A.K.; data curation, K.U. and A.D.; formal analysis, K.U., M.K. (Michał Kubrak) and A.K.; investigation, K.U., M.K. (Michał Kubrak) and A.K.; resources, M.K. (Mykola Karpenko), A.D. and M.S.; writing—original draft preparation, K.U. and A.B.; writing—review and editing, K.U., A.B., M.S. and M.K. (Michał Kubrak); visualisation, K.U., A.D. and M.K. (Mykola Karpenko); supervision, K.U., A.K. and A.B.; project administration, K.U. and M.S.; funding acquisition, K.U., M.S. and A.D. All authors have read and agreed to the published version of the manuscript.

Funding: A. Bergant gratefully acknowledges the support of Slovenian Research Agency (ARRS) conducted through the research project L2-1825 and the programme P2-0162.

Data Availability Statement: The code generated during the study and experimental data are available from the corresponding author by request.

Conflicts of Interest: The authors declare no conflict of interest.

Nomenclature

A_i, B_i and C_i	unsteady friction coefficients (-)
c	pressure wave speed (m/s)
D	pipe internal diameter (m)
Ep and Et	pressure and time compliance parameters (%)
e	pipe-wall thickness (m)
f	transient friction factor (-)
f_q	Darcy–Weisbach friction factor (-)
g	acceleration due to gravity (m/s^2)
j	imaginary unit (-)
k	empirical unsteady friction coefficient of the IAB model (-)
L	pipe length (m)
m_i and n_i	frictional weighting function coefficients (-)

N	number of computational reaches (-)
p	pressure (Pa)
p_R	reservoir pressure (Pa)
R	pipe internal radius (m)
Re_0	initial Reynolds number (-)
s	Laplace parameter (1/s)
T	temperature in Celsius degrees (°C)
t	time (s)
u	dummy variable (s)
Wh	water hammer number (-)
w	weighting function of unsteady friction (-)
v	average flow velocity (m/s)
v_0	initial liquid velocity (m/s)
x	space coordinate (m)
y_i	time dependent historical velocity effect (m/s)
Δt	numerical time step (s)
$\Delta\hat{t}$	dimensionless time step (-)
Δx	numerical spatial step (m)
Δv	velocity change at the valve (m/s)
ε	pipe-wall roughness (m)
η	correction factor of unsteady friction (-)
κ_n	n^{th} zeros of the Bessel function of type J_2 (-)
μ	dynamic viscosity (Pa·s)
μ'	second viscosity coefficient (Pa·s)
ν	kinematic viscosity of liquid (m^2/s)
ρ	liquid density (kg/m^3)
τ	wall shear stress (Pa)
Acronyms	
CBM	convolution-based model
CFM	Courant–Friedrichs–Lewy condition
CORR	corrected
EXP	experimental
FULL CONV	ineffective solution of the convolutional integral
HDPE	high-density polyethylene
IAB	instantaneous acceleration-based model
LFM	lumped friction method
MOC	method of characteristics
PVC	polyvinyl chloride
SM	standard method

Appendix A. Estimation of the Weighting Function Coefficients

The estimation of the weighting function coefficients is performed at the initial stage of transient simulations (set-up of the initial conditions) by the following procedure:

I. First calculate the constant time step Δt, next the dimensionless time step:

$$\Delta\hat{t} = \Delta t \cdot \frac{\nu}{R^2} = \frac{Wh}{N}, \tag{A1}$$

where: $Wh = \frac{\nu L}{cR^2}$ is a water hammer number.

II. When a dimensionless time step is known, calculate efficient weighting function coefficients m_1, m_2 and n_1, n_2 (for a simplified two-term function):

(a) m_1 calculation when $\Delta\hat{t} \leq 10^{-4}$:

$$m_1 = 0.03234 \cdot \Delta\hat{t}^{-0.5} + 48.35 \cdot \Delta\hat{t}^{0.5437} + 9.717 \cdot \Delta\hat{t}^{3.85} - 1.318, \tag{A2}$$

m_1 calculation when $\Delta\hat{t} > 10^{-4}$:

$$m_1 = 0.148 \cdot \exp(-\Delta\hat{t} \cdot 188.8) + 0.3227 \cdot \exp(-\Delta\hat{t} \cdot 1316) + 0.8039 \cdot \exp(-\Delta\hat{t} \cdot 5728) + 2.458 \cdot \exp(-\Delta\hat{t} \cdot 19{,}270) + 1, \quad \text{(A3)}$$

(b) m_2 calculation when $\Delta\hat{t} \leq 10^{-4}$:

$$m_2 = 0.1963 \cdot \Delta\hat{t}^{-0.5} + 2.88 \cdot \Delta\hat{t}^{3.575} - 0.2661 \cdot \Delta\hat{t}^{5.276} - 0.2351, \quad \text{(A4)}$$

m_2 calculation when $\Delta\hat{t} > 10^{-4}$:

$$m_2 = 2\,214 \cdot \exp(-\Delta\hat{t} \cdot 62.02) + 4.155 \cdot \exp(-\Delta\hat{t} \cdot 386.6) + 7.929 \cdot \exp(-\Delta\hat{t} \cdot 2191) + 20.485 \cdot \exp(-\Delta\hat{t} \cdot 12{,}570) + 1, \quad \text{(A5)}$$

(c) n_1 calculation when $\Delta\hat{t} \leq 10^{-5}$:

$$n_1 = 0.001476 \cdot \Delta\hat{t}^{-1} + 0.1203 \cdot \Delta\hat{t}^{-0.5} + 526.7 \cdot \Delta\hat{t}^{0.5567} + 6.091, \quad \text{(A6)}$$

n_1 calculation when $\Delta\hat{t} > 10^{-5}$:

$$n_1 = 9.317 \ \exp(-\Delta\hat{t} \cdot 4459) + 87 \cdot \exp(-\Delta\hat{t} \cdot 29{,}320) + 188.1 \cdot \exp(-\Delta\hat{t} \cdot 104{,}300) + 477.43 \cdot \exp(-\Delta\hat{t} \cdot 290{,}500) + 26.3744, \quad \text{(A7)}$$

(d) n_2 calculation when $\Delta\hat{t} \leq 10^{-4}$:

$$n_2 = 0.09021 \cdot \Delta\hat{t}^{-1} + 0.382 \cdot \Delta\hat{t}^{-0.4592} + 218.1 \cdot \Delta\hat{t}^{0.2615}, \quad \text{(A8)}$$

n_2 calculation when $\Delta\hat{t} > 10^{-4}$:

$$n_2 = 56.56 \cdot \exp(-\Delta\hat{t} \cdot 79.71) + 136.5 \cdot \exp(-\Delta\hat{t} \cdot 489.6) + 396.7 \cdot \exp(-\Delta\hat{t} \cdot 2880) + 1903.3 \cdot \exp(-\Delta\hat{t} \cdot 15{,}760) + 70.8493. \quad \text{(A9)}$$

III. Calculate correction coefficient η:

(a) For laminar flow when $\Delta\hat{t} \leq 0.02$:

$$\eta = \frac{\left[2 \cdot m_{1z} \cdot \Delta\hat{t}^{0.5} + m_{2z} \cdot \Delta\hat{t}^{1} + \left(\frac{2}{3}\right) \cdot m_{3z} \cdot \Delta\hat{t}^{1.5} + \left(\frac{1}{2}\right) \cdot m_{4z} \cdot \Delta\hat{t}^{2} + \left(\frac{2}{5}\right) \cdot m_{5z} \cdot \Delta\hat{t}^{2.5} + \left(\frac{1}{3}\right) \cdot m_{6z} \cdot \Delta\hat{t}^{3}\right]}{\sum_{i=1}^{2} \frac{m_i}{n_i} \cdot \left(1 - e^{-n_i \cdot \Delta\hat{t}}\right)}, \quad \text{(A10)}$$

where: m_{1z} = 0.282095; m_{2z} = −1.25; m_{3z} = 1.057855; m_{4z} = 0.9375; m_{5z} = 0.396696; and m_{6z} = −0.351563.

For laminar flow when $\Delta\hat{t} > 0.02$:

$$\eta = \frac{C1 + C2}{\sum_{i=1}^{2} \frac{m_i}{n_i} \cdot \left(1 - e^{-n_i \cdot \Delta\hat{t}}\right)}, \quad \text{(A11)}$$

where:

$$C1 = 2 \cdot m_{1z} \cdot 0.02^{0.5} + m_{2z} \cdot 0.02^{1} + \left(\frac{2}{3}\right) \cdot m_{3z} \cdot 0.02^{1.5} + \left(\frac{1}{2}\right) \cdot m_{4z} \cdot 0.02^{2} + \left(\frac{2}{5}\right) \cdot m_{5z} \cdot 0.02^{2.5} + \left(\frac{1}{3}\right) \cdot m_{6z} \cdot 0.02^{3}, \quad \text{(A12)}$$

$$C2 = \sum_{i=1}^{5} \frac{\left(1 - e^{-n_{iz} \cdot \Delta\hat{t}}\right)}{n_{iz}} - \sum_{i=1}^{5} \frac{\left(1 - e^{-n_{iz} \cdot 0.02}\right)}{n_{iz}}, \quad \text{(A13)}$$

and: n_{1z} = 26.3744; n_{2z} = 70.8493; n_{3z} = 135.0198; n_{4z} = 218.9216; and n_{5z} = 322.5544;

(b) For turbulent flow (Re > 2320):

$$\eta = \frac{A^* \cdot \sqrt{\frac{\pi}{B^*}} \cdot erf\left(\sqrt{\Delta\hat{t} \cdot B^*}\right)}{\sum_{i=1}^{2} \frac{m_i}{n_{is}} \cdot \left(1 - e^{-n_{is} \cdot \Delta\hat{t}}\right)}, \tag{A14}$$

where:

$$A^* = \sqrt{\frac{1}{4\pi}};\ B^* = \frac{Re^{\kappa}}{12.86};\ \kappa = \log_{10}\left(15.29/Re^{0.0567}\right), \tag{A15}$$

n_{is} is scaled coefficient using a universal scaling procedure:

$$n_{1s} = n_1 - 171.6545 + B^*;\ n_{2s} = n_2 - 171.6545 + B^*. \tag{A16}$$

IV. Calculate the constants in the efficient solution of convolution integral

$$A_1 = e^{-n_1 \cdot \Delta\hat{t}};\quad B_1 = \frac{m_1}{\Delta\hat{t} \cdot n_1} \cdot [1 - A_1];\quad C_1 = A_1 \cdot B_1, \tag{A17}$$

$$A_2 = e^{-n_2 \cdot \Delta\hat{t}};\quad B_2 = \frac{m_2}{\Delta\hat{t} \cdot n_2} \cdot [1 - A_2];\quad C_2 = A_2 \cdot B_2\,. \tag{A18}$$

Finally, the temporary unsteady friction factor during simulations is calculated by the following equation:

$$f_{(t+\Delta t)} = f_{q,(t+\Delta t)} + \frac{32\nu}{D\left|v_{(t+\Delta t)}\right| v_{(t+\Delta t)}} \cdot \sum_{i=1}^{2} \underbrace{\left[y_i(t) \cdot A_i + \eta \cdot B_i \cdot \left(v_{(t+\Delta t)} - v_{(t)}\right) + (1-\eta) \cdot C_i \cdot \left(v_{(t)} - v_{(t-\Delta t)}\right)\right]}_{y_i(t+\Delta t)} \tag{A19}$$

Note that:

- when calculated velocity is in range $-10^{-5} < v < 10^{-5}$, assume $v = -10^{-5}$ if it has a minus sign and $v = 10^{-5}$ when it has a positive sign (to avoid division by zero);
- select optimal number of grid points through the pipe axis; it should generally not exceed $N = 52$;
- set $y_i(t) = 0$ as an initial condition (for steady flow).

References

1. Pothof, I.; Karney, B. Guidelines for Transient Analysis in Water Transmission and Distribution Systems. In *Water Supply System Analysis—Selected Topics*; IntechOpen: London, UK, 2012. [CrossRef]
2. Jansson, M.; Andersson, M.; Karlsson, M. High-speed imaging of water hammer cavitation in oil–hydraulic pipe flow. *Fluids* **2022**, *7*, 102. [CrossRef]
3. Mousavifard, M.; Norooz, R. Numerical analysis of transient cavitating pipe flow by Quasi 2D and 1D models. *J. Hydraul. Res.* **2022**, *60*, 295–310. [CrossRef]
4. Warda, H.A.; Wahba, E.M.; Salah El-Din, M. Computational Fluid Dynamics (CFD) simulation of liquid column separation in pipe transients. *Alex. Eng. J.* **2020**, *59*, 3451–3462. [CrossRef]
5. Zhou, L.; Li, Y.; Zhao, Y.; Ou, C.; Zhao, Y. An accurate and efficient scheme involving unsteady friction for transient pipe flow. *J. Hydroinform.* **2021**, *23*, 879–896. [CrossRef]
6. Andrade, D.M.; Rachid, F.B.F. A versatile friction model for Newtonian liquids flowing under unsteady regimes in pipes. *Meccanica* **2022**, *57*, 43–72. [CrossRef]
7. Guerrero, B.; Lambert, M.F.; Chin, R.C. Extension of the 1D Unsteady Friction Model for Rapidly Accelerating and Decelerating Turbulent Pipe Flows. *J. Hydraul. Eng.* **2022**, *148*, 04022014. [CrossRef]
8. Santos, J.D.B.; Anjos, G.R.; Savi, M.A. An investigation of fluid-structure interaction in pipe conveying flow using reduced-order models. *Meccanica* **2022**. [CrossRef]
9. Cherian, R.M.; Sajikumar, N.; Sumam, K.S. Influence of Fluid–Structure Interaction on Pressure Fluctuations in Transient Flow. *J. Pipeline Syst. Eng. Pract.* **2021**, *12*, 04021002. [CrossRef]
10. Henclik, S. Application of the shock response spectrum method to severity assessment of water hammer loads. *Mech. Syst. Signal Process.* **2021**, *157*, 107649. [CrossRef]
11. Daily, J.W.; Hankey, W.L.; Olive, R.W.; Jordaan, J.M. Resistance coefficients for accelerated and decelerated flows through smooth tubes and orifices. *Trans. ASME* **1956**, *78*, 1071–1077. [CrossRef]

12. Carstens, M.R.; Roller, J.E. Boundary-shear stress in unsteady turbulent pipe flow. *J. Hydraul. Div. ASCE* **1959**, *95*, 67–813. [CrossRef]
13. Safwat, H.H.; van den Polder, J. Experimental and analytic data correlation study of water column separation. *J. Fluids Eng.* **1973**, *95*, 91–97. [CrossRef]
14. Brunone, B.; Golia, U.M.; Greco, M. Some remarks on the momentum equations for fast transients. In Proceedings of the Hydraulic Transients with Column Separation (9th and Last Round Table of the IAHR Group), IAHR, Valencia, Spain, 4–6 September 1991; pp. 201–209.
15. Vítkovský, J.; Lambert, M.; Simpson, A.; Bergant, A. Advances in unsteady friction modelling in transient pipe flow. In Proceedings of the 8th International Conference on Pressure Surges, The Hague, The Netherlands, 12–14 April 2000; pp. 471–482.
16. Ramos, H.; Covas, D.; Borga, A.; Loureiro, D. Surge damping analysis in pipe systems: Modelling and experiments. *J. Hydraul. Res.* **2004**, *42*, 413–425. [CrossRef]
17. Reddy, H.P.; Silva-Araya, W.F.; Chaudhry, M.H. Estimation of decay coefficients for unsteady friction for instantaneous, acceleration-based models. *J. Hydraul. Eng.* **2012**, *138*, 260–271. [CrossRef]
18. Cao, Z.; Wang, Z.; Deng, J.; Guo, X.; Lu, L. Unsteady friction model modified with compression–expansion effects in transient pipe flow. *J. Water Supply Res. Technol.-Aqua* **2022**, *71*, 330–344. [CrossRef]
19. Hu, Y.; Zhou, L.; Pan, T.; Fang, H.; Li, Y.; Liu, D. Godunov-type solutions for free surface transient flow in pipeline incorporating unsteady friction. *J. Water Supply: Res. Technol.-Aqua* **2022**, *71*, 546–562. [CrossRef]
20. Pan, T.; Zhou, L.; Ou, C.; Wang, P.; Liu, D. Smoothed particle hydrodynamics with unsteady friction model for water hammer pipe flow. *J. Hydraul. Eng.* **2022**, *148*, 04021057. [CrossRef]
21. Zhou, L.; Li, Y.; Karney, B.; Cheng, Y. Godunov-type solutions for transient pipe flow implicitly incorporating Brunone unsteady friction. *J. Hydraul. Eng.* **2021**, *147*, 04021021. [CrossRef]
22. Abdeldayem, O.M.; Ferràs, D.; van der Zwan, S.; Kennedy, M. Analysis of unsteady friction models used in engineering software for water hammer analysis: Implementation Case in WANDA. *Water* **2021**, *13*, 495. [CrossRef]
23. Wan, W.; Mehmood, K. Instantaneous acceleration-based modeling of pumping systems response under transient events. *Int. J. Mech. Sci.* **2022**, *224*, 107354. [CrossRef]
24. Zielke, W. Frequency-dependent friction in transient pipe flow. *J. ASME* **1968**, *90*, 109–115. [CrossRef]
25. Trikha, A.K. An efficient method for simulating frequency-dependent friction in transient liquid flow. *J. Fluids Eng. ASME* **1975**, *97*, 97–105. [CrossRef]
26. Kagawa, T.; Lee, I.; Kitagawa, A.; Takenaka, T. High speed and accurate computing method of frequency-dependent friction in laminar pipe flow for characteristics method. *Trans. Jpn. Soc. Mech. Eng. Part A* **1983**, *49*, 2638–2644. [CrossRef]
27. Schohl, G.A. Improved approximate method for simulating frequency—Dependent friction in transient laminar flow. *J. Fluids Eng. ASME* **1993**, *115*, 420–424. [CrossRef]
28. Urbanowicz, K. Fast and accurate modelling of frictional transient pipe flow. *Z. Angew. Math. Mech.* **2018**, *98*, 802–823. [CrossRef]
29. Adamkowski, A.; Lewandowski, M. Experimental examination of unsteady friction models for transient pipe flow simulation. *J. Fluids Eng.* **2006**, *128*, 1351–1363. [CrossRef]
30. Vardy, A.E.; Brown, J.M.B. Transient turbulent friction in smooth pipe flows. *J. Sound Vib.* **2003**, *259*, 1011–1036. [CrossRef]
31. Urbanowicz, K. Modern modeling of water hammer. *Pol. Marit. Res.* **2017**, *24*, 68–77. [CrossRef]
32. Johnston, D.N. Efficient methods for numerical modelling of laminar friction in fluid lines. *J. Dyn. Syst. Meas. Control ASME* **2006**, *128*, 829–834. [CrossRef]
33. Wylie, E.B.; Streeter, V.L.; Suo, L. *Fluid Transients in Systems*; Prentice-Hall Inc.: Englewood Cliffs, NJ, USA, 1993.
34. Bergant, A.; Simpson, A.R.; Vítkovský, J.P. Developments in unsteady pipe flow friction modelling. *J. Hydraul. Res.* **2001**, *39*, 249–257. [CrossRef]
35. Vítkovský, J.P.; Bergant, A.; Simpson, A.R.; Lambert, M.F. Systematic evaluation of one-dimensional unsteady friction models in simple pipelines. *J. Hydraul. Eng.* **2006**, *132*, 696–708. [CrossRef]
36. Zielke, W. Frequency-Dependent Friction in Transient Pipe Flow. Doctoral Thesis, University of Michigan, Ann Arbor, MI, USA, 1966.
37. Zarzycki, Z. On weighting function for wall shear stress during unsteady turbulent pipe flow. In Proceedings of the 8th International Conference on Pressure Surges, BHR Group, The Hague, The Netherlands, 12–14 April 2000; pp. 529–543.
38. Vardy, A.E.; Brown, J.M.B. Transient turbulent flow in fully-rough pipes. *J. Sound Vib.* **2004**, *270*, 233–257. [CrossRef]
39. Urbanowicz, K.; Bergant, A.; Karadzić, U.; Jing, H.; Kodura, A. Numerical investigation of the cavitating flow for constant water hammer number. *J. Phys. Conf. Ser.* **2021**, *1736*, 012040. [CrossRef]
40. Vardy, A.E.; Brown, J.M.B. Evaluation of unsteady wall shear stress by Zielke's method. *J. Hydraul. Eng.* **2010**, *136*, 453–456. [CrossRef]
41. Xu, Y.; Jiao, Z.; Zhao, L. Fast meshless solution with lumped friction for water hammer. In Proceedings of the BATH/ASME 2020 Symposium on Fluid Power and Motion Control, Virtual, 9–11 September 2020. FPMC2020-2789, V001T01A043. [CrossRef]
42. Urbanowicz, K. Analytical expressions for effective weighting functions used during simulations of water hammer. *J. Theor. Appl. Mech.* **2017**, *55*, 1029–1040. [CrossRef]
43. Bergant, A.; Karadžić, U.; Vítkovský, J.P.; Vušanović, I.; Simpson, A.R. A discrete gas cavity model that considers the frictional effects of unsteady pipe flow. *Stroj. Vestn.-J. Mech. Eng.* **2005**, *51*, 692–710.

44. Adamkowski, A.; Henclik, S.; Janicki, W.; Lewandowski, M. The influence of pipeline supports stiffness onto the water hammer run. *Eur. J. Mech. B/Fluids* **2016**, *61*, 297–303. [CrossRef]
45. Henclik, S. Numerical modeling of water hammer with fluid–structure interaction in a pipeline with viscoelastic supports. *J. Fluids Struct.* **2018**, *76*, 469–487. [CrossRef]
46. Holmboe, E.L. Viscous Distortion in Wave Propagation as Applied to Waterhammer and Short Pulses. Doctoral Thesis, Carnegie Institute of Technology, Pittsburgh, PA, USA, 1964.
47. Covas, D. Inverse Transient Analysis for Leak Detection and Calibration of Water Pipe Systems Modelling Special Dynamic Effects. Doctoral Thesis, Imperial College London (University of London), London, UK, 2003.

water

MDPI

Article

Water Hammer in Steel–Plastic Pipes Connected in Series

Michał Kubrak [1,*], Apoloniusz Kodura [1], Agnieszka Malesińska [1] and Kamil Urbanowicz [2]

[1] Faculty of Building Services, Hydro and Environmental Engineering, Warsaw University of Technology, 00-653 Warsaw, Poland
[2] Faculty of Mechanical Engineering and Mechatronics, West Pomeranian University of Technology in Szczecin, 70-310 Szczecin, Poland
* Correspondence: michal.kubrak@pw.edu.pl

Abstract: This paper experimentally and numerically investigates the water hammer phenomenon in serially connected steel and HDPE pipes with different diameters. The aim of the laboratory tests was to obtain the time history of the pressure head at the downstream end of the pipeline system. Transient tests were conducted on seven different pipeline system configurations. The experimental results show that despite the significantly smaller diameter of the HDPE pipe compared to the steel pipe, introducing an HDPE section makes it possible to suppress the valve-induced pressure surge. By referring to the results of the experimental tests conducted, the comparative numerical calculations were performed using the fixed-grid method of characteristics. To reproduce pressure wave attenuation in a steel pipe, Brunone-Vitkovský instant acceleration-based model of unsteady friction was used. To include the viscoelastic behavior of the HDPE pipe wall, the one-element Kelvin–Voigt model was applied. By calibrating the unsteady friction coefficient and creep parameters, satisfactory agreement between the calculated and observed data was obtained. The calibrated values of parameters for a single experimental test were introduced in a numerical model to simulate the remaining water hammer runs. It was demonstrated that using the same unsteady friction coefficient and creep parameters in slightly different configurations of pipe lengths can be effective. However, this approach fails to reliably reproduce the pressure oscillations in pipeline systems with sections of significantly different lengths.

Keywords: water hammer; hydraulic transients; serially connected pipes; unsteady friction; viscoelasticity

Citation: Kubrak, M.; Kodura, A.; Malesińska, A.; Urbanowicz, K. Water Hammer in Steel–Plastic Pipes Connected in Series. *Water* **2022**, *14*, 3107. https://doi.org/10.3390/w14193107

Academic Editor: Francesco Napolitano

Received: 31 August 2022
Accepted: 27 September 2022
Published: 2 October 2022

Publisher's Note: MDPI stays neutral with regard to jurisdictional claims in published maps and institutional affiliations.

Copyright: © 2022 by the authors. Licensee MDPI, Basel, Switzerland. This article is an open access article distributed under the terms and conditions of the Creative Commons Attribution (CC BY) license (https://creativecommons.org/licenses/by/4.0/).

1. Introduction

Any rapid alteration in the fluid flow velocity results in a pressure wave that propagates through the pipeline system. Such events, known as hydraulic transients or water hammer, can cause serious problems. It is well known that violent pressure surges may lead to pipe bursts, hydraulic device failures or water contamination [1,2]. Due to the fact that pipeline systems are subjected to a wide variety of operating conditions, transient flows commonly occur. For this reason, in recent decades, numerous studies have been conducted on this topic. In practice, pressurized fluid distribution systems often consist of serially connected pipes with different properties, i.e., different pipe-wall materials and different diameters. Nevertheless, to date, only a few researchers have addressed the problem of hydraulic transients in complex multi-pipe systems.

Gong et al. [3] investigated the possibility of replacing a metallic pipe with a plastic section in order to attenuate the pressure waves. They demonstrated that it is feasible to mitigate pressure surges in water distribution systems if the branch that links the transient source and the water main is connected to a plastic section instead of steel pipe. Garg and Kumar [4] studied the water hammer in a pipeline with two different materials and their combined configuration and reported that pipes made of viscoelastic material can be used to renovate existing water pipelines to effectively control the water hammer. Triki [5] proposed a protection technique based on replacing a short section of the transient sensitive regions of the existing piping system by another one made of polymeric material. Ferrante [6] investigated the effect of the junction between two pipes with different polymeric materials on the transient pressure wave and proved that the reflection coefficient at the junction between the two pipes depends on the viscoelastic parameters and on the transient time-history. Bettaieb et al. [7] studied the performance of metallic–polymeric pipe configurations under transient conditions caused by pump failure and demonstrated that wave reflections from junction connections significantly affect the pressure wave in water distribution systems.

Despite this interest, no one—to the best of our knowledge—has studied the water hammer phenomenon in a serially connected steel–plastic pipeline system with significantly different inner diameters and for several different lengths of each section. The present work aimed to experimentally and numerically investigate a valve-induced water hammer phenomenon in serially connected pipes with different properties, in terms of both the pipe dimensions and the pipe-wall material. The measurements were taken on a setup from which it was possible to collect transient data in a pipeline that consisted of steel and high-density polyethylene (HDPE) sections connected in series. Moreover, the experiments were carried out for various lengths of steel and HDPE sections, while maintaining a constant total length of the pipeline system. By referring to the results of the conducted experimental tests, the comparative numerical calculations were performed using the fixed-grid method of characteristics.

This paper is divided into six sections. In the second section, some theoretical aspects of hydraulic transients in pressurized pipes are presented. Experimental tests are outlined in the third section. Numerical calculations are addressed in Sections 4 and 5. Conclusions are drawn in the final section.

2. Mathematical Description of Hydraulic Transients in Pressure Pipes

The 1D unsteady pressure pipe flow is governed by the system of first-order hyperbolic partial differential equations resulting from mass and momentum conservation [8]:

$$\frac{\partial V}{\partial t} + V\frac{\partial V}{\partial x} + g\frac{\partial H}{\partial x} + f\frac{V|V|}{2D} = 0 \quad (1)$$

$$\frac{\partial H}{\partial t} + V\frac{\partial H}{\partial x} + \frac{c^2}{g}\frac{\partial V}{\partial x} = 0 \quad (2)$$

where f is the friction factor $(-)$, c is the pressure wave velocity (m/s), H is the piezometric head (m), V is the mean flow velocity (m/s), g is the gravity acceleration (m/s^2), x is the space coordinate (m), t is time (s), and D is the internal pipe diameter (m).

Equation (1) is the momentum equation and Equation (2) is called the continuity equation. In classic (elastic) water hammer theory, the pressure wave velocity is defined by:

$$c = \sqrt{\frac{\frac{K}{\rho}}{1 + \alpha\frac{KD}{Es}}} \quad (3)$$

where α is the parameter dependent on the pipe diameter and constraints $(-)$, K is the bulk modulus of elasticity of the liquid (Pa), ρ is the fluid density (m^3/s), E is the modulus of elasticity of the pipe-wall material (Pa) and s is the pipe-wall thickness (m).

Traditionally, transient analyses were carried out by incorporating a steady or quasi-steady friction factor into the momentum equation. It is well known that this approach overestimates pressure oscillations during fast transient events [9]. For this reason, the efforts of many researchers have been focused on identifying the dynamic effects accounting for pressure wave damping. Generally, two types of models have been proposed in the literature to reduce discrepancies between the measurements and calculation results:

unsteady friction models and viscoelastic models. In the following subsections, the models used in this paper were briefly described.

2.1. Unsteady Friction

In general, two approaches are widely used to take into account unsteady friction: the weighting function-based (WFB) method and instantaneous acceleration-based (IAB) method. The WFB method involves including the friction term in the form of convolution related to the instantaneous flow velocity and to the weighted past velocity changes. This approach, first proposed by Zielke [10] for unsteady laminar flow, was later adapted for transient turbulent flow [11,12]. In the IAB method, the unsteady friction coefficient depends on the local and instantaneous convective acceleration of flowing liquid. Due to its simplicity, probably the best-known example of this method is the model proposed by Brunone [13,14] and later improved by Vítkovský [15] by including an additional sign operator in the convection term. To date, Brunone-Vítkovský model is the only unsteady friction model used in some of the available commercial software tools for simulation of hydraulic transients [16]. In this approach, the friction factor is given by the following equation:

$$f = f_q + \frac{k_V}{V|V|}\left(\frac{\partial V}{\partial t} + c\,\mathrm{sign}(V)\left|\frac{\partial V}{\partial x}\right|\right) \quad (4)$$

where f_q is the quasi-steady friction factor $(-)$, k_V is the unsteady friction coefficient $(-)$ and sign(V) = (+ 1 for $V > 0$ or -1 for $V < 0$).

The sign term in Equation (4) gives the correct sign of the convective term for all possible flow [17]. The presented approach combines local inertia and wall friction unsteadiness. In order to analytically determine k_V, the Vardy and Brown shear decay coefficient can be used [18]:

$$k_V = \frac{\sqrt{C*}}{2}, \quad (5)$$

where for laminar flow, C^* is equal to:

$$C* = 0.00476 \quad (6)$$

and for turbulent flow, C^* is given by:

$$C* = \frac{7.41}{\mathrm{Re}^{\log(14.3/\mathrm{Re}^{0.05})}}, \quad (7)$$

where Re—Reynolds number $(-)$.

Issues regarding the compliance of the measurement data and the calculation results using this particular unsteady friction model are discussed in Section 5. More details on unsteady friction modeling can be found in [17,19].

2.2. Viscoelastic Behavior of Polymer Pipes

In contrast to elastic materials, viscoelastic materials exhibit a retarded strain after initial instantaneous strain upon applied loading [20]. For this reason, the classic formulation of the continuity equation is no longer suitable for describing unsteady flow in polymer pipes. Polymers exhibit both elastic and viscous characteristics. In order to correctly mathematically describe this property, the response of viscoelastic material is often represented by the mechanical response of an elastic spring connected to the viscous dashpot [21]. This approach, known as "mechanical analogs", is widely used in simulating hydraulic transients in plastic pipes. Figure 1 presents the scheme of the generalized Kelvin–Voigt model.

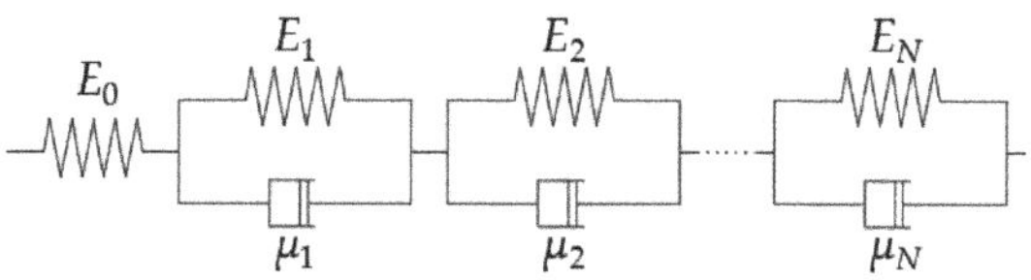

Figure 1. Generalized Kelvin–Voigt model.

Figure 1 shows a series of N Kelvin–Voigt elements with a modulus of elasticity E_k (subscript k denotes the number of a particular element), viscosity of μ_k, creep compliance $J_k = 1/E_k$ and compliance $J_0 = 1/E_0$. The creep function corresponding to the generalized Kelvin–Voigt model is given by:

$$J(t) = J_0 + \sum_{k=1}^{N} J_k \left(1 - e^{\frac{-t}{\tau_k}}\right), \quad (8)$$

where τ is the retardation time of the k-th Kelvin–Voigt element (s).

The viscoelasticity of plastic pipes can be incorporated by adding a retarded strain term to the continuity equation [22,23]:

$$\frac{\partial H}{\partial t} + V\frac{\partial H}{\partial x} + \frac{c^2}{g}\frac{\partial V}{\partial x} + \frac{2c^2}{g}\sum_{k=1}^{N}\frac{\partial \varepsilon_k}{\partial t} = 0, \quad (9)$$

where ε—radial strain of the k-th Kelvin–Voigt element (−).

Equation (9) and Equation (1) describe the unsteady fluid flow in polymer pipes. Further detailed mathematical description of the 1D viscoelastic behavior of polymer pipes during transient flow can be found, e.g., in [23–25]. The analysis of transients in viscoelastic pipelines by means of 2D Kelvin-Voigt model is detailed in [26,27].

3. Laboratory Tests

The aim of the experiments presented in this paper was to record pressure changes during a valve-induced water hammer in a horizontal pipeline system that consists of serially connected pipes with different pipe-wall materials, with different inner diameters and with different lengths. Transient data were collected on a laboratory setup that consisted of three essential elements: a pressure tank, pipeline with fittings, and data acquisition system. Two different conduits were used—steel pipe with an inner diameter of D_S = 0.053 m and HDPE pipe with an inner diameter of D_P = 0.032 m and wall thickness of s_P = 0.0024 m (D_S/D_P = 1.65). In order to capture the influence of the length of each section on pressure changes, tests were performed for different proportions of the length of both pipes, keeping the total length of the system constant. To reduce complexity, the order in which the pipes were fitted was always the same, i.e., the steel pipeline was always connected to the pressure tank that supplied the system. The scheme of the laboratory setup is presented in Figure 2.

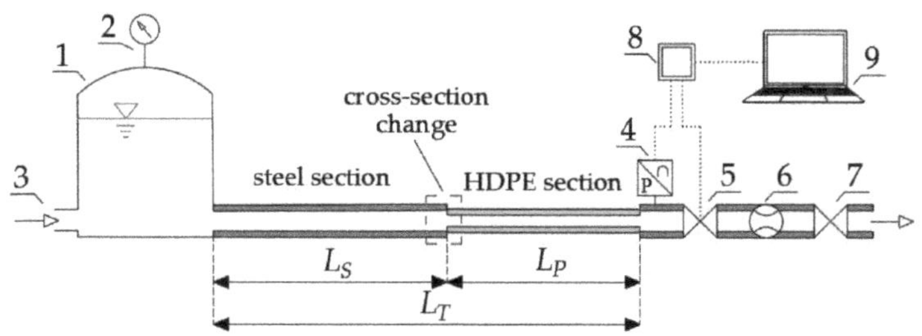

Figure 2. Laboratory setup: 1—pressure tank; 2—Bourdon pressure gauge; 3—water supply system; 4—pressure sensor; 5—valve initiating water hammer; 6—inductive flowmeter; 7—regulating valve; 8—data logger; 9—laptop; L_S—length of the steel section (m); L_P—length of the plastic (HDPE) section (m); L_T—total length of the pipeline system (m).

The valve that initiated the water hammer as well as the pressure sensor were connected to the short steel section at the downstream end of the pipeline system with the same diameter as the main steel pipe. The distances between the end of the HDPE pipe and pressure sensor (element 4 in Figure 2) and between the end of the HDPE and the valve (element 5 in Figure 2) were 10 cm and 30 cm, respectively. All pipeline sections were rigidly fixed to the floor. The data acquisition system consisted of an inductive flowmeter, a piezoresistive pressure sensor with the range of −0.1–1.2 MPa, and a valve closing timer connected to the laptop through a data logger. Pressure samples were recorded with a frequency of 500 Hz. As mentioned before, the main goal was to perform water hammer tests for different lengths of steel and HDPE sections. Six different steel–plastic configurations were considered. An additional water hammer run was conducted in a straight steel pipeline (without any HDPE section). Overall, seven different configurations of the pipeline systems were considered. In order to be able to compare individual experiments, for each pipeline system configuration, the water hammer phenomenon was initiated for a constant steady-state discharge of approximately 9.167×10^{-4} m^3/s. Due to the fact that the water temperature significantly affects the pressure wave during transient flow in plastic pipes [28], the experiments were conducted, taking care to maintain a constant water temperature (19 °C) for each system configuration. The water hammer was initiated by a manual valve closure. For each run, the registered valve closing time was between 0.04–0.11 s. The main parameters of the pipeline system along with the values of the mean flow velocity during steady flow conditions are listed in Table 1.

Table 1. Dimensions of the pipeline system in each configuration and steady state flow parameters.

System Config.	L_S (m)	L_P (m)	L_T (m)	L_S/L_T (%)	V_{0S} (m/s)	V_{0P} (m/s)
#1	15.75	32.35	48.10	33	0.43	0.97
#2	18.90	29.20	48.10	39	0.42	0.95
#3	25.15	22.95	48.10	52	0.41	0.93
#4	31.40	16.70	48.10	65	0.41	0.93
#5	39.15	8.95	48.10	81	0.42	0.95
#6	45.40	2.70	48.10	94	0.41	0.93
#7	48.10	0.00	48.10	100	0.42	0.95

Note: where V_{0S}—initial steady flow velocity in the steel pipe (m/s); V_{0P}—initial steady flow velocity in the HDPE pipe (m/s).

It is worth highlighting that the internal diameters of the pipes used in this setup differ significantly. Hence, for a given initial discharge, the mean flow velocity in the plastic section is much higher than that in the steel section. Since the plastic section is located at the downstream side of the pipeline system and connected near the valve that initiates unsteady flow, this consequently translates into a higher pressure increase during a water hammer event. In order to compare the pressure signals obtained in each water hammer run, the collected transient data are summarized in Figure 3.

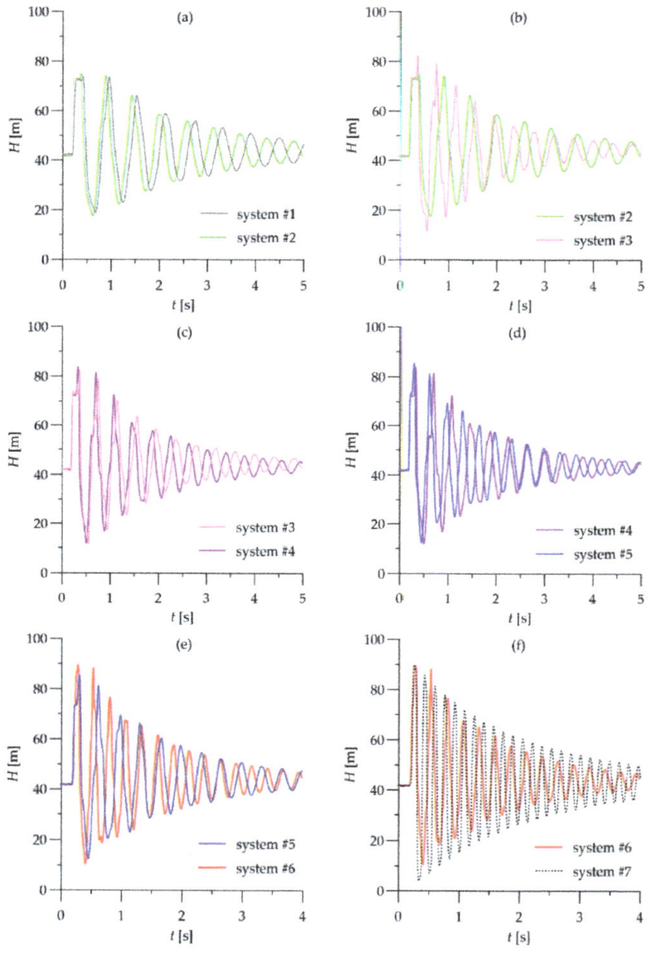

Figure 3. Comparison between experimental transient data: (**a**) System #1 vs. system #2; (**b**) System #2 vs. system #3; (**c**) System #3 vs. system #4; (**d**) System #4 vs. system #5; (**e**) System #5 vs. system #6; and (**f**) System #6 vs. system #7.

As expected, it is apparent from Figure 3 that, with the increase in the share of the steel section in the total length of the pipeline system, the mean pressure wave velocity of the entire pipeline system increases. This can be observed as each subsequent pressure signal shifts to the left, i.e., the return time of the wave reflected from the tank shortens. Interestingly, even despite the much higher initial flow velocity in the HDPE section than in the steel section (during steady state conditions $V_{0P}/V_{0S} = 2.26$), the maximum pressure increase in each system configuration with a plastic section (systems from #1 to #6) is lower than that in the steel pipeline (system #7). This is due to the viscoelastic properties of the HDPE pipe, which result in less severe transients for the same flow change compared to steel pipe. It can be observed in Figure 3e,f that the second positive pressure peak recorded in system #6 (red line) is practically the same as the first one. Apparently, for this particular L_S/L_T ratio, the overlapping of pressure waves due to sudden pipeline contraction occurs. Furthermore, similarly to the results of water hammer experiments in viscoelastic pipeline systems with sudden cross-section changes [29,30], the effect of the diameter change is only visible in the first phases of pressure oscillations. Unlike in the case of water hammer in a steel pipeline with sudden contractions and expansions [31], after a couple of consecutive amplitudes, the pressure wave becomes smoothed and resembles oscillations in a simple, single-diameter pipeline. In order to compare differences in the first pressure peak, transient data obtained from all runs are compiled in Figure 4.

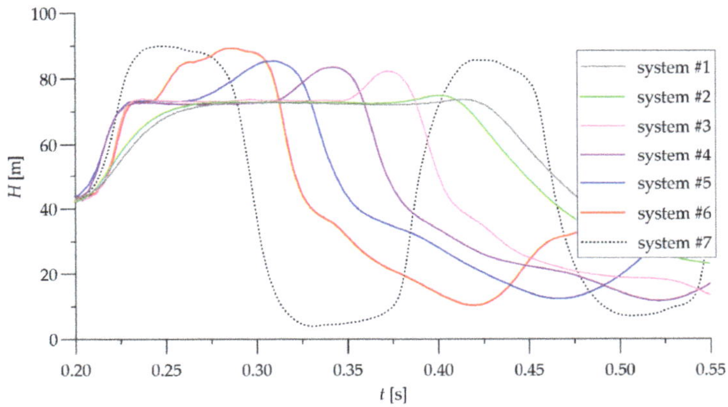

Figure 4. First phase of pressure oscillations registered during all water hammer runs.

In Figure 4, it can be observed that, in the case of water hammer in systems #1 and #2 (gray and green lines), the $H(t)$ curve becomes almost horizontal for the first half period of the reflection wave. This is typical for viscoelastic pipeline systems and indicates that the pipe-wall material creeps greatly at this time interval [32]. In each case, the interference of pressure waves resulting from the cross-section change is visible. After the first pressure peak, an increase in the pressure level can be observed, which results from the transmission of the pressure wave from a steel section (large diameter) through an HDPE section (small diameter). This effect becomes more significant as the length of the steel section increases, which also affects the maximum pressure rise more. In the pipeline system #1, which has the longest HDPE section, the observed maximum pressure increase is ΔH_{max} = 31.62 m, whereas in the steel pipeline (system #7), ΔH_{max} = 48.28 m (where $\Delta H_{max} = H_{max} - H_0$). Changes in the dimensionless maximum pressure increase (H_{max}/H_0) as a function of the L_S/L_T ratio are presented in Figure 5.

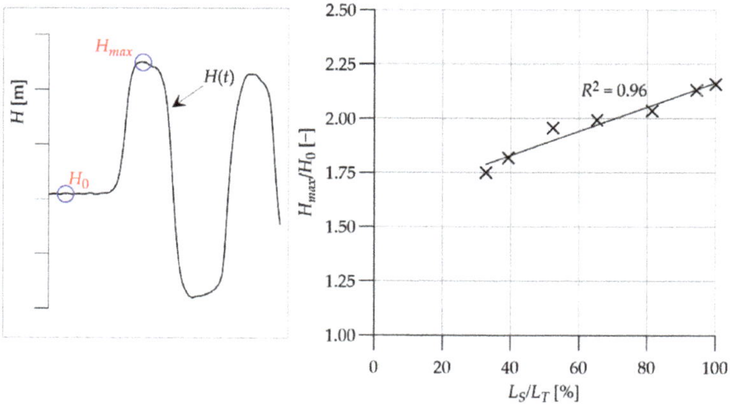

Figure 5. Values of dimensionless maximum pressure increase as a function of L_S/L_T.

Figure 5 shows a clear trend in the increase in the maximum pressure caused by the lengthening of the steel section. This result indicates that using an in-line polymeric pipe section in order to suppress pressure surges is also possible for pipes with significantly smaller internal diameter compared to the diameter of the original pipe. Evidently, the larger the diameter and longer the length of the polymeric section, the better the pressure wave damping that can be obtained. However, in some fluid distribution systems, increasing the mean flow velocity may be advantageous (e.g., hydraulic transportation systems). In order to analyze the influence of the length of each section on the pressure wave velocity, an attempt to calculate its value was made. The pressure signal obtained from steel–HDPE pipelines is distorted due to sudden cross-section changes. Pressure wave reflections are especially visible in the initial phases of water hammer phenomenon. To minimize the error in the calculation of c, the time difference between the occurrence of the fifth and sixth pressure peaks was taken into account. The results of the calculations are presented as a function of the L_S/L_T ratio in Figure 6.

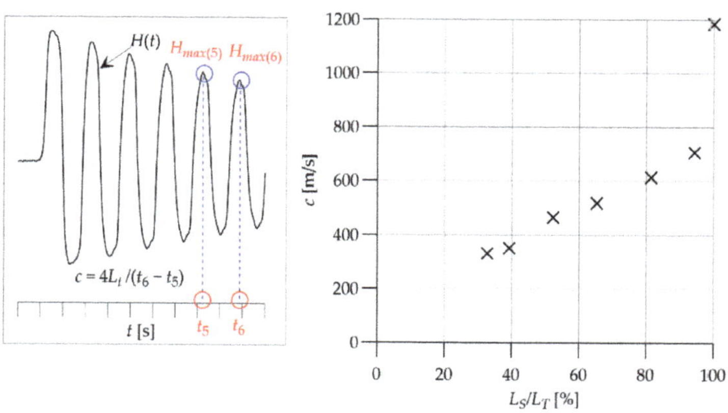

Figure 6. Values of mean pressure wave velocities in tested systems as a function of L_S/L_T.

As can be seen in Figure 6, for all steel–HDPE pipeline configurations, the pressure wave velocity increases linearly with the increase in the L_S/L_T ratio. However, for a pipeline system without any HDPE section (system #7; L_S/L_T = 100%), there is a clear sudden increase in the value of pressure wave velocity. This effect is attributed to the method of determining the value of c presented on the right side of the plot. In principle, the experimentally obtained $H(t)$ signal is a harmonic function with a period equal to the pipeline period of reflection time. In order to calculate the pressure wave velocity, the time interval between successive maximum pressure peaks was taken into account. In fact, this is a measure of the phase speed, not a direct measure of the speed of sound. Despite the limitations of this method of calculating c, Figure 6 to some extent illustrates the changes of pressure wave velocity due to the change in lengths of each section. Detailed discussion on the pressure wave velocity in pipeline systems can be found in [33].

Further analysis of Figure 3 shows that, despite the differences in the values of ΔH_{max} and c in each water hammer run, the observed pressure wave decays at a similar rate for each steel–HDPE configuration. In order to quantify the damping rate of each particular pipeline system, the logarithmic damping decrement can be calculated as [34]:

$$\delta = \frac{1}{n} \ln\left(\frac{H_m}{H_{m+n}}\right) \tag{10}$$

where n is the number of subsequent pressure peaks used to calculate the damping decrement and m is the number of the first pressure peak taken into account.

Calculated values of the logarithmic damping decrement for $m = 1$ and $n = 10$ are presented in Figure 7.

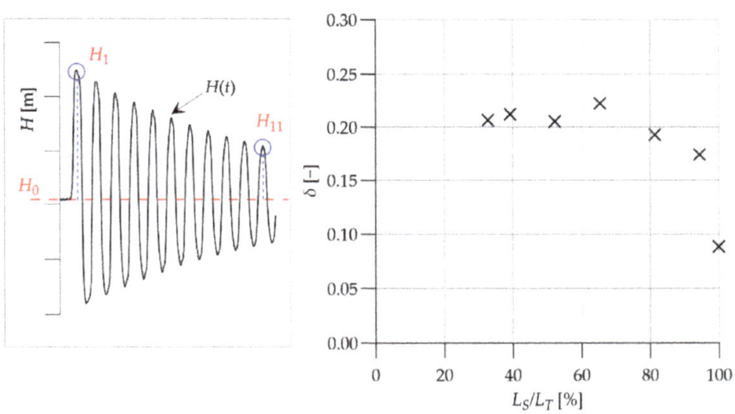

Figure 7. Values of logarithmic damping decrement as a function of L_S/L_T.

Figure 7 shows that the pressure magnitudes damped at similar rates for each steel–HDPE pipeline system configuration as the calculated values of δ are between 0.174 and 0.221. As anticipated, the weakest damping of the pressure wave was obtained for the pipeline without any HDPE section (δ = 0.087). It should be noted that no clear correlation between the length of each section and the logarithmic damping decrement was revealed. This is probably due to the cross-section changes, which cause the initial pressure amplitudes not to be damped regularly.

4. Water Hammer Solver

To numerically model the water hammer events obtained during the laboratory tests, a hydraulic transient solver that is able to take into account both the pipe-wall viscoelasticity and unsteady friction presented in [35,36] was used. Here, for brevity, only its simplified description is given.

In most practical cases, the convective acceleration terms $V(\partial V/\partial x)$ and $V(\partial H/\partial x)$ are negligibly small [37]. Hence, simplified governing equations are usually used to numerically simulate hydraulic transients:

$$\frac{\partial H}{\partial t} + \frac{c^2}{gA}\frac{\partial Q}{\partial x} + \frac{2c^2}{g}\sum_{k=1}^{N}\frac{\partial \varepsilon_k}{\partial t} = 0 \tag{11}$$

$$\frac{\partial Q}{\partial t} + gA\frac{\partial H}{\partial x} + \frac{fQ|Q|}{2DA} = 0 \tag{12}$$

where Q is the volumetric flow rate (m^3/s).

This set of differential equations was solved using the fixed-grid method of characteristics with a specified time interval. A fragment of the uniformly spaced computational grid is presented in Figure 8.

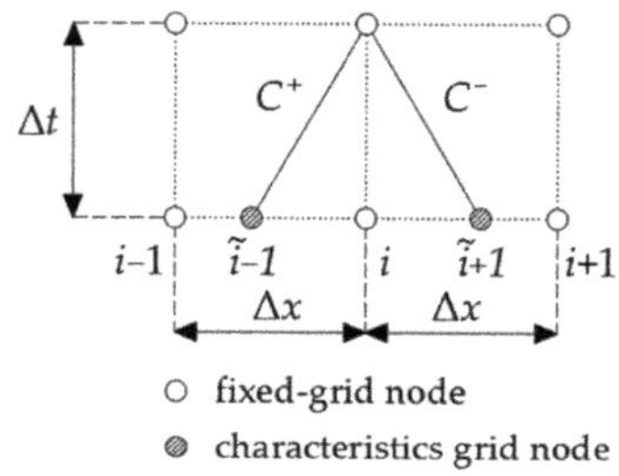

Figure 8. Numerical grid in fixed-grid method of characteristics.

The unknown values of pressure and discharge are calculated from positive and negative characteristics equations:

$$C^+: \quad Q_{i,t} = C_P - C_{a+}H_{i,t} \tag{13}$$

$$C^-: \quad Q_{i,t} = C_N + C_{a-}H_{i,t} \tag{14}$$

where C_N, C_{a+}, C_P, and C_{a-} are known coefficients used to describe the steady-state friction, the unsteady friction, and viscoelasticity of the pipe-wall material, respectively.

In general form, these coefficients are defined as:

$$C_P = \frac{Q_{i-1,t-1} + BH_{i-1,t-1} + C'_{P1} + C''_{P1} + C'''_{P1}}{1 + C'_{P2} + C''_{P2}} \tag{15}$$

$$C_N = \frac{Q_{i-1,t-1} + BH_{i-1,t-1} + C'_{P1} + C''_{P1} + C'''_{P1}}{1 + C'_{P2} + C''_{P2}} \tag{16}$$

$$C_{a+} = \frac{B + C'''_{P2}}{1 + C'_{P2} + C''_{P2}} \tag{17}$$

$$C_{a-} = \frac{B + C'''_{N2}}{1 + C'_{N2} + C''_{N2}} \tag{18}$$

where parameter B, defined as $B = gA/c$, depends on the fluid and pipe properties.

In Equations (15)–(17), indexes $'$, $''$ and $'''$ refer to steady-state friction, unsteady friction, and viscoelasticity, respectively. In order to model the pipeline systems used in the experiments, for each section (steel and HDPE), individual computational grids connected with series junction were created. For both sections, steady-state friction was taken into account. For first-order accuracy calculations, the values of C'_{P1} and C'_{N1} are equal to:

$$C'_{P1} = -R\Delta t|Q_{i-1,t-1}|Q_{i-1,t-1}, \tag{19}$$

$$C'_{N1} = -R\Delta t|Q_{i+1,t-1}|Q_{i+1,t-1}. \tag{20}$$

Apart from the steady-state friction term, in each section, different dynamic effects were taken into account. In the steel pipe, Brunone-Vitkovský IAB unsteady friction model described in Section 2.1 was applied. For the HDPE section, unsteady friction was neglected and only the viscoelastic behavior of the pipe-wall material was taken into account. Descriptions of the remaining C_N, C_{a+}, C_P, and C_{a-} coefficients for each section of the pipeline are listed in Table 2.

Table 2. Coefficients C_N, C_{a+}, C_P, and C_{a-} are used to model each section of the pipeline.

Steel Section	
$C''_{P1} = k_v\theta Q_{i,t-1} - k_v(1-\theta)(Q_{i-1,t-1} - Q_{i-1,t-2}) - k_v\text{sgn}(Q_{i-1,t-1})\left\|Q_{i,t-1} - Q_{i-1,t-1}\right\|$	(21)
$C''_{N1} = k_v\theta Q_{i,t-1} - k_v(1-\theta)(Q_{i+1,t-1} - Q_{i-1,t-2}) - k_v\text{sgn}(Q_{i+1,t-1})\left\|Q_{i,t-1} - Q_{i+1,t-1}\right\|$	(22)
$C''_{P2} = C''_{N2} = k_v\theta$	(23)
$C'''_{P1} = C'''_{N1} = C'''_{P2} = C'''_{N2} = 0$	(24)
HDPE Section	
$C''_{P1} = C''_{P2} = C''_{N1} = C''_{N2} = 0$	(25)
$C'''_{P1} = -C'''_{N1} = -2cA\Delta t \sum_{k=1}^{N} \left\{ -C_0 \frac{J_k}{\tau_k} H_{i,0} + C_0 \frac{J_k}{\tau_k} \left[1 - \left(1 - e^{-\Delta t/\tau_k}\right) \frac{\tau_k}{\Delta t} \right] H_{i,0} + -C_0 \frac{J_k}{\tau_k} \left[\left(1 - e^{-\Delta t/\tau_k}\right) \frac{\tau_k}{\Delta t} - e^{-\Delta t/\tau_k} \right] (H_{i,t-1} - H_{i,0}) - \frac{e^{-\Delta t/\tau_k}}{\tau_k} \varepsilon_{k(i,t-1)}$	(26)
$C'''_{P2} = C'''_{N2} = 2cA\Delta t C_0 \sum_{k=1}^{N} \left[\frac{J_k}{\tau_k} \left(1 - e^{-\Delta t/\tau_k}\right) \right]$	(27)
$C_0 = \frac{\alpha D_S \gamma}{2 s_P},$	(28)

where γ—volumetric weight of the fluid (kg/m^3); θ—relaxation coefficient dependent on a numerical scheme (in our calculations, θ = 1).

Boundary conditions were applied according to [2]. In the first node, a constant level upstream reservoir was assumed. As mentioned before, the connection node between two pipeline sections was defined as a series junction. In the last node, the downstream valve boundary was used, with the relative valve closing function defined as:

$$\tau = \left(1 - \frac{t}{t_c}\right)^2 \tag{29}$$

where t_c—valve closing time (s).

5. Numerical Calculations

The aim of the numerical calculations was to simulate the water hammer runs that were conducted during the experimental study. In all cases, the experimental values of the flow rate, initial pressure, and valve closing time were used as input. The observed pressure signals were used to calibrate the numerical model, i.e., to determine the values of the unsteady friction coefficient and creep parameters that reproduce the observed pressure oscillations. Calibrated values of the parameters for a single experimental test were later introduced in a numerical model to simulate the remaining runs.

The procedure of the numerical calculations was as follows: in the first step, the experiment concerning the water hammer in a steel pipeline without an HDPE section was analyzed (system #7). At this stage, the unsteady friction coefficient k_V was calibrated. In the second step of numerical analysis, the pipeline system configuration with the longest

HDPE section was considered (system #1). In order to simulate the dynamic effect of pressure wave attenuation in the steel section, the earlier calibrated value of k_V was used as an input. The calibration of the model at this stage consisted of determining the creep parameters of the HDPE section which provided good compliance between the observed and calculated $H(t)$ curve. In the third step, the simulations of the remaining experimental configurations were made (systems #2–#6) by using the k_V coefficient determined in the first step and the creep parameters calibrated in the second step of the numerical calculations.

While calibrating the numerical model simulating system #7 (steel pipeline without HDPE section), the lowest discrepancy between the experimental data and the results of numerical calculations was obtained for the pressure wave velocity c_S = 1195 m/s and unsteady friction coefficient k_V = 0.042. To obtain a stable solution, the CFL condition had to be satisfied. Calculations were made for the space interval in the steel section of Δx_S = 1 m. In order to obtain the Courant number as close as possible to unity, the time interval Δt = 0.00083 s was applied (Ca = 0.99). For comparison, calculations were also performed for k_v determined using analytically derived shear decay coefficients (Equations (5) and (7)). The results are presented in Figure 9.

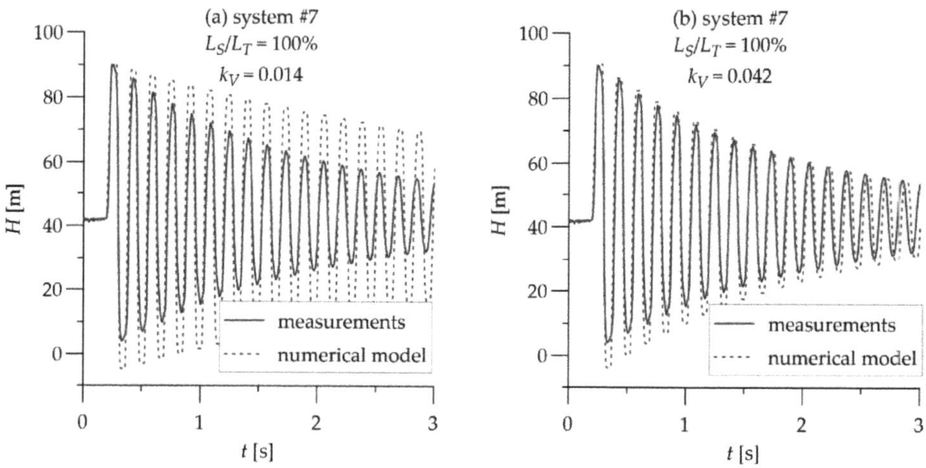

Figure 9. Comparison between calculated and measured pressure oscillations: (**a**) Unsteady friction coefficient calculated using Equations (5) and (7); and (**b**) Calibrated unsteady friction coefficient.

As can be seen in Figure 9a, the calculated pressure oscillations using k_v determined with the use of the shear decay coefficient over-predict successive amplitudes of the pressure wave compared to the experimental data. The calibrated value of k_v is exactly three times higher than predicted with theoretical formulas. From Figure 9b, it is apparent that for k_v = 0.042, satisfactory agreement between the experimental and simulated $H(t)$ curves is obtained.

Turning now to simulating the transient response in the steel–HDPE configuration with the longest plastic section (system #1), the calibration of creep parameters was performed assuming the total number of Kelvin–Voigt elements equal to one. Although in [23] it was indicated that the results obtained for $N = 1$ are poor, in other works, it was reported that it is possible to obtain satisfactory agreement between the simulated and observed pressure signal with the one-element model [38,39]. As the calibration was performed with a simple "trial and error" method to minimize the square error between the calculated and observed transient data, such simplification considerably facilitates the whole process. The best compliance between the calculations results and measurements in system #1 was obtained for $J = 2.20 \times 10^{-10}$ Pa and τ = 0.034 s. The value of the pressure wave velocity in the HDPE section (also determined by "trial and error" method) used as an input was equal to c_P = 300 m/s. To avoid numerical diffusion and dispersion, the space interval applied for the HDPE section was Δx_P = 0.25 m (Ca = 0.99). In numerical calculations simulating hydraulic transients in all steel–plastic configurations, the same value of space interval in the steel section of Δx_S = 1 m and the time interval Δt = 0.00083 s was applied.

In order to check the suitability of the previously calibrated parameters for pipeline systems with different lengths of each section, calculations simulating tests in the remaining configurations (systems #2–#6) were conducted. A comparison between the calculated and measured pressure oscillations for all steel–HDPE configurations is presented in Figure 10.

The results in Figure 10a show that the approach adopted to simulate a valve-induced water hammer event in the steel–HDPE pipeline can lead to satisfactory results in terms of the simulated pressure oscillations compared to the measured data. It is worth noting that although the phase shift and extreme pressure peaks are reasonably well reproduced, the shape of the calculated first peak differs from the observed one (zoom in Figure 10). This is probably due to the short steel section at the downstream end of the laboratory setup in which the pressure sensor was connected, which was not taken into account in the numerical model. Despite the fact that the distance from the end of the plastic pipe to the pressure sensor was not long, the cross-section change at this section could distort the observed pressure signal. Figure 10b shows that using the same unsteady friction coefficient and creep parameters in a slightly different configuration of pipe section lengths can be effective. The calculated pressure oscillations are similar to the experimental transient data in terms of both the extreme pressures and the effective pressure wave velocity of the entire pipeline system. However, systems #1 and #2 differ only slightly, as the proportion of the length of the steel section to the total pipeline length (L_S/L_T) is equal to 33% and 39%, respectively.

Figure 10c–f demonstrate that introducing parameters calibrated on a particular pipeline system to a numerical model with significantly different lengths of each section fails to predict the experimental data. Although the maximum and minimum pressure peaks are well reproduced, the shape of the pressure signal differs distinctly. In Figure 10c,d, the calculated effective pressure wave velocity is lower than the observed one as the simulated signal shifts to the right. Interestingly, as can be seen in Figure 10e for L_S/L_T = 81%, the calculated and experimental $H(t)$ curves have the same pressure wave period time. In the pipeline system with the longest steel section (Figure 10f), the calculated effective pressure wave velocity is higher than the observed one.

Figure 10 shows that, as the share of the steel section in the total length of the pipeline increases, the difference in the damping rate of the calculated and experimental pressure oscillations becomes clearer, i.e., the calculated pressure wave decays faster than the observed one. This is probably related to the fact that the viscoelastic pipe wall's behavior is more significant in a low-frequency pipeline system [40]. The calibrated values of the creep parameters referred to the pipeline system with the longest HDPE section, in which the pressure wave propagated at a relatively low speed. In such a pipeline system, the plastic pipe has more time to dissipate the energy due to the viscous resistance of HDPE. As observed in the experimental study, the elongation of the steel section causes the pressure wave velocity to increase, and thus, the creep process of the plastic material is disturbed by the returning (reflected from the tank) pressure wave sooner. As a consequence, there is less time for the HDPE section to absorb the energy of the water. This result also supports previous findings presented by Mitosek and Chorzelski [41] who experimentally demonstrated that the pressure wave velocity in viscoelastic pipes strongly depends on their length.

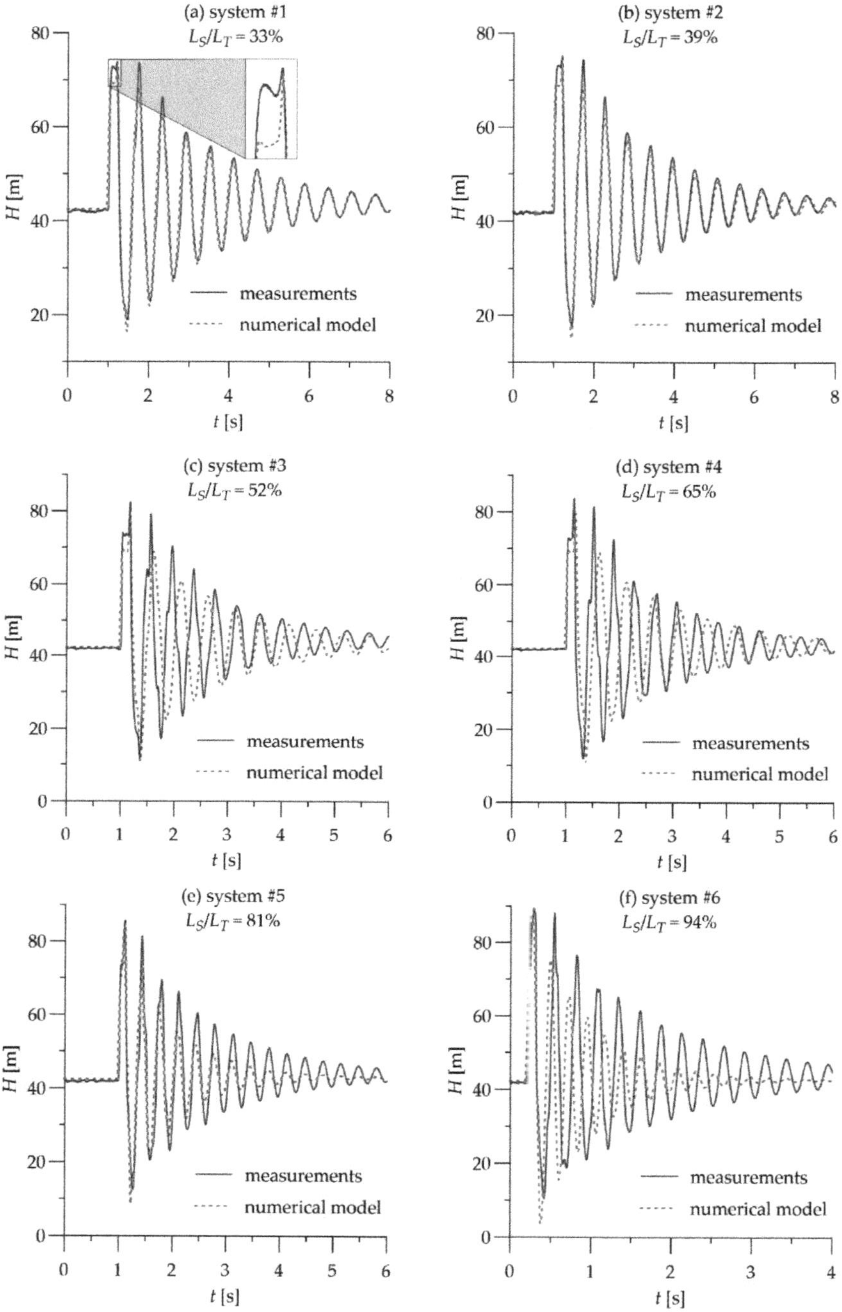

Figure 10. Comparison between calculated and measured pressure oscillations: (**a**) System #1; (**b**) System #2; (**c**) System #3; (**d**) System #4; (**e**) System #5; and (**f**) System #6.

In order to assess the compliance between simulated and measured pressure oscillations, the pressure wave parameters used earlier (dimensionless pressure increase, mean pressure wave velocity, and the logarithmic damping decrement), calculated for both experimental and numerical data, are compiled in Figure 11.

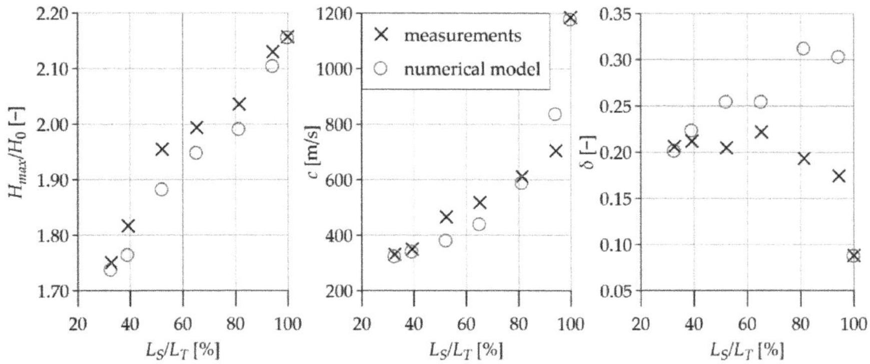

Figure 11. Comparison between calculated and measured values of the dimensionless pressure increase, mean pressure wave velocity, and the logarithmic damping decrement.

Comparison of the pressure wave parameters shown in Figure 11 confirms the conclusions drawn from the visual analysis of simulated and observed pressure signals. It should be noted that, in all cases, values of the dimensionless pressure increase obtained on the basis of experimental data are similar, although slightly lower than those obtained with the use of the numerical model. Figure 11 also shows that, as the L_S/L_T ratio increases, the mean velocity of the pressure wave increases at a different rate for the calculated and observed $H(t)$ curves. The greatest discrepancy can be seen in the logarithmic damping decrement. Figure 11 illustrates the opposite trend of the change of δ with the increase in the length of the steel section. As expected, experimental data indicate that as the L_S/L_T ratio increases, the pressure wave decays slower. However, pressure oscillations calculated for steel-pipeline systems tend to attenuate faster with the increase in the L_S/L_T ratio. This indicates that the calibrated values of creep compliance and retardation time which referred to the system with the longest plastic section are overestimated in terms of describing a transient response in higher-frequency pipeline systems. The obtained results show that calibrated creep parameters do not have any global character. Any alteration in the pipeline system (e.g., shortening or lengthening of the plastic section in multi-pipe system) requires these parameters to be recalibrated.

6. Conclusions

In this study, the water hammer phenomenon in serially connected steel and HDPE pipes with different inner diameters was analyzed from both an experimental and a numerical point of view. Laboratory tests were conducted on a simple tank–pipeline–valve model and included different configurations of lengths of each pipe. The collected transient data revealed that the maximum pressure increase linearly depended on the share of the steel section in the total pipeline length. Despite the significantly smaller diameter of the HDPE pipe compared to the steel pipe, introducing an in-line plastic section into the pipeline system suppressed the valve-induced pressure surges.

In order to numerically simulate experimental water hammer events, a transient solver which made it possible to take into account unsteady friction and the viscoelastic behavior of the pipe wall was used. The combination of Brunone-Vitkovský IAB model applied to the steel section and the one-element Kelvin–Voigt model accounting for the viscoelastic properties of the HDPE pipe made it possible to obtain satisfactory agreement between the calculated and measured pressure signals.

Numerical parameters calibrated on a single pipeline system were introduced in a numerical model to simulate the water hammer in different configurations of lengths of steel and HDPE sections. It was demonstrated that this approach fails to reproduce the observed $H(t)$ curve for a pipeline system with significantly different lengths of each section. From a practical standpoint, any change in the configuration of the pipeline system causes requires the creep parameters to be recalibrated.

This research has raised many questions in need of further investigation. Future work will concentrate on the performance of the WFB model of unsteady friction in numerical models simulating the water hammer in steel–plastic pipeline systems. Moreover, the influence of applying a different number of Kelvin–Voigt elements on simulating hydraulic transients in multi-pipe systems will be examined.

Author Contributions: Conceptualization, M.K. and A.K.; methodology, M.K., A.K., and A.M.; software, M.K. and K.U.; validation, M.K.; formal analysis, M.K., K.U., and A.M.; investigation, M.K. and K.U.; resources, M.K., A.K., and K.U.; writing—original draft preparation, M.K.; writing—review and editing, A.K., K.U., and A.M.; visualization, M.K.; supervision, M.K.; project administration, M.K.; funding acquisition, M.K. All authors have read and agreed to the published version of the manuscript.

Funding: This work was supported by the Faculty of Building Services, Hydro and Environmental Engineering, Warsaw University of Technology (Grant no. 504/04608).

Data Availability Statement: The code generated during the study and experimental data are available from the corresponding author by request.

Conflicts of Interest: The authors declare no conflict of interest.

Abbreviations

Nomenclature

A	cross-sectional area of water stream (m^2)
c	pressure wave velocity (m/s)
D	pipe internal diameter (m)
E_0	elastic bulk modulus of the spring (Pa)
E_k	elastic bulk modulus of the k-th Kelvin–Voigt element (Pa)
f	friction factor (−)
g	gravity acceleration (m/s^2)
H	pressure head (m)
ΔH_{max}	maximum pressure head increase (m)
J_0	instantaneous creep compliance (Pa^{-1})
J_k	creep compliance of the k-th Kelvin–Voigt element (Pa^{-1})
K	bulk modulus of elasticity of the liquid (Pa)
k_V	unsteady friction coefficient (−)
L	length of the individual pipe section (m)
m	the number of the first pressure peak taken into account to calculate logarithmic damping decrement (−)
n	the number of subsequent pressure peaks used to calculate damping decrement (−)
N	total number of Kelvin–Voigt elements (−)
Q	volumetric flow rate (m^3/s)
s	pipe-wall thickness (m)
t	time (s)
t_c	valve closing time (s)
V	mean flow velocity (m/s)
X	space coordinate (m)

Δt	time interval (s)
Δx	space interval (m)
α	parameter dependent on pipe diameter and constraints (−)
γ	volumetric weight of the fluid (kg/m^3)
δ	logarithmic damping decrement (−)
ε_0	instantaneous (elastic) strain (−)
ε_ρ	retarded strain (−)
θ	relaxation coefficient for flow-time derivative calculation in IAB model numerical scheme (−)
μ_K	viscosity of the k-th dashpot (kg/sm)
τ_K	the retardation time of k-th Kelvin–Voigt element (s)
Acronyms	
IAB	instantaneous acceleration-based model of unsteady friction
HDPE	high-density polyethylene
WFB	weighting function-based model of unsteady friction
Subscripts	
i	number of computational node
k	number of Kelvin–Voigt element
S	steel section
P	HDPE section
0	steady-state flow parameter

References

1. Wylie, E.B.; Streeter, V.L.; Suo, L. *Fluid Transients in Systems*; Prentice Hall, Prentice-Hall International: Upper Saddle River, NJ, USA; London, UK, 1997; ISBN 978-0-13-934423-7.
2. Chaudhry, M.H. *Applied Hydraulic Transients*; Springer: New York, NY, USA, 2014; ISBN 978-1-4614-8538-4.
3. Gong, J.; Stephens, M.L.; Lambert, M.F.; Zecchin, A.C.; Simpson, A.R. Pressure Surge Suppression Using a Metallic-Plastic-Metallic Pipe Configuration. *J. Hydraul. Eng.* **2018**, *144*, 04018025. [CrossRef]
4. Garg, R.K.; Kumar, A. Experimental and Numerical Investigations of Water Hammer Analysis in Pipeline with Two Different Materials and Their Combined Configuration. *Int. J. Pres. Ves. Pip.* **2020**, *188*, 104219. [CrossRef]
5. Triki, A. Water-Hammer Control in Pressurized-Pipe Flow Using an in-Line Polymeric Short-Section. *Acta Mech.* **2016**, *227*, 777–793. [CrossRef]
6. Ferrante, M. Transients in a Series of Two Polymeric Pipes of Different Materials. *J. Hydraul. Res.* **2021**, *59*, 810–819. [CrossRef]
7. Bettaieb, N.; Guidara, M.A.; Haj Taieb, E. Impact of Metallic-Plastic Pipe Configurations on Transient Pressure Response in Branched WDN. *Int. J. Pres. Ves. Pip.* **2020**, *188*, 104204. [CrossRef]
8. Parmakian, J. *Waterhammer Analysis*; Prentice Hall: New York, NY, USA, 1955.
9. Landry, C.; Nicolet, C.; Bergant, A.; Müller, A.; Avellan, F. Modeling of Unsteady Friction and Viscoelastic Damping in Piping Systems. *IOP Conf. Ser. Earth Environ. Sci.* **2012**, *15*, 052030. [CrossRef]
10. Zielke, W. Frequency-Dependent Friction in Transient Pipe Flow. *J. Basic Eng.* **1968**, *90*, 109–115. [CrossRef]
11. Vardy, A.E.; Brown, J.M.B. Transient, Turbulent, Smooth Pipe Friction. *J. Hydraul. Res.* **1995**, *33*, 435–456. [CrossRef]
12. Vardy, A.E.; Brown, J.M.B. Transient Turbulent Friction in Fully Rough Pipe Flows. *J. Sound Vib.* **2004**, *270*, 233–257. [CrossRef]
13. Brunone, B.; Golia, U.M.; Greco, M. Effects of Two-Dimensionality on Pipe Transients Modeling. *J. Hydraul.* **1995**, *121*, 906–912. [CrossRef]
14. Brunone, B.; Golia, U.M. Discussion of "Systematic Evaluation of One-Dimensional Unsteady Friction Models in Simple Pipelines" by J. P. Vitkovsky, A. Bergant, A. R. Simpson, and M. F. Lambert. *J. Hydraul. Eng.* **2008**, *134*, 282–284. [CrossRef]
15. Vítkovský, J.P.; Bergant, A.; Simpson, A.R.; Lambert, M.F. Systematic Evaluation of One-Dimensional Unsteady Friction Models in Simple Pipelines. *J. Hydraul. Eng.* **2006**, *132*, 696–708. [CrossRef]
16. Abdeldayem, O.; Ferràs, D.; van der Zwan, S.; Kennedy, M. Analysis of Unsteady Friction Models Used in Engineering Software for Water Hammer Analysis: Implementation Case in WANDA. *Water* **2021**, *13*, 495. [CrossRef]
17. Bergant, A.; Ross Simpson, A.; Vitkovsky, J. Developments in Unsteady Pipe Flow Friction Modelling. *J. Hydraul. Res.* **2001**, *39*, 249–257. [CrossRef]
18. Vardy, A.E.; Brown, J.M.B. Transient turbulent friction in smooth pipe flows. *J. Sound Vib.* **2003**, *259*, 1011–1036. [CrossRef]
19. Ghidaoui, M.S.; Zhao, M.; McInnis, D.A.; Axworthy, D.H. A Review of Water Hammer Theory and Practice. *App. Mech. Rev.* **2005**, *58*, 49–76. [CrossRef]
20. Shaw, M.T. *Introduction to Polymer Viscoelasticity*, 4th ed.; Wiley: Hoboken, NJ, USA, 2018; ISBN 978-1-119-18182-8.
21. Keramat, A.; Tijsseling, A.S.; Hou, Q.; Ahmadi, A. Fluid–Structure Interaction with Pipe-Wall Viscoelasticity during Water Hammer. *J. Fluid. Struct.* **2012**, *28*, 434–455. [CrossRef]

22. Pezzinga, G.; Scandura, P. Unsteady Flow in Installations with Polymeric Additional Pipe. *J. Hydraul.* **1995**, *121*, 802–811. [CrossRef]
23. Covas, D.; Stoianov, I.; Mano, J.F.; Ramos, H.; Graham, N.; Maksimovic, C. The Dynamic Effect of Pipe-Wall Viscoelasticity in Hydraulic Transients. Part II—Model Development, Calibration and Verification. *J. Hydraul. Res.* **2005**, *43*, 56–70. [CrossRef]
24. Covas, D.; Stoianov, I.; Ramos, H.; Graham, N.; Maksimovic, C. The Dynamic Effect of Pipe-Wall Viscoelasticity in Hydraulic Transients. Part I—Experimental Analysis and Creep Characterization. *J. Hydraul. Res.* **2004**, *42*, 517–532. [CrossRef]
25. Weinerowska-Bords, K. Viscoelastic Model of Waterhammer in Single Pipeline – Problems and Questions. *Archives Hydro-Eng. Environ. Mech.* **2006**, *53*, 21.
26. Pezzinga, G.; Brunone, B.; Cannizzaro, D.; Ferrante, M.; Meniconi, S.; Berni, A. Two-Dimensional Features of Viscoelastic Models of Pipe Transients. *J. Hydraul. Eng.* **2014**, *140*, 04014036. [CrossRef]
27. Pezzinga, G.; Brunone, B.; Meniconi, S. Relevance of Pipe Period on Kelvin-Voigt Viscoelastic Parameters: 1D and 2D Inverse Transient Analysis. *J. Hydraul. Eng.* **2016**, *142*, 04016063. [CrossRef]
28. Sun, Q.; Zhang, Z.; Wu, Y.; Xu, Y.; Liang, H. Numerical Analysis of Transient Pressure Damping in Viscoelastic Pipes at Different Water Temperatures. *Materials* **2022**, *15*, 4904. [CrossRef] [PubMed]
29. Meniconi, S.; Brunone, B.; Ferrante, M. Water-Hammer Pressure Waves Interaction at Cross-Section Changes in Series in Viscoelastic Pipes. *J. Fluid. Struct.* **2012**, *33*, 44–58. [CrossRef]
30. Kubrak, M.; Malesińska, A.; Kodura, A.; Urbanowicz, K.; Stosiak, M. Hydraulic Transients in Viscoelastic Pipeline System with Sudden Cross-Section Changes. *Energies* **2021**, *14*, 4071. [CrossRef]
31. Malesinska, A.; Kubrak, M.; Rogulski, M.; Puntorieri, P.; Fiamma, V.; Barbaro, G. Water Hammer Simulation in a Steel Pipeline System with a Sudden Cross-Section Change. *J. Fluid. Eng.* **2021**, *143*, 091204. [CrossRef]
32. Keramat, A.; Haghighi, A. Straightforward Transient-Based Approach for the Creep Function Determination in Viscoelastic Pipes. *J. Hydraul. Eng.* **2014**, *140*, 04014058. [CrossRef]
33. Tijsseling, A.S.; Vardy, A.E. What Is Wave Speed? In Proceedings of the 12th International Conference on Pressure Surges, Dublin, Ireland, 18–20 November 2015; pp. 343–360.
34. Adamkowski, A.; Henclik, S.; Janicki, W.; Lewandowski, M. The Influence of Pipeline Supports Stiffness onto the Water Hammer Run. *Eur. J. Mech.-B/Fluid.* **2017**, *61*, 297–303. [CrossRef]
35. Soares, A.K.; Covas, D.I.; Reis, L.F. Analysis of PVC Pipe-Wall Viscoelasticity during Water Hammer. *J. Hydraul. Eng.* **2008**, *134*, 1389–1394. [CrossRef]
36. Covas, D. Inverse Transient Analysis for Leak Detection and Calibration of Water Pipe Systems—Modelling Special Dynamic Effects. Ph.D. Thesis, University of London, London, UK, 2003.
37. Sarker, S.; Sarker, T. Spectral Properties of Water Hammer Wave. *App. Mech.* **2022**, *3*, 799–814. [CrossRef]
38. Weinerowska-Bords, K. Alternative Approach to Convolution Term of Viscoelasticity in Equations of Unsteady Pipe Flow. *J. Fluid. Eng.* **2015**, *137*, 054501. [CrossRef]
39. Soares, A.K.; Covas, D.I.C.; Reis, L.F.R. Leak Detection by Inverse Transient Analysis in an Experimental PVC Pipe System. *J. Hydroinformatics* **2011**, *13*, 153–166. [CrossRef]
40. Duan, H.-F.; Ghidaoui, M.; Lee, P.J.; Tung, Y.-K. Unsteady Friction and Visco-Elasticity in Pipe Fluid Transients. *J. Hydrau. Res.* **2010**, *48*, 354–362. [CrossRef]
41. Mitosek, M.; Chorzelski, M. Influence of Visco-Elasticity on Pressure Wave Velocity in Polyethylene MDPE Pipe. *Archives Hydro-Eng. Environ. Mech.* **2003**, *50*, 127–140.

water

MDPI

Review

On Sources of Damping in Water-Hammer

Alan E. Vardy

University of Dundee and Dundee Tunnel Research, Dundee DD1 4HN, UK; a.e.vardy@dundee.ac.uk

Abstract: Various potential causes of damping of pressure waves in water-hammer-like flows are discussed, with special attention being paid to their qualitative influences on measured pressure histories. A particular purpose is to highlight complications encountered when attempting to interpret causes of unexpected behaviour in pipe systems. For clarity, each potential cause of damping is considered in isolation even though two or more could exist simultaneously in real systems and could even interact. The main phenomena considered herein are skin friction, visco-elasticity, bubbly flows and porous pipe linings. All of these cause dispersive behaviour that can lead to continual reductions in pressure amplitudes. However, not all are dissipative and, in such cases, the possibility of pressure amplification also exists. A similar issue is discussed in the context of fluid–structure interactions. Consideration is also given to wavefront superpositions that can have a strong influence on pressure histories, especially in relatively short pipes that are commonly necessary in laboratory experiments. For completeness, attention is drawn towards numerical damping in simulations and to a physical phenomenon that has previously been wrongly cited as a cause of significant damping.

Keywords: damping; attenuation; water-hammer; pressure waves; wavespeed; unsteady friction; fluid–structure interaction; wave superposition; valve closure

Citation: Vardy, A.E. On Sources of Damping in Water-Hammer. *Water* **2023**, *15*, 385. https://doi.org/10.3390/w15030385

Academic Editors: Helena M. Ramos and Kamil Urbanowicz

Received: 5 December 2022
Revised: 30 December 2022
Accepted: 3 January 2023
Published: 18 January 2023

Copyright: © 2023 by the author. Licensee MDPI, Basel, Switzerland. This article is an open access article distributed under the terms and conditions of the Creative Commons Attribution (CC BY) license (https://creativecommons.org/licenses/by/4.0/).

1. Introduction

The root causes of pressure surges in pipeline systems are well understood and the analysis of them is a relatively mature subject. At first sight, it might be expected that this would imply that systems can always be designed with high confidence whereas, in practice, this is not the case. Likewise, it might be expected that, if a system does not behave in the originally intended manner, it should be relatively straightforward to infer the reason why. Again, however, this is not always the case; indeed, it is rarely so. One reason for this is that many different phenomena can occur simultaneously and interact with one another. Even if an engineer is well familiar with each individual phenomenon, it may be difficult to predict the consequences of interactions or to recognise their relative importance—or even their existence—in systems that are behaving in an unexpected manner.

In both cases, the task can be especially challenging when damping phenomena cause significant changes in the amplitudes and/or shapes of the wavefronts propagating through the systems. Accordingly, it is beneficial to have a clear understanding of the potential sources of damping and their likely importance in any particular situation. This paper is an attempt to summarise the common causes and to give a degree of guidance on when they are likely to be significant. It is not intended as a review, and neither is it intended to be comprehensive. These objectives would not be practicable, partly because of space limitations, but also because of the author's own ignorance, which has frequently been exposed when attempting to unravel secrets that nature has hidden in unforeseen behaviour. Indeed, identifying the causes of unwanted behaviour can be highly challenging even for persons with much wider experience than the author.

1.1. Damping

The focus of this paper is on the so-called damping of pressure waves propagating in an essentially one-dimensional (1-D) manner. For simplicity, most of the descriptions refer to pipes and these are assumed to be of a circular cross-section. However, many of the phenomena are also observed in other contexts, notably free-surface flows. The term 'damping' is used generically. It usually implies a continuing reduction in amplitudes as time evolves and it often also implies changes in the shapes of waves as they propagate. Commonly, it is a consequence of phenomena that cause significant dissipation, but this is not always the case. The examples given below also include cases where it results from dispersive behaviour that, at least as a first approximation, simply redistributes energy. This is not a trivial distinction. Dissipative phenomena always cause damping, and dispersive phenomena often do so, but there are exceptions in which re-distributions of energy cause superpositions that would not otherwise have occurred. Examples of this are given below.

In the context of water-hammer in pipelines, damping is usually beneficial. It can enable strong disturbances to decay before the arrival of subsequent disturbances, thereby reducing the potential consequences of superpositions. However, it is not always helpful. For example, it complicates the application of leak-detection techniques based on intentionally created pressure waves. Even when the damping is not strong enough to prevent good estimates of the locations of leaks or restrictions, etc., it may downgrade the reliability of estimations of their sizes.

1.2. Outline of Paper

Most of this paper is devoted to presenting the causes of damping and to summarising their consequences. Thereafter, attention is focussed on the use of this knowledge in the assessment of measured pressures. In such cases, the outcomes are known, and the challenge is to deduce what caused them. To avoid unnecessary complications, the discussion centres on the flows in two simple geometrical configurations. One is the almost trivial case of a single pipe along which a wave is travelling. The other is the classical reservoir-pipe-valve system typically chosen in many studies of water-hammer. The first of these is adequate for studies of phenomena in which significant damping occurs during the propagation of individual waves. The second is more representative of studies undertaken in laboratories. Typical practical applications include additional complications, of course, such as networks, pumps, air valves, etc., but their inclusion herein would complicate and extend this paper unnecessarily. Likewise, attention is focussed exclusively on individual causes of damping, not on the many possible combinations thereof. This is important for clarity of presentation, and it is also realistic. Although it will be common for two or more causes of damping to exist simultaneously, it will be usual for one of these to be dominant and the others to be of secondary importance.

For completeness, it is declared that the remainder of this paper focusses exclusively on waves that may be regarded as travelling in a plane-wave manner. That is, they can be represented with sufficient accuracy without taking account of the variations in pressure over the pipe cross-section (even though lateral variations in axial velocity will exist). This limitation will rarely, if ever, have significant consequences for persons assessing waves caused by strong flow disturbances of the type that are usual in water-hammer-like flows, i.e., resulting from events such as valve closure or pump trip. In such cases, low-frequency components of waves are dominant. In some other types of application, however, high-frequency components can be dominant. This is commonly the case in studies of acoustics, and it is also important in specialist applications such as leak detection in pipelines. When waves are of a sufficiently high frequency for their wavelengths to be shorter than a few pipe diameters, cross-sectional variations in pressure become significant. Indeed, in the case of waves with frequencies greater than well-known cut-off frequencies, radial disturbances dominate axial ones. Many possible modes of vibration exist simultaneously at any frequency, but plane-wave behaviour is strongly dominant at frequencies smaller than, say, a quarter of the lowest cut-off frequency. Thereafter, radial modes rapidly increase in importance and plane-wave components decay rapidly.

The scope of this paper is also limited to interpretations in the time domain. Again, this is appropriate for the water-hammer-like flows at which this paper is targeted and in which low frequencies are dominant. In principle, however, it is possible to convert time-domain histories into frequency-domain spectra and to interpret everything in the context of frequency-dependence. This can be advantageous in particular applications, such as leak and blockage detection, even when attention is directed primarily at frequencies that are well below cut-off frequencies. It would be useful for another paper to discuss damping in applications in which frequency-domain interpretations are more instructive than time-domain ones. However, such a paper would need to be written by a different author with a sufficiently deep knowledge of the assessment of frequency-dependent damping in relevant applications.

Throughout this paper, the focus is on readily observable flow behaviour or, more strictly, on readily measurable properties of the flow, especially on pressure histories. These can be measured with a good accuracy by surface-mounted sensors, whereas measurements of parameters such as velocity and shear stress are not normally available in practical engineering applications with rapidly varying flows. Indeed, even if they were, it is unlikely that they would simplify interpretations greatly. This is because strong variations can exist even in individual cross-sections at a single instant in time. For them to add significantly to the usefulness of pressure measurements, it would be necessary to be able to deduce their average values (e.g., cross-sectional flow rates, not just point velocities). For analogous reasons, this paper does not attempt to use energy-based methods to assess damping even though they can be used to good effect in theoretical modelling (e.g., Karney [1], Axworthy et al. [2]).

2. Skin Friction

The most well-known cause of damping is skin friction. Its influence can be illustrated in a simple manner by considering an instantaneous step change in pressure at one end of a semi-infinite pipe. In the absence of friction (and all other complications), the wavefront would propagate in an unchanged form, illustrated by the dotted line in Figure 1. If, instead, the existence of skin friction is acknowledged, the pressure will decrease downstream because a longitudinal pressure gradient is needed to counter the resistance at the wall. The greater the distance travelled by the wavefront, the greater the overall resistance and so the pressure at the wavefront reduces continually. This effect was discussed by Leslie and Tijsseling [3] in the context of a wavefront generated by a valve closure. The reducing amplitude of the step at the leading edge of the wave implies a reducing change in the velocity. That is, the flow is decelerating, and the magnitude of the pressure gradient is influenced by this.

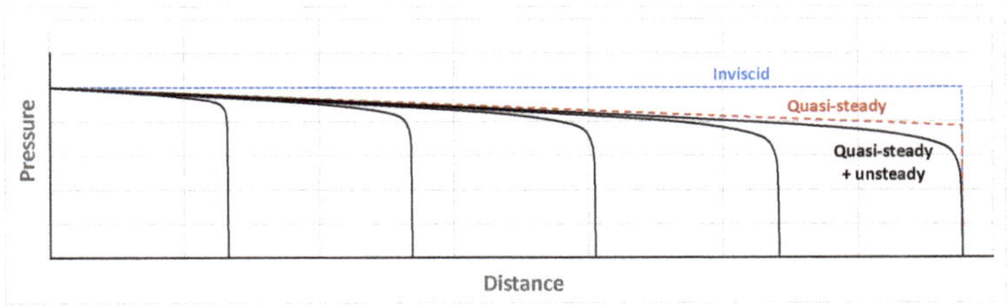

Figure 1. Influence of skin friction on a compressive wavefront in pipe flow.

A first approximation to the magnitude of the evolving shear stress at any location can be obtained by assuming that, at any instant, the velocity distribution over the cross-section is identical to that in a steady flow. In this case, the wall shear stress will be deduced from the steady-flow relationships and the resulting pressure distribution will be as indicated in Figure 1 by the broken line labelled 'quasi-steady' friction. Strictly, even for the single wavefront under consideration, the assumption of quasi-steady behaviour does not imply that the velocity behind the wavefront is axially uniform (although it will often be nearly so). The reason for this is that the continual increase in resistance resulting from the increasing distance travelled by the wavefront causes a continual reduction in the pressure change at the wavefront and, hence, also of the velocity change. Such changes are transmitted back to the upstream boundary at the speed of sound and their reflections at the boundary travel back downstream, albeit at almost the same speed as the wavefront itself. For completeness, it is noted that even the speed of the propagation of the wavefront itself does not remain exactly constant, although the variation is of negligible consequence in almost all liquid flows.

The assumption that skin friction behaves in a quasi-steady manner is, at best, an approximation. In reality, at any location, the response of the velocity profile to a change in the flow rate is time-dependent. The outermost regions of the cross-section respond first and vorticity diffusion gradually causes consequential changes in the core region of the flow. The times required for the radial vorticity diffusion to cause the necessary adjustments of the velocity profiles are far greater than those required for the wavefronts to propagate over similar distances. This leads to pressure distributions resembling the continuous lines in Figure 1. The greater the distance of any particular location behind the step, the greater the elapsed time since the step passed that location. Therefore, the greater the time during which the velocity profile has been adjusting, the closer its approximation to a quasi-steady form. The processes of vorticity diffusion are somewhat different in laminar and turbulent flows and the consequences of this can be significant. Ghidaoui and Kolyshkin [4] discuss this in detail and show that it is responsible for differences between the small-time effects seen in Figure 2 below and longer-term effects measured in some other experiments, e.g., Shuy [5].

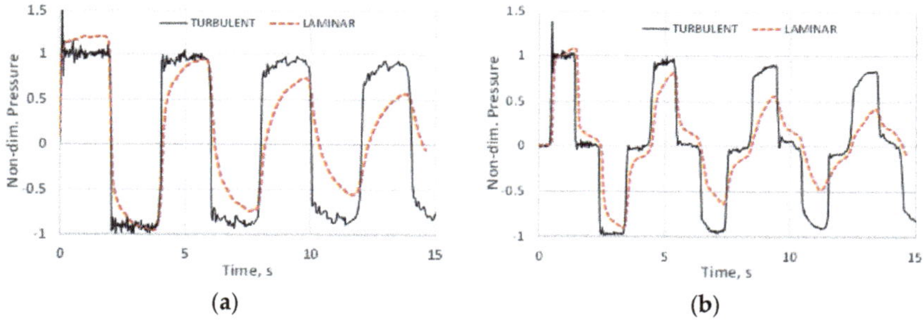

Figure 2. Measured pressures following rapid valve closure in a laboratory pipe (after Holmboe, [6]). (**a**) At the valve. (**b**) At mid-pipe.

The strength of this unsteady effect at any instant depends upon the rates of change in the flow rate, not only at the current instant, but also before this instant because of the time needed for vorticity diffusion. These are strong during the passage of a wavefront and much weaker after its passage and this is the root cause of the characteristic shape of the wavefront seen in Figure 1. This shape is commonly seen in measurements made in laboratory studies of the phenomenon, but it is usually less pronounced in full-scale measurements. There is a simple reason for this. Although wavefronts generated by events such as a valve closure or pump-start-up can be quite compact, no such generalisation can be made about the time intervals between them because these depend on the overall system, not on local phenomena. As pointed out by Duan et al. [7], pipe length:diameter (L/D) ratios are important in this respect. These tend to be much longer in full-scale engineering applications than in laboratory experiments, so the time intervals between the successive reflections of any particular wavefront are correspondingly larger. Overall, therefore, the relative importance of unsteady-friction effects is usually much greater in laboratory investigations than in most practical applications. It is somewhat ironic that the need for researchers to understand the consequences of unsteady friction exists, in part, because their experiments cannot reproduce full-scale L/D ratios.

Measured Pressures

Quasi-steady friction in pipe flows is well understood and widely available representations of it are adequate for most engineering purposes. The position with unsteady friction is less secure, but it is not the purpose of this paper to review analytical models of it. Instead, the key is to be able to recognise it in physical measurements. In practice, this has to be done at second-hand through measurements of pressures because shear stresses are almost never measured directly except in highly specialised laboratory experiments designed expressly for this purpose (e.g., Shuy [5], Vardy et al. [8]).

Figure 2 shows well-known experimental measurements obtained by Holmboe [6] in a reservoir-pipe-valve system. These experiments were reported by Holmboe and Rouleau [9] and used by Zielke [10] to assess his ground-breaking model of unsteady friction in laminar flows. Because the experiments were designed explicitly to assess the influences of skin friction, the pipe was encased in concrete, thereby eliminating any risk of significant influences of pipe movement. The pressure disturbances were caused by sudden closures of the valve following a period of steady flow. This caused steep pressure wavefronts to travel upstream towards the reservoir and the figure shows such a wavefront together with successive wavefronts after reflections at both ends of the pipe. The left-hand box shows pressures just upstream of the valve and the right-hand box shows pressures at the mid-length of the pipe. Both laminar flow and turbulent flow cases are shown and, to facilitate comparisons, the pressures are scaled by the Joukowsky pressure.

Consider first the turbulent flow case. Temporarily disregarding the sharp peak, the initial pressure rise is a close approximation to a step increase and the subsequent reflections after the wavefront has propagated back and forth along the pipe are almost equally steep. During the early periods, the pressure between successive reflections is nearly constant, but it becomes progressively less so with increasing time. The reason for this can be inferred from the corresponding measurements for the laminar flow case. The first reflection clearly exhibits the behaviour described above in relation to unsteady friction and the distortion becomes increasingly pronounced with each subsequent reflection. Indeed, the distortion is so strong that it visibly affects the shape of the whole of the wavefront, even in the first reflection. With this basis for comparison, it is possible to infer with high confidence that the similar, but much less pronounced, trend in the turbulent flow measurements is also attributable to unsteady friction. In both cases, the pressures between successive passes of the leading tips of the wavefronts can be interpreted as approaching quasi-steady values asymptotically.

For completeness, brief attention should be paid to the slightly noisy nature of the turbulent flow measurements. Although other explanations are possible, it seems likely that these are attributable to the pressure sensors, not to the flow itself. This inference is supported by the fact that the noise is consistently greater at the valve than at the mid-pipe. The strong initial overshoot in the turbulent flow traces is also assumed to be due to this cause. The only other possibility envisaged by the present author, namely a very strong axial vibration of the valve, seems implausible.

In both the laminar and turbulent cases, the peak positive and negative values decrease monotonically in time. This is a natural consequence of the dissipative influences of viscous phenomena, both quasi-steady and unsteady. However, it is also seen that the decay in amplitudes is stronger at the mid-pipe than at the valve, especially in the laminar flow case. At first sight, this might seem counter-intuitive, but it is a natural consequence of the ever-increasing length in the wavefront that has been influenced significantly by the unsteady component. When this length exceeds the length of the pipe, the superposition of the wavefronts propagating in opposite directions prevents the development of the complete pressure changes associated with the individual wavefronts. This effect is discussed more fully in Section 8 below.

Holmboe's experiments were close to ideal for researchers studying unsteady friction. However, it is emphasised that the influence of unsteady components of skin friction in practical applications is less strong than that seen in Figure 2. There are three reasons for this. First, the measurements were made in a pipe with a very small length:diameter ratio so the time intervals between successive steps are short. Second, the valve closure was as rapid as the experimenters could achieve. Third, the Reynolds numbers of the initial steady flows were smaller than will be usual in large-scale engineering. The second and third of these factors are highly desirable for researchers studying unsteady friction and even the first has advantages. Nevertheless, it must not be inferred that the overall influence of unsteady friction in practical pipe systems is so strong. It is indeed strong close to sudden wavefronts, but highly abrupt wavefronts are less common in practice and the time intervals between successive wavefronts are usually much greater than those required for vorticity diffusion over a pipe's cross-section.

Figure 3 shows theoretical predictions of pressure histories from a study described by Vítkovský et al. [11] caused by rapid and slow valve closures in a turbulent pipe flow. This enables the relative influences of the quasi-steady (QS) and unsteady (US) contributions to be assessed. Using the discussion on Figure 1 above, it can be inferred that the damping is caused primarily by the quasi-steady component, whereas the distortion of the leading part of the wavefront is caused primarily by the unsteady component. Moreover, in common with Figure 2 above, successive wavefront reflections arrive at intervals that are shorter than those needed for vorticity diffusion to enable cross-sectional velocity profiles to approach quasi-steady conditions closely. With the slower valve closure, the differences between predictions with and without unsteady friction are less obvious visually than for the rapid closure. This is consistent with expectations discussed above in relation to timescales, but these histories are also strongly influenced by the superpositions of successive wavefronts.

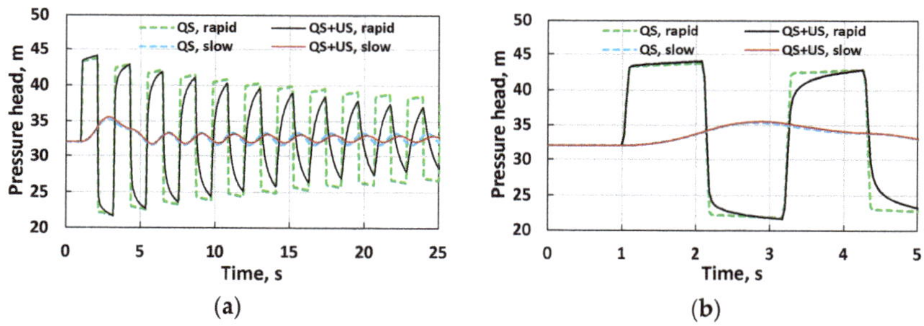

Figure 3. Influence of skin friction on simulated pressures in an unsteady, turbulent, pipe flow induced by rapid and slow valve closures. (**a**) Progressive decay. (**b**) Detailed form.

3. Pipe Wall Properties

In the preceding section, no account was taken of pipe wall flexibility. This is sometimes a close approximation to reality—tunnels, pipes encased in concrete, etc. In such cases, the wavespeed is controlled almost exclusively by the fluid properties. For water at normal pressures and temperatures, this is of the order of 1450 m/s. At the other extreme, e.g., in blood vessels, the wavespeed may be controlled almost exclusively by the vessel wall properties. In this section, attention is focussed on pipes for which the wavespeeds are determined mostly by the fluid properties, but the influence of pipe flexibility is nevertheless significant. Nevertheless, it is assumed that the wall stresses are in radial equilibrium with the fluid pressure at all times. That is, radial inertia is disregarded. This is a good approximation except in the case of very high frequency fluctuations.

The simplest case is characterised by steel pipes that are thin-walled, but of sufficient thickness to withstand all water-hammer pressures that could occur. In this case, when a sudden event such as rapid valve closure occurs, the principal water-hammer wavefront is somewhat smaller than in a rigid pipe because the increased pressure causes a small increase in the cross-sectional area. In effect, the expansion reduces the overall stiffness, thereby reducing the wavespeed and, hence, also the pressure rise needed to cause the required change in the velocity. In most applications, it is sufficient to analyse such flows in the same manner as for a nominally rigid pipe except for using a reduced, but constant, wavespeed. The expected pressure histories then differ from those in rigid pipes only by reduced amplitudes and by increased intervals between successive reflections.

Although usually unnecessary, a more rigorous assessment of the influence of pipe elasticity would acknowledge that, because of Poisson's Ratio effects, the circumferential stresses that resist radial expansion must give rise to axial stresses. This topic is addressed in Section 4. First, however, consideration is given to the waves in pipes that respond gradually to changes in the fluid pressure.

Inelastic Pipe Walls

Steel pipes (and other metal pipes) are conventionally treated as linearly elastic in studies of water-hammer. PVC pipes, however, are far from linearly elastic and yet they are commonplace in practical applications such as water supply. The wall material of such pipes can behave in a visco-elastic manner. That is, the pipe diameter responds relatively slowly to changes in the fluid pressure. Imagine, for instance, a sudden, sustained increase in the fluid pressure. The apparent stiffness associated with the initial pipe response to this is greater than the subsequent apparent stiffness, which falls asymptotically to a value that, in the absence of a further pressure change, is maintained indefinitely. The timescales associated with this relaxation are typically much greater than those associated with vorticity diffusion over a pipe cross-section, so the distortion of the pressure wavefronts is also much greater.

The consequences of this for water-hammer phenomena are illustrated in Figure 4, which again uses data obtained in the study reported by Vítkovský et al. [11]. The graphs showing the influence of quasi-steady (QS) and unsteady (US) components of skin friction are the same as those in Figure 4. By inspection, these are much less important than the influence of the assumed visco-elastic (VE) properties of the pipe wall. Indeed, the latter is so dominant that little useful purpose would be served by taking account of unsteady components of skin friction in analyses of water-hammer in pipes with wall properties such as these (although quasi-steady friction will still be influential in long pipes). In a research context, it will be rare for reliable deductions to be made about unsteady friction from measurements made in such pipes. This is the case even though the prevailing shear stresses will be comparable to those in nominally rigid or linearly elastic pipes. However, it is also true that the existence of unsteady friction will reduce somewhat the accuracy with which the visco-elastic effects themselves can be deduced.

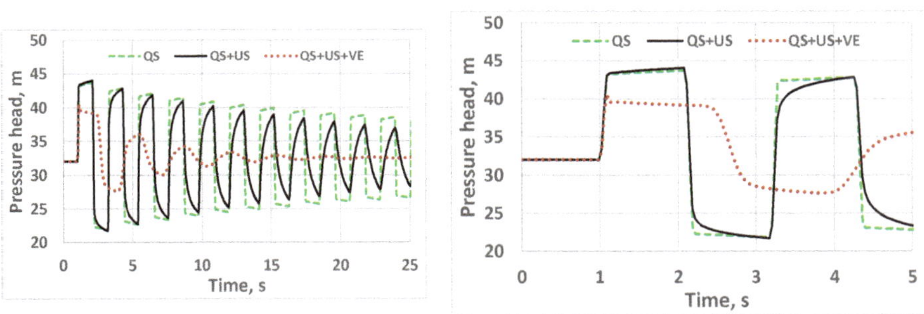

(a) (b)

Figure 4. Influence of visco-elasticity on simulated pressures in an unsteady, turbulent, pipe flow induced by rapid valve closure. (**a**) Progressive decay. (**b**) Detailed form.

In passing, it is informative to consider the implications of figures such as Figure 4 for researchers attempting to calibrate numerical models of the consequences of visco-elasticity from measurements in relatively short pipes used typically in laboratory experiments. As seen in the preceding section, even with the shorter delays caused by unsteady friction, wavefronts can soon begin to overlap with their own reflections from boundaries. As a consequence, only the early stages of the phenomenon can be calibrated directly. The increased timescales associated with visco-elasticity can be a major complication in this respect.

A related difficulty illustrated in Figure 4 arises when wavespeeds need to be inferred from experimental measurements. In this context, however, the appropriate interpretation of the term 'wavespeed' tends to be context-dependent. As discussed by Tijsseling and Vardy [12], it is important to distinguish between speeds at which small disturbances travel (i.e., the speed of sound relative to the fluid) and the speed at which large disturbances appear to travel (e.g., distances moved by wave crests in a given time). Using the latter, the intervals between successive reflections in the elastic-walled pipe case are clearly distinguishable from the instants when the pressures begin to fall. The same is not true, however, in the case of visco-elastic pipes. The interval between the start of the first two maxima is reasonably distinct, but the intervals between the succeeding maxima are less distinct. Moreover, the intervals between the successive maxima increase with time and the extent of the overlap between each wave and its reflection also increases. As a consequence, if physical measurements were of this form, it would not be safe to infer the wavespeeds from the time intervals between the maxima. When such behaviour is interpreted in the frequency domain instead of the time domain, the wavespeed is said to be frequency-dependent (e.g., Aliabadi et al. [13]

4. Axial and Lateral Structural Movement

Attention now turns to the consequences of Poisson's Ratio effects in pipe walls, and linearly elastic behaviour is again assumed. In the event of a sudden disturbance to the flow, e.g., valve closure, the change in the internal pressure causes a change in the circumferential stress in the pipe wall and, hence, also causes a change in the axial stress in the wall. This change propagates axially in the wall at the speed of sound in the solid material. Typically, as illustrated in Figure 5, it travels much faster than the pressure wave in the fluid. In turn, the Poisson's Ratio effects at the axial-stress wavefront cause a change in the circumferential stress that changes the pipe diameter and, hence, induces a pressure change in the fluid. This form of fluid–structure interaction (FSI) is an example of a dispersive process that is not necessarily dissipative (at least, not strongly so).

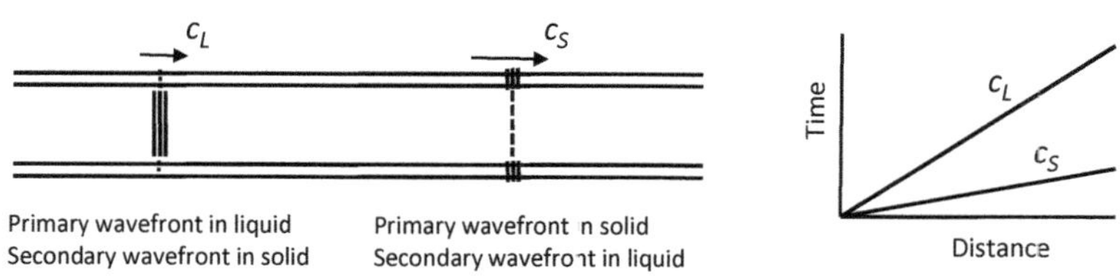

Figure 5. Wave pair in fluid–structure interaction.

In the context of water-hammer, the pressure wavefront associated with the axial stress wave is usually known as a precursor wave and the pressure change that accompanies it is typically much smaller than that in the primary wavefront. Usually, it is of secondary importance, but, as illustrated below, this is not always the case. This depends upon how the two wavefronts behave when they meet a pipe boundary. Whenever either type of wavefront reaches an end of a pipe, it causes two reflected wavefronts, namely one of each type. Likewise, at a junction of two or more pipes, it causes two wavefronts to propagate along each pipe connected to the junction. It is possible for this to cause cumulative reductions in the amplitudes of the principal wavefronts that, especially in networks, can have similar consequences to conventional damping. However, it is also possible for the effect to cause greater changes in the pressure in some locations than those in the initial disturbance. Therefore, it is best to regard it as a fragmentation process that will commonly lead to reduced peak-to-peak amplitudes rather than to regard it as damping per se.

Phenomena associated with changes in pipe cross-sectional areas are far from being the only potential structural influences on pressure histories in pipes. Other possibilities include, for instance, the non-rigid behaviour of supposedly rigid restraints and strong vibrations in suspended pipes. In most applications, the freedom of pipes to move is relatively small and, in this case, the influence of the movements is likely to be significant only relatively close to the wavefronts. Nevertheless, the consequences of even small movements can be difficult to unravel from other, longer-lasting effects. It is, therefore, useful to remember that supports designed to exercise restraint as a consequence of their stiffness do not exert any restraint at all until movement actually occurs. In principle, movements are possible either normal to a pipe axis or parallel to it (or both). Flexural motion is usually easier to restrain than axial motion, but this is not always true. In any case, supports that are capable of effectively preventing significant lateral movements might be much less effective in preventing axial movements, e.g., Tijsseling and Vardy [14]. Both types of movement are readily initiated when the pressure waves meet changes in the axial orientation of a pipe, at bends and T-junctions, for instance. This can have especially strong consequences in suspended pipe systems, but it is also relevant to pipes supported on the ground or even buried in the ground.

So far, it has been implicitly assumed that the initiating event primarily causes a disturbance in the fluid and that the consequential response of the pipe is of secondary importance. However, this is not always the case. Externally induced shaking of pipes, e.g., in earthquakes, causes structural movements that can have strong consequences for the pressure change. Figure 6 shows an extreme example described by Vardy et al. [15] in which the initiating disturbance was induced structurally, not hydraulically, by the axial impact at a closed end of a pipe. The main experiments were undertaken with the T-piece configuration shown in Figure 6a, but preliminary measurements were made with only a single pipe, with the junction replaced by a closed end. Figure 6b shows the pressure measurements at the mid-length of the single pipe. Waves in the liquid and solid are denoted by 'L' and 'S', respectively. By inspection, the initial axial stress wave (S1) arrives long before the pressure wave (L1), and so does the reflection (S2) of the stress wave from the remote end. Its subsequent reflection (S3) from the impact end arrives shortly afterwards. The pressure wavefront L2 is a disturbance moving upstream, having been generated when the stress wave S1 reflected at the remote end. It is noteworthy that the magnitude of this wavefront exceeds that of L1. A further important feature of the figure is that the maximum and minimum pressures occur long after the initial pressure wave and its first reflection. This illustrates the potential for a dispersive effect to cause amplification, not damping.

Figure 6c shows the pressure at the main junction in the full T-piece configurations. The first event, labelled L2*, is the response of the pressure to the reflection that, in the axial-only configuration, gave rise to L2 and S2 on arrival of the first stress wave (S1). Likewise, the event labelled L2** corresponds to the arrival of the stress waves from the remote ends of the branches. This is followed closely by the arrival of the first pressure wavefront (L1) from the initial impact. The gradual reduction in pressure during the intervening period is a consequence of flexural movements in the branches. As this develops, the lateral movement of the branches at the junction corresponds to an elongation of the axial pipe. As stated above, this is an extreme example of the possible consequences of FSI, because it is caused by an external structural force. Nevertheless, it demonstrates that structural movement can cause behaviour that will complicate attempts to make reliable inferences from the measurements of pressure alone.

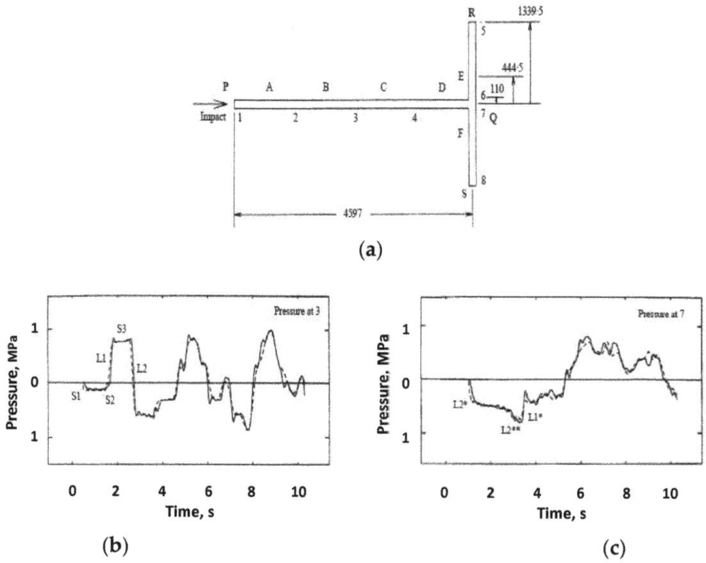

Figure 6. Influence of axial and flexural stresses in pipe walls (after Vardy et al. [15]). (**a**) T-piece experiment. (**b**) Axial pipe only. (**c**) Full T-piece configuration.

5. Variable Wavespeed

As is well known, the relationship between changes in the pressure and velocity associated with pressure waves is strongly dependent on the speed of sound. An obvious consequence of this is that any phenomenon that can cause changes in the speed of sound will also cause dispersive behaviour and, hence, influence damping (either increasing or countering it). An obvious example arises when free gas bubbles exist in a liquid. Even very small proportions of gas can have a strong influence because they can hugely increase the overall compressibility of the fluid whilst having a minimal effect on its bulk density.

Consider first a system in which the gas is distributed uniformly. In this case, pressure waves causing only small proportional changes in the absolute pressure would have little influence on the wavespeed so this would be almost uniform throughout. The dominant difference from the same system with a wholly liquid flow would be a reduction in the pressure changes needed to cause any particular changes in the flow. However, in neglecting other effects, such as skin friction, there would be neither dissipation nor significant dispersion. A different picture exists when the pressure changes are large enough to cause significant changes in the bubble volume because this changes the local compressibility and, hence, also the local wavespeeds. As a consequence, it is not possible for a strong wavefront to travel along a pipe in an unchanged form.

This effect is illustrated in Figure 7 for an almost equivalent compressible phenomenon, namely pressure wave propagation in an ideal gas. This is a convenient choice, partly because of the author's background, but also because the relationships between the density, sound speed and pressure are especially simple. To highlight the effect under discussion, all other causes of change except quasi-steady friction are suppressed; e.g., the pipe is rigid, unsteady skin friction is neglected and there are no heat transfers between the fluid and the pipe. The figure shows the pressure and velocity histories at regularly spaced locations along a pipe when the pressure at its upstream end is increased suddenly to a new value and is then held constant. The continuous and broken lines show conditions with upstream pressure increases of 1% and 10% of ambient pressure, respectively. In both cases, the velocity reaches a maximum immediately behind the wavefront and then gradually decays. In contrast, the pressure increases continuously after the wavefront passes. This behaviour exists in both cases, and it is especially pronounced in the case with the larger increase in the pressure. Even in this case, however, the proportional increase in the speed of sound across the wavefront is only about 1.3%. Much greater changes can be expected in the case of liquid flows with free gas bubbles.

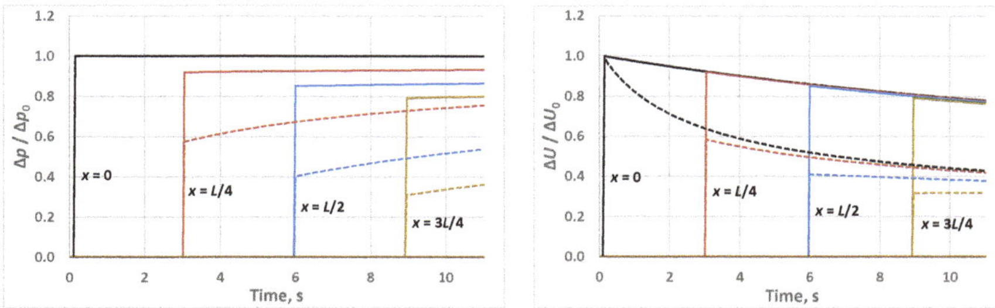

(a) (b)

Figure 7. Influence of fluid compressibility on wavefront propagation. (Continuous lines: $\Delta p_0 = 1$ kPa; Broken lines: $\Delta p_0 = 10$ kPa.). (**a**) Pressure. (**b**) Velocity.

In the case of bubbly flows, it will rarely be reasonable to expect the gas to be distributed uniformly throughout a system. When this is not the case, spatial variations in the speed of sound will exist even in ambient conditions and these will further complicate the propagation of waves. As an extreme example, consider a local region of uniformly distributed bubbles in an otherwise gas-free system. When a pressure wave arrives at an interface between the different zones, it will partially reflect in a manner that resembles the corresponding behaviour on arrival at a junction between two pipes with different diameters. The equivalent behaviour will exist continuously along regions with gradual variations in the bubble content. The wavefronts propagating into regions of increasing gas will flatten and the wavefronts propagating in the other direction will steepen. In both cases, however, the process will cause continuous reflections that will influence conditions elsewhere in the system. In some locations, this will appear to be damping, whereas, in others, it will have the opposite effect. Such behaviour will inevitably complicate the task of anyone attempting to interpret the measured pressure signals without prior knowledge of the processes giving rise to them.

Even greater complexity exists when spatially varied distributions of bubbles convect through a system because this changes the system response to identical initiating disturbances. All effects, including the strength of the damping behaviour, are affected. However, the possible variations are too great to consider in an overview paper such as this. Likewise, no attention is paid to even more extreme conditions associated with phenomena such as cavitation, column separation or pressurised free-surface flows, all of which can also cause other behaviours that are significantly more important than the damping effects that they might induce.

6. Porous Surfaces

Strong damping can exist when fluid can discharge laterally through pipe walls. It will be rare for this to be desirable in the case of liquid flows, but it can be beneficial in some gas flows. Well-known examples are gun silencers and vehicle exhausts. Usually, however, both of these exist over only short lengths of pipe. To add variety, the author takes this opportunity to indulge himself by addressing a topic that has been important in his own career, namely pressure wavefronts generated when trains enter tunnels at high speed. Such wavefronts can be characterised as a relatively steep ramp; e.g., a pressure rise of, say, 2 kPa in less than 10 tunnel diameters followed by an extended, more gentle ramp. The initial steep ramp is generated during the short period when a train nose suddenly causes a partial blockage of the tunnel portal. The subsequent, more gradual rise develops as the length of train inside the tunnel increases, causing increasing frictional resistance.

In tunnels of slab-track construction, i.e., with relatively smooth and impervious surfaces over the whole cross-sectional perimeter (including the track bed), a wave-steepening effect seen in Figure 8a exists and, in the absence of effective countermeasures, it can lead to the emission of unacceptable sonic boom-like disturbances from tunnel exit portals. Methods of countering this behaviour are outwith the scope of this paper and, instead, attention is focussed on the corresponding outcome in ballast-track tunnels (tracks mounted on deep layers of coarse gravel). In such tunnels, the wavefronts initially steepen, but after a sufficient distance, they begin to flatten, as seen in Figure 8b. The mechanism causing the strong influence of ballast is only partially understood, but two important facts are known. First, the behaviour is not primarily attributable to enhanced skin friction, either quasi-steady or unsteady. Second, the greater the volume of air cavities within the gravel, the greater the damping effect.

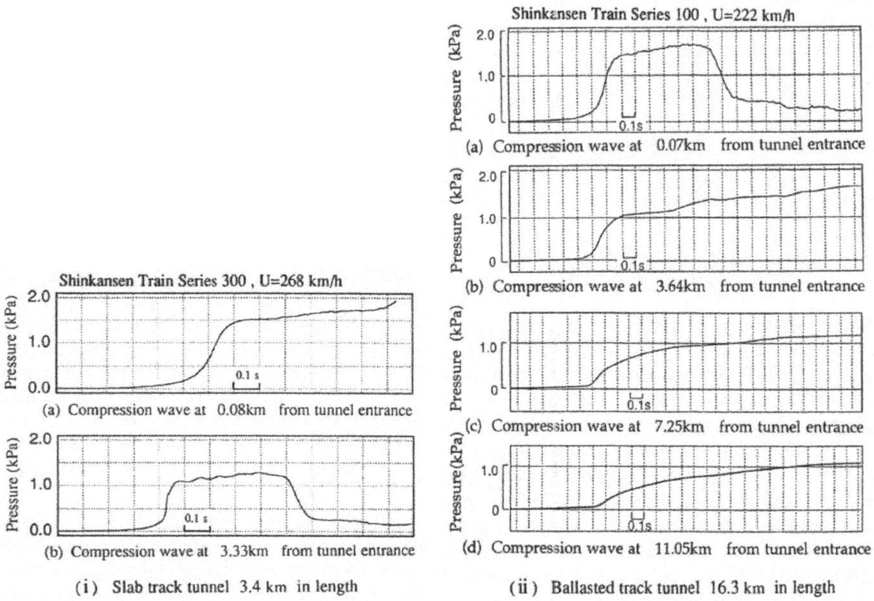

Figure 8. Compression waves in long, ballast-track and slab-track tunnels (after Ozawa et al. [16]. N.B.: The strong pressure reduction at 3.33 km in the slab-track tunnel is a reflection from the tunnel exit portal whereas the strong reduction at 0.07 km in the ballast-track tunnel occurred when the train nose passed the sensor).

A partial explanation for this behaviour can be inferred by focussing on what happens at any typical location along the tunnel. As the wavefront passes the location, the pressure above the gravel increases rapidly and begins to force air laterally into the bed, thereby compressing the air trapped in the cavities in the gravel. Resistance to the flow into the ballast layer causes delays in the pressurisation process so the equilibrium of the pressure does not exist until some distance behind the wavefront. The rate of the flow into the ballast depends upon the pressure difference between the air above and within the bed. This difference is negligible at the leading and trailing edges of the ramp. It reaches a maximum where the overlying pressure is approximately half of the value just after the steep nose-entry ramp. Although incomplete, this explanation of one cause of the ballast behaviour has recently been used to propose a method of reproducing ballast-like behaviour in slab-track tunnels (Liu et al. [17]. That is, research targeted at explaining why a problem does not exist in one type of tunnel has led to a method of minimising the problem in tunnels where it does exist.

As an historical aside, it may be of interest to know that the risk of sonic boom-like disturbances being emitted from railway tunnels was not recognised until it had actually happened. It was first encountered during the commissioning of the first phase of the Shinkansen railway in Japan around half a century ago. At that stage, it was immediately observed that the problem existed only for slab-track tunnels, not for ballast-track tunnels. Of additional interest, the problem with the slab-track tunnels was overcome by building short tunnel extensions with porous walls. Readers might enjoy the challenge of figuring out why, at least so far, it has always been more beneficial to provide these at tunnel entrances than at their exits, where they would act as silencers.

7. Delayed Reflections

The reflection process of a pressure wave at an open end of a pipe—connected to a large reservoir, say—is rapid, but it is not instantaneous (Rudinger [18]). Instead, differences between the impedances in the pipe and the reservoir cause interactions that need time to decay. Disturbances propagating from the pipe into the reservoir radiate in a spherical-like manner whereas the reflections along the pipe approximate closely to planar. The time required for the changes in pressure at the outlet to die away is short—typically in the order of the time required for a wave to travel one pipe diameter—so the phenomenon is justifiably neglected in many practical applications. However, this is not always the case, notably when the incident wavefront has significant high-frequency components with wavelengths that are not much longer than the pipe diameter, as is the case in some musical instruments, for instance.

The upper row of Figure 9 depicts the pressure histories at distances of 5, 4, 3, 2 and 1 diameter from the outlet plane of a duct connected to a large reservoir. The rising limbs of the curves show the progress of a steep (near-shock-like) wavefront approaching the reservoir and the falling limbs show its reflection back along the duct. In case (a), the reflection process is assumed to be instantaneous whereas, in case (b), allowance is made for the amplitudes of the induced disturbances in the reservoir to decay. To a close approximation, the curves showing the progress of the reflected wave in case (a) are reversed and inverted images of those showing the incident wavefront. In contrast, the reflected shapes in case (b) are extended in time, strongly so during the later stages of the reflection. As a rough guide, the delays during the first two-thirds of the reflection are less than the time required to travel on the duct diameter, but they increase strongly during the later stages of the reflection.

The lower row of the figure shows the corresponding behaviour for a wavefront that is approximately 1.5 duct diameters long. This is again very steep, albeit less so than the first wavefront. Case (c) shows predictions based on an instantaneous reflection and it is seen that the maximum pressure at a distance of one diameter from the outlet plane is smaller than at greater distances. This is a simple consequence of an overlapping of the later stages of the incident wavefront with the early stages of the reflection. The same overlapping effect necessarily also occurs in case (d) which allows for the delayed response caused by the reservoir. However, the delays allow the pressures close to the outlet to exceed those predicted on the assumption of instantaneous reflection processes.

Tijsseling [19] used an example such as this to disprove a hypothesis put forward by other authors who had postulated much stronger delays and had argued that they could be a strong cause of damping observed in the measurements of water-hammer in pipe flows. Tijsseling correctly reasoned that the timescales of the reflection process were far smaller than those of the discrepancies that the original authors were attempting to explain. In a nutshell, the reflection process does cause delays and damping, but the amplitudes of the delays are very small in comparison with those caused by the other effects discussed above.

For completeness, it is noted that a similar behaviour exists when pressure waves encounter any change in the cross-sectional area; e.g., at a junction of a pipe with a downstream pipe of a larger diameter. However, the differences between the impedances of the two pipes are even smaller than those at a pipe outlet into a reservoir so the induced reflections are smaller. Likewise, the geometrical discontinuity is smaller and so the timescales of the reflection process are smaller. It is almost never useful to take account of this form of damping in the case of liquid-filled pipes.

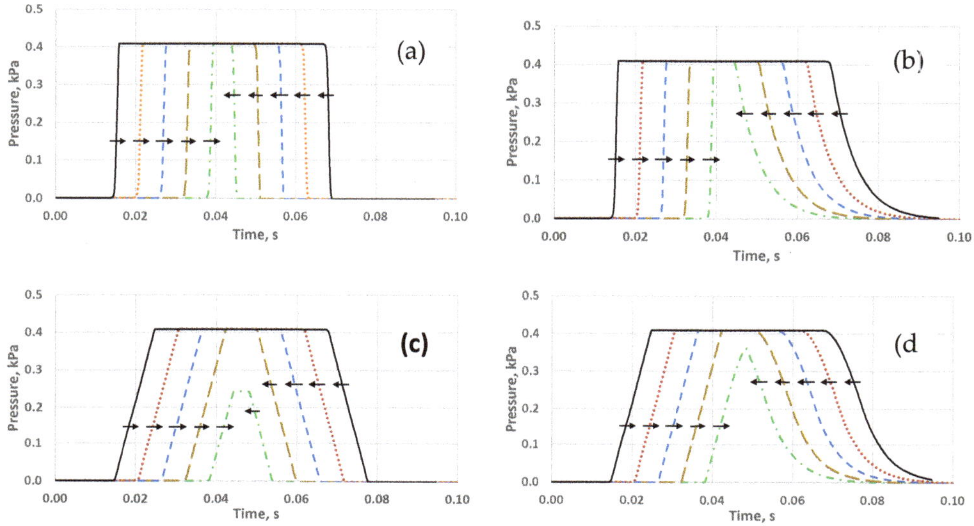

Figure 9. Pressures at successive locations as a wavefront approaches and reflects from a reservoir. (**a**) Steep ramp, assuming instantaneous reflections. (**b**) Steep ramp, allowing for true reflection delays. (**c**) More gentle ramp, assuming instantaneous reflections. (**d**) More gentle ramp, allowing for true reflection delays. (The arrows indicate the direction of travel of the wavefronts along the duct as they pass the locations at which the graphs are drawn.

8. Experimental Measurements

Attention now turns to consideration of the implications of the above causes of damping for persons attempting to interpret physical measurements. Such persons include, for example, (i) engineers seeking to understand unexpected behaviour, (ii) researchers seeking to isolate particular phenomena and (iii) developers of new techniques for such purposes as leak detection. In all cases, it is necessary (or, at least, helpful) to have a good understanding of the extent to which nominally secondary effects might complicate the task. When damping is seen to be present, it may be necessary to identify its cause so that its implications for other operating regimes can be assessed. Suppose, for instance, that one particular effect is dominant in the measurements, but another one is also exerting some influence. Without a good understanding of both causes, it would be unsafe to assume that the relative importance of the two effects will be the same in other contexts. As a simple example, skin friction is a secondary effect in many laboratory studies of water-hammer, but it can be the dominant cause of the pressure increase in the period after a valve in a long pipeline is closed rapidly. Indeed, for long pipelines, even the correct interpretation of 'rapid' can surprise the unwary.

Against this background, it is informative to revisit some of the figures presented above, attempting as far as practicable to approach them from the standpoint of persons seeing the traces for the first time and without the benefit of theoretical comparisons. First, however, it is useful to discuss briefly the influence of overlapping wavefronts resulting from, say, the finite times required for valve closure. Figure 10 shows the pressure histories in a classical reservoir-pipe-valve system when the valve is closed rapidly, but not instantaneously, causing a linear reduction in the flow rate to zero in the time required for a sound wave to travel the length of the pipe. To simplify the interpretations, no damping phenomena are simulated; e.g., the flow is treated as inviscid, and no account is taken of local losses or of time delays at the reservoir. The pressure histories are shown at the ends and mid-point of the pipe and also at its quarter points. However, since the pressure at the reservoir never changes, only four curves are seen. Considering each of these in turn:

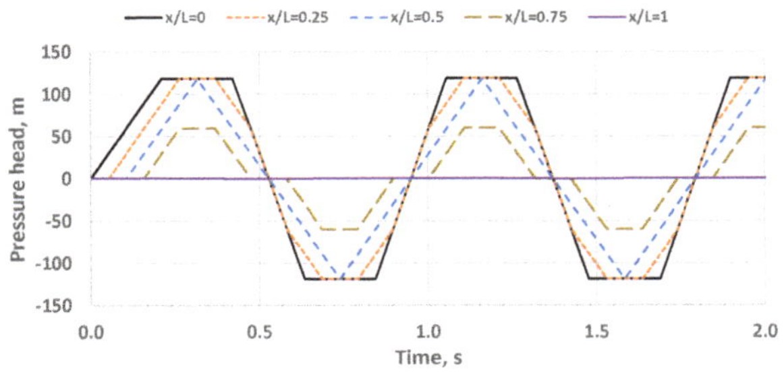

Figure 10. Influence of wave superpositions on pressure histories. (Pressures relative to the reservoir pressure.)

- The pressure at the valve rises linearly to a maximum in 0.21 s and then remains constant for a further 0.21 s until the reflection from the reservoir begins to arrive. In the following 0.21 s, it falls linearly to a minimum before remaining constant until the arrival of the next reflection. The amplitudes of the maximum and minimum are equal so the rates of the change in the pressure during the reflections at the valve are double the rate of the change during the closure itself.
- The pressure at $3L/4$ (i.e., at a distance of $L/4$ from the valve) follows a broadly similar pattern except that the changes between the maximum and minimum values occur in three distinct periods. During each reduction, the first period begins when a reflection begins to arrive from the reservoir, and it ends when the subsequent reflection from the valve begins to arrive. During the next period, the pressure change is caused jointly by the second half of the reflection from the reservoir and the first half of its subsequent reflection from the valve. The third period begins at the end of the reflection from the reservoir and continues until the end of the reflection from the valve.
- Next, consider the pressure at the mid-pipe length. At that location, the beginning of any particular reflection coincides with the end of a reflection travelling in the opposite direction. As a consequence, there is no intermediate period of constant pressure, and all rates of change are equal to that in the original closure.
- Finally, consider the pressure history at a distance of $L/4$ from the reservoir. Here, the later parts of each ramp wave travelling towards the reservoir coincide with the earlier parts after they have reflected at the reservoir and, as a consequence, the maximum pressure is smaller than at locations further from the reservoir. Conversely, it is greater than that at locations closer to the reservoir.

A simple message to be taken from this description is that wave superpositions can have a strong influence on pressure histories at any particular location. In this example, each ramp is linear, so the effects are easily seen. In practical applications, however, wavefronts rarely approximate closely to linear and the consequences of superpositions can be less obvious. Another important observation from Figure 10 is that the magnitudes of successive maxima and minima are not affected by the superpositions, and neither are the rates of the change in the pressure of the individual wavefronts. Therefore, although the superpositions can complicate the interpretation of the physical measurements exhibiting damping, they are not, in themselves, a cause of the damping of individual waves even though, at one location illustrated in Figure 10, they cause a big reduction in the amplitudes.

8.1. Re-Assessment of above Figures

Figure 2 is now revisited from the standpoint of an engineer who has only the physical measurements and a description of the reservoir-pipe-valve system. Even though no theoretical curves are given to show the predictions in the absence of damping, it can readily be deduced that strong damping exists. The gradual increase in pressure at the valve during the period of $2L/c$ between the valve closure and the beginning of the first reflection is consistent with the normal expectations of the consequence of a frictional pressure gradient along the pipe in the original steady flow. However, the strongly curved shape in the laminar flow case during the subsequent period of $2L/c$ cannot be attributable to quasi-steady friction. In principle, neither can it be attributed with total certainty to any other particular cause. Nevertheless, since the pipe was encased in concrete, the effect would probably be correctly expected to be caused by unsteady fiction. The measured pressures in the turbulent flow case exhibit the same generic behaviour so the same inferences can be drawn, although the effect is much weaker, so it is obvious until after several reflections have occurred.

A simple comparison of the measurements at the valve and mid-pipe reveals a continual increase in the damping effect as the main wavefront propagates. The most obvious effect is the continual reduction in the amplitudes of the pressure changes. A more important observation, however, is that the individual changes occur successively more slowly. This is clear evidence of a dispersive behaviour. However, it is not easy to assess the rate of the dispersion because the measured values are influenced by the superposition of the reflections such as that described above (Figure 10). This is apparent from the differences between the pressure amplitudes at the valve and at the mid-length of the pipe (n.b., compared with the pressures at $L/4$ in the above example with the inviscid flow). The effect can be seen in both flow cases, although it is less strong in the turbulent flow case than in the laminar flow case.

In summary, quite a lot of valuable information can be deduced from these particular measurements even without the benefit of theoretical predictions with which to make comparisons. However, it is far from common for such clear deductions to be drawn. To illustrate this, consider again Figure 3 and, for this purpose, imagine that the continuous lines had been obtained by physical measurements. In the case of the rapid valve closure, the curve exhibits similar features to those in Figure 2 so it is likely that similar deductions would be made from it, and reasonably so. Nevertheless, the degree of confidence that would be justified in the deductions would be less strong. One reason for this is that measurements are available only at one location. As a consequence, the distortion of successive cycles cannot be proven to increase during propagation along the pipe (even though other possibilities might seem implausible). Likewise, although the existence of superposition effects seems highly probable, the confirmation that would be provided by additional measurements at one or more locations along the pipe is not available.

Notwithstanding the reduced evidence, it is likely that correct inferences would be drawn from measurements showing as much detail as those for the rapid valve-closure case in Figure 3. The same is not true, however, for the slower valve-closure case. For this, it would not even be possible to deduce details of the valve-closure process itself, including, for instance, its start time and, more importantly, its duration. As a consequence, it would not be safe to use the measurements as a basis for predicting, say, what would happen in the event of more rapid closures. In contrast, deductions from measurements made with closure times smaller than $2L/c$ could, with care, be used to estimate likely outcomes for either slower or faster closures. It is somewhat ironic that the case that appears to exhibit greater damping is the one that gives fewer clues about the causes of the damping and, hence, about its wider implications.

Next, consider Figure 4. Once again, all of these pressure histories are theoretical predictions, but, for present purposes, imagine that the physical measurements of the pressure in a reservoir-pipe-valve system yielded the dotted (red) curve labelled QS + US + VE and that it were the only available history. Temporarily disregarding the first $2L/c$; i.e., before the arrival of the first reflection from the reservoir, it could be tempting to attribute the subsequent behaviour to strong unsteady friction. The detailed shape of any particular cycle differs from that displayed in Figure 2, for instance, but this could be imagined to be a consequence of unknown factors such as the characteristics of the valve and the closure sequence. However, consideration of the first $2L/c$ should swiftly dispel this direction of thinking. During this period, the pressure reduces continually and yet, as seen above, if skin friction were the cause of the change, the pressure would increase, not decrease. It would not be plausible to attribute the measured decrease to a negative pre-existing gradient, so it has to be due to some other cause.

This is a strong clue that the pipe itself is responding in a gradual manner to the pressure increase caused by the valve closure. That is, the cross-sectional area is increasing and is continuing to do so even though the pressure is reducing. This implies that the pipe material is behaving in a manner such as visco-elastic or visco-plastic. The direct evidence of this behaviour ceases at $2L/c$, but it would be inferred from the figure that it would have continued for a longer time if the reflection from the reservoir had not interrupted the process. However, there is insufficient evidence to be confident that the strongly dispersive behaviour is attributable to a delayed response of the pipe wall material to imposed pressure changes. In principle, other changes such as, for example, a non-uniform axial distribution of gas bubbles could be responsible. Therefore, anyone needing to identify the true cause of the strong dispersion would need to seek additional information to supplement the measured pressure history.

As a final example, consider again Figure 6, which shows more complex pressure histories than those in other figures. As indicated above, the influence of FSI in this example is far greater than in typical practical situations encountered by either researchers or most practising engineers. Nevertheless, the figure does serve to illustrate the influence that can be exerted by phenomena that are overlooked when assessing measured pressure histories. Recall that the largest sudden pressure change seen in Figure 6b (i.e., L2) exists because the remote end of the pipe was free to move axially. A very different outcome would have resulted if the end had been restrained, but even if it were possible to prevent all axial motion, however small, there would still have been a small reflection because of radial elasticity. FSI will rarely have a sustained influence in most practical applications, but it certainly has an influence when movement is possible, and this will inevitably complicate inferences about other aspects of pressure histories. In practical assessments, it is likely that FSI effects will not be recognised and will either be interpreted as noise or be attributed to other causes. The real point here is that multiple effects coexist in real pressure measurements, and each will contribute to outcomes, especially close to the wavefronts. One consequence is that the interpretation of such measurements is a specialist skill. Another is that researchers need to think broadly when designing experiments targeted at the investigation of specific phenomena. The author wishes all of them much satisfaction in the exercise of their skills.

8.2. Unidentified Oscillations

Experimental measurements of conditions following rapid valve closure are commonly used for the assessment of theoretical models of water-hammer phenomena, as in Figure 6 above, for example. In such cases, it is necessary to decide how to represent the valve boundary in the associated theoretical comparisons. One option is to use the measured pressure history directly. Another is to attempt to model the valve itself. Yet another is to idealise the measured history in some way, perhaps using a smoothing technique to reduce the signal noise. Each of these methods has its proponents, but none is strictly rigorous. For instance, the direct use of measured histories should not continue after significant reflections have begun to arrive from along the pipe. This is not necessarily a significant restriction, however, because, after closure, a valve can be modelled by stipulating zero flow (unless it is itself able to move). A particular concern arises when the measured signal immediately after closure exhibits high-frequency oscillations. It might not be possible to deduce with confidence whether these are attributable to a hydraulic cause induced by the vibration of the valve or to a measurement error in the pressure sensor. Either way, the analyst needs to decide whether and/or how to allow for them.

Suppose that such oscillations exist but are not seen in subsequent reflections. In principle, this could be because (a) physical processes have damped them out or (b) they were due to sensor errors. Now, further suppose that the oscillations are input faithfully to a numerical simulation but are still not seen in the predicted reflections. This could be an indication that a correctly modelled physical phenomenon has damped them out. Alternatively, it could be a consequence of non-physical numerical damping (see the next section). Equally, it could be because they have become obscured by some other phenomenon. The point being made here is that it can be very difficult indeed to infer the validity of a theoretical method when the initial triggering event is subject to uncertainty. This is one reason why the present author considers that descriptions such as 'good agreement' or even 'excellent agreement' can convey misleading messages to unwary readers.

9. Numerical Damping

So far, only physical causes of damping have been considered. However, it would be remiss to close the discussion without also commenting on non-physical damping in numerical simulations, which can be important supplementary tools in the interpretation of measured physical behaviour. Almost all numerical simulations are affected to some degree by this problem. For example, it can be a consequence of using numerical grids with regions that are too coarse to propagate higher frequencies correctly. Alternatively, it can be caused by interpolation algorithms or by incorrect estimations of flux across interfaces between adjacent cells. Numerical damping can also be introduced intentionally for special reasons; for instance, by the use of 'artificial viscosity' to suppress unrealistic oscillations close to locations of especially rapid change.

Numerical damping is especially significant when its existence is not recognised. This is a particular issue when simulations are used to provide guidance on the interpretation of experimental measurements. It is all too easy for numerical damping to cause behaviour that enhances apparent agreement between predicted and measured results. This has the potential to cause true physical damping to be interpreted as a consequence of some other effect that a numerical model has been designed to simulate. However, potentially adverse consequences of numerical damping are certainly not limited to the interpretation of measurements. In one especially severe case, an editor over-ruled several reviewers' objections to a paper that presented results from a large parametric study designed to assess the influence of a range of parameters. The paper was published even though all predictions at higher frequencies tended to asymptotic conditions that were demonstratively physically impossible. This example is mentioned in the hope that it will serve as a warning to readers of this present paper. It was especially unfortunate because a key objective of the paper was to enable practising engineers to assess when they needed to take account of the various parameters under study. It is to be hoped that the paper was not widely read.

10. Summary and Conclusions

A number of possible causes of the damping of pressure waves in pipelines have been described, with special attention paid to characteristics that researchers and designers might encounter in assessments of measured pressure histories. This can be important when attempting to understand unexpected behaviour, perhaps as a prelude to implementing mitigating countermeasures. It is also important when there is a need to know whether it is safe to use measured results in one pipe system as a basis for predicting conditions to be expected in other systems.

Attention has been paid to damping caused by the response of skin friction on pipe walls to pressure waves causing rapid changes in the axial mean velocity. It has been shown that the time required for vorticity diffusion over a pipe cross-section can cause large differences between the wall stresses and their corresponding values in steady flows. However, the differences reduce rapidly after a wavefront has passed and have little impact on sustained damping. Nevertheless, the effect can have a strong influence on the interpretation of laboratory measurements. Commonly, pipe lengths in laboratories are such that the influence of unsteady skin friction on measured pressure histories is much stronger than in longer pipes typical in engineering practice. This is because the superpositions of elongated wavefronts with their own reflections travelling in the opposite direction can cause continual reductions in the amplitude that would not be seen in most regions of long pipes.

Similar behaviour can arise as a consequence of the visco-elastic nature of many 'plastic' pipes. Sufficiently close to a wavefront, the visual effect can resemble that caused by unsteady friction, although its magnitude is greater. In common with unsteady friction, complications can arise as a consequence of wave superpositions when measurements are made in short pipes. Moreover, the decay times are longer and so the periods of significant overlap are greater. In addition, visco-elastic pipes tend to be much more flexible than metal pipes and this complicates comparisons between the two because of its influence on effective wavespeeds even after the non-linear contributions of the phenomena have decayed.

Another potential cause of damping considered above is bubbly flows. Even low concentrations of bubbles can be influential because they cause wavespeeds to be pressure-dependent and, hence, cause dispersion. This can be especially strong if the gas concentration is not uniform along the pipe. Nevertheless, if the gas is not soluble in the liquid, the process is not inherently dissipative and, although the pressure amplitudes may decrease in some locations, they could increase in others. Therefore, it is debatable whether decreased amplitudes arising from this cause should be regarded as damping per se. Similar comments apply to the case of gas flow in ducts with porous walls. Again, the influence is especially strong when the porosity varies along the duct.

A slightly more complex case arises when pipes have significant freedom to move axially or laterally, as is the case, for instance, with suspended pipe systems. Then, FSI phenomena exist, and their importance will depend strongly on the interactions between internal pressure forces and forces due to structural movements. Once again, it is possible for this to cause decreased pressure amplitudes in some locations, but increased amplitudes in others. In addition, however, the interactions can lead to pressure histories that are much more complex than those in immovable pipes and this inevitably complicates the identification of any truly damping phenomena that might also be present.

For completeness, attention has also been paid briefly to one phenomenon that has been cited as having a significant influence on damping even though it is physically incapable of doing so. This is the reflection process at an open end of a duct; e.g., at the junction of a pipe and a reservoir. Time delays are indeed inevitable as a consequence of the time required for the reservoir to exercise its dominance in the sustainable pressure at the pipe outlet, but the delays are typically shorter than that required for a wavefront to travel one pipe diameter. This is a much shorter delay than those associated with any other of the phenomena considered above.

Funding: This research received no external funding.

Institutional Review Board Statement: Not applicable.

Informed Consent Statement: Not applicable.

Data Availability Statement: Not applicable.

Acknowledgments: The author wishes to express special thanks to Anton Bergant, Emeritus Head of Applied Research and Computation Department, Litostroj Power, Slovenia and John Vítkovský, Senior Hydrologist, Department of Environment and Science, Queensland Government, Australia, for searching their records to provide experimental and theoretical data illustrating unsteady friction and visco-elasticity.

Conflicts of Interest: The author declares no conflict of interest.

Nomenclature

c	speed of sound
D	diameter
FSI	fluid–structure interaction
L	Length
p	pressure
QS	quasi-steady skin friction
t	time coordinate
US	unsteady component of skin friction
VE	visco-elastic
x	axial coordinate

References

1. Karney, B.W. Energy relations in transient closed-conduit flow. *J. Hydraul. Eng. ASCE* **1990**, *116*, 1180–1196. [CrossRef]
2. Axworthy, D.H.; Ghidaoui, M.S.; McInnes, D.A. Extended thermodynamics derivation of energy dissipation in unsteady pipe flow. *J. Hydraul. Eng. ASCE* **2000**, *126*, 276–287. [CrossRef]
3. Leslie, D.J.; Tijsseling, A.S. Travelling discontinuities in waterhammer theory—Attenuation due to friction. In Proceedings of the 8th BHR Group International Conference on Pressure Surges, The Hague, The Netherlands, 12–14 April 2000; Professional Engineering Publishing: London, UK; pp. 323–335.
4. Ghihaoui, M.S.; Kolyshkin, A.A. Stability analysis of velocity profiles in water-hammer flows. *J. Hydraul. Eng. ASCE* **2001**, *127*, 499–512. [CrossRef]
5. Shuy, E.B. Wall shear stress in accelerating and decelerating turbulent pipe flows. *J. Hydraul. Res. IAHR* **1996**, *34*, 173–183. [CrossRef]
6. Holmboe, E.L. Viscous Distortion in Wave Propagation as Applied to Water Hammer and Short Pulses. Ph.D. Thesis, Carnegie Institute of Technology, Pittsburgh, PA, USA, 1964.
7. Duan, H.F.; Ghidaoui, M.S.; Lee, P.J.; Tung, Y.K. Relevance of unsteady friction with to pipe size and length in pipe fluid transients. *J. Hydraul. Eng. ASCE* **2012**, *138*, 154–166. [CrossRef]
8. Vardy, A.E.; Bergant, A.; He, S.; Ariyaratne, C.; Koppel, T.; Annus, I.; Tijsseling, A.S.; Hou, Q. Unsteady skin friction experimentation in a large diameter pipe. In Proceedings of the 3rd IAHR International Meeting of the Workgroup on Cavitation and Dynamic Problems in Hydraulic Machinery and Systems, Brno, Czech Republic, 14–16 October 2009; Rudolf, P., Ed.; Eindhoven University of Technology: Eindhoven, The Netherlands; pp. 593–602.
9. Holmboe, E.L.; Rouleau, W.T. The effect of viscous shear on transients in liquid lines. *J. Basic Eng. ASME* **1967**, *89*, 174–180. [CrossRef]
10. Zielke, W. Frequency-dependent friction in transient pipe flow. *J. Basic Eng. ASME* **1968**, *90*, 109–115. [CrossRef]
11. Vitkovski, J.; Stephens, M.; Bergant, A.; Simpson, A.; Lambert, M. Numerical Error in Weighting Function-Based Unsteady Friction Models for Pipe Transients. *J. Hydraul. Eng.* **2006**, *132*, 709–721. [CrossRef]
12. Tijsseling, A.S.; Vardy, A.E. What is wavespeed? In Proceedings of the 12th International Conference on Pressures Surges, Dublin, Ireland, 18–20 November 2015; Tijsseling, A.S., Ed.; BHR Group: London, UK; pp. 343–360.
13. Aliabadi, H.K.; Ahmadi, A.; Keramat, A. Frequency response of water hammer with fluid-structure interaction in a viscoelastic pipe. *Mech. Syst. Signal Proc.* **2020**, *144*, 106848. [CrossRef]
14. Tijsseling, A.S.; Vardy, A.E. Axial modelling and testing of a pipe rack. In Proceedings of the 7th International Conference on Pressure Surges and Fluid Transients in Pipelines and Open Channels, Harrogate, UK, 16–18 April 1996; BHR Group: London, UK; pp. 363–383.
15. Vardy, A.E.; Fan, D.; Tijsseling, A.S. Fluid-structure interaction in a T-piece pipe. *J. Fluids Struct.* **1996**, *10*, 763–786. [CrossRef]
16. Ozawa, S.; Maeda, T.; Matsumura, T.; Nakatani, K.; Uchida, K. Distortion of compression wave during propagation along Shinkansen tunnel. In Proceedings of the 8th International Symposium on Aerodynamics and Ventilation in Vehicle Tunnels, Liverpool, UK, 6–8 July 1994; BHR Group: London, UK; pp. 211–226.
17. Liu, F.; Vardy AEPokrajac, D. Influence of air chambers on wavefront steepening in railway tunnels. *Tunn. Undergr. Sp. Technol.* **2021**, *117*, 104120. [CrossRef]

18. Rudinger, G. The reflection of pressure waves of finite amplitude from an open end of a duct. *J. Fluid Mech.* **1957**, *3*, 48–66. [CrossRef]
19. Tijsseling, A.S. Discussion of "Effect of boundary on water hammer wave attenuation and shape" by Huade Cao, Ioan Nistor and Magdi Mohareb. *J. Hydraul. Eng. ASCE* **2021**, *147*, 07021011. [CrossRef]

Disclaimer/Publisher's Note: The statements, opinions and data contained in all publications are solely those of the individual author(s) and contributor(s) and not of MDPI and/or the editor(s). MDPI and/or the editor(s) disclaim responsibility for any injury to people or property resulting from any ideas, methods, instructions or products referred to in the content.

water

MDPI

Article

Finite Volume Method for Modeling the Load-Rejection Process of a Hydropower Plant with an Air Cushion Surge Chamber

Jianwei Lu [1], Guoying Wu [2], Ling Zhou [1,3,*] and Jinyuan Wu [1,4]

1 College of Water Conservancy and Hydropower Engineering, Hohai University, Nanjing 210098, China
2 China Water Resources Pearl River Planning, Surveying and Designing Co., Ltd., Guangzhou 510610, China
3 Yangtze Institute for Conservation and Development, Nanjing 210098, China
4 Shanghai Municipal Engineering Design Institute (Group) Co., Ltd., Shanghai 200092, China
* Correspondence: zlhhu@163.com

Abstract: The pipe systems of hydropower plants are complex and feature special pipe types and various devices. When the Method of Characteristics (MOC) is used, interpolation or wave velocity adjustment is required, which may introduce calculation errors. The second-order Finite Volume Method (FVM) was presented to simulate water hammer and the load-rejection process of a hydropower plant with an air cushion surge chamber, which has rarely been considered before. First, the governing equations were discretized by FVM and the flux was calculated by a Riemann solver. A MINMOD slope limiter was introduced to avoid false oscillation caused by data reconstruction. The virtual boundary strategy was proposed to simply and effectively handle the complicated boundary problems between the pipe and the various devices, and to unify the internal pipeline and boundary calculations. FVM results were compared with MOC results, exact solutions, and measured values, and the sensitivity analysis was conducted. When the Courant number was equal to 1, the results of FVM and MOC were consistent with the exact solution. When the Courant number was less than 1, compared with MOC, the second-order FVM results were more accurate with less numerical dissipation. As the Courant number gradually decreased, the second-order FVM simulations were more stable. For the given numerical accuracy, second-order FVM had higher computational efficiency. The simulations of load rejection showed that compared with the MOC results, the second-order FVM calculations were closer to the measured values. For hydropower plants with complex pipe systems, wave velocity or the Courant number should be adjusted during MOC calculation, resulting in calculation error, and the error value is related to the parameters of the air cushion surge chamber (initial water depth, air cushion height, etc.). The second-order FVM can more accurately, stably, and efficiently simulate the load-rejection process of hydropower plants compared with MOC.

Keywords: air cushion surge chamber; method of characteristics; finite volume method; load rejection process

Citation: Lu, J.; Wu, G.; Zhou, L.; Wu, J. Finite Volume Method for Modeling the Load-Rejection Process of a Hydropower Plant with an Air Cushion Surge Chamber. *Water* **2023**, *15*, 682. https://doi.org/10.3390/w15040682

Academic Editor: Kamil Urbanowicz

Received: 23 January 2023
Revised: 4 February 2023
Accepted: 7 February 2023
Published: 9 February 2023

Copyright: © 2023 by the authors. Licensee MDPI, Basel, Switzerland. This article is an open access article distributed under the terms and conditions of the Creative Commons Attribution (CC BY) license (https://creativecommons.org/licenses/by/4.0/).

1. Introduction

In hydropower plants, complicated hydraulic transients often occur during startup, shutdown, or load change of the power generation unit. Dangerous water hammer events are caused by some inappropriate operations in the water system and likely produce abnormally high pressures, which may induce pipe rupture and damage other hydraulic devices. Due to the advantages of low construction cost and ecological environment impact, air cushion surge chambers have been widely used in water hammer protection in hydropower stations, ensuring the safety of the hydraulic operation [1]. However, water hammer protection devices also increase the operational complexity of the hydraulic system. Therefore, accurate and efficient numerical simulations of water hammer events become more important for the proper design and safe operation of hydropower plants.

The Method of Characteristics (MOC) is widely used in the simulation of hydraulic transition processes in hydropower plants [2]. However, in the actual complex water transmission system, there are certain short pipes, T-pipes, and tandem pipes. In the calculation of simulation with MOC modeling, interpolation calculation or wave speed adjustment is required; the former reduces the calculation efficiency and calculation accuracy and the latter often introduces calculation errors, which may lead to poor simulation results [1–3].

In recent years, many scholars have tried to use the Finite Volume Method (FVM) for the simulation of water hammer problems. Based on the system's mass and energy conservation, FVM can solve the discontinuous problem well without causing spurious numerical oscillations. Guinot [4] was the first to introduce the first-order FVM numerical method into the solution of the water hammer problem, and its calculation results are basically consistent with those of the MOC calculation. Zhao [5] developed the first-order and second-order FVM Godunov-type scheme (GTS) to simulate water-hammer problems in a simple horizontal pipeline with an instantaneously closed valve.

Zhou et al. [6,7] firstly developed a GTS approach to simulate transient cavitating pipe flow. Elong [8] solved the two-dimensional shallow water equations using the first-order finite volume method (FVM), the Harten Lax and van Leer (HLL) scheme, and the monotone upwind scheme for conservation laws (MUSCL) to simulate floods. Zhou [9] and Xue [10] conducted a simulation study of water–gas two-phase homogeneous flow using the FVM format and found that second-order FVM can effectively avoid spurious numerical oscillations and can improve the stability and accuracy of the calculation results.

Zhou et al. [2,3] developed one explicit solution source item approach for second-order FVM GTS to easily incorporate various forms of the existing unsteady friction models, including original convolution-based models, simplified convolution-based models, and Brunone instantaneous acceleration-based models. They pointed out that the first-order Godunov scheme and fixed-grid MOC scheme can achieve the same accuracy, but both display strong numerical damping once the Courant number is less than one. In contrast, the second-order GTS is more robust, even for Courant numbers significantly less than those for simple water hammer events.

Overall, the motivation and reason of attempting the second-order FVM GTS to simulate the load-rejection process of hydropower plants with an air cushion surge chamber are as follows. The fixed-grid MOC scheme is of first-order accuracy and has been widely used for solving water hammer equations in the simulation of hydraulic transition processes in hydropower plants. Since the real water pipe systems are usually complicated and made of pipe sections with different lengths, diameters, and materials, it is impossible to make the Courant number exactly equal to one in every pipe of such a complex pipe system; thus, the MOC scheme has to be implemented either via interpolation or wavespeed adjustment, which may induce large accumulated numerical errors. Importantly, as the Courant number is less than one and decreases, compared with MOC, the second-order GTS results are more accurate and more stable with less numerical dissipation. The previous work mainly focuses on the FVM GTS simulating the water hammer problem in a simple reservoir–pipe–valve system. However, it is necessary to further investigate the feasibility of GTS for more complicated hydraulic transient problems in a real hydraulic system with more complicated pipe components and devices, and to explore the possible computation error caused by the classic MOC scheme in a hydropower plant with a complex pipe system.

The main aim of this paper was to develop an accurate and efficient water hammer numerical model, which is significant for the proper design and safe operation of hydropower plants. A second-order FVM GTS fully considering the various pipe components and devices was developed to simulate the hydraulic transients and load-rejection process of the hydropower plant with an air cushion surge chamber, which has rarely been involved in previous published works. Importantly, the virtual boundary strategy was proposed to simply and effectively handle the complicated boundary problems between the pipe and various devices. Namely, virtual boundaries were introduced at the upstream and downstream boundaries and at the connection between the hydraulic components (air chamber and unit) and the pipeline to achieve uniformity in the calculation of the control cells inside the pipeline and at the boundaries. The results calculated by the proposed second-order FVM GTS models were compared with the exact solution and the measured values as well as predictions by the MOC scheme. The accuracy and efficiency of the proposed approach were discussed. Another important purpose is that the proposed accurate model was used to explore the possible computation error caused by the MOC scheme in a hydropower plant with a complex pipe system.

2. Numerical Models of Hydropower Plant Hydraulic Transients

2.1. Water Hammer Control Equations

The classic water hammer equations for pipe flow are as follows [1]:

$$g\frac{\partial H}{\partial x}+V\frac{\partial V}{\partial x}+\frac{\partial V}{\partial t}+(J-gS_0)=0 \tag{1}$$

$$V\frac{\partial H}{\partial x}+\frac{\partial H}{\partial t}+\frac{a^2}{g}\frac{\partial V}{\partial x}=0 \tag{2}$$

where H is the piezometric head; V is the average flow velocity; a is the water hammer wave speed; g is the gravitational acceleration; x is the coordinate distance along the tube axis; t is time; J is the steady-state friction coefficient of the pipe; and S_0 is the slope of the pipe.

2.2. Control Equations of Air Cushion Surge Chamber

In the analysis of air chamber shown in Figure 1, the pressure at any instant was assumed to be the same throughout the volume. The compressibility of the water in the air chamber was considered negligible compared with air compressibility. Inertia and friction were neglected. The air was assumed to follow the reversible polytropic relation

$$H_aV_a^k=H_{a0}V_{a0}^k=Constant \tag{3}$$

where H_a and V_a are the absolute pressure head and volume of the air within the air chamber, and their initial values are H_{a0} and V_{a0}, respectively; and k is the polytropic exponent. The adiabatic process with k = 1.4 is often used for the fast transients whereas the isothermal with k = 1.0 is often presented for the slower compressions; an intermediate polytropic case with k = 1.2 is often suggested as a reasonable compromise.

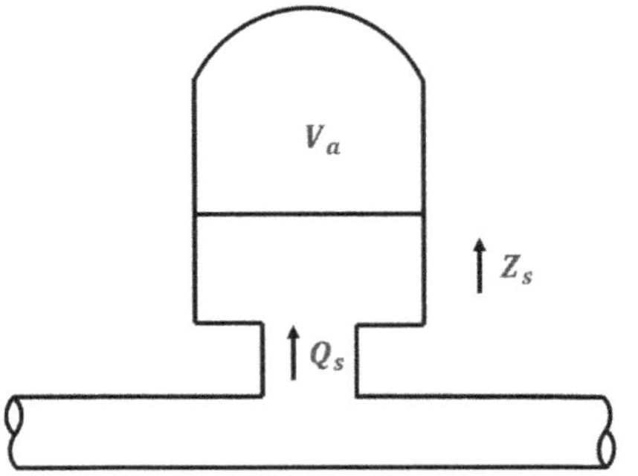

Figure 1. Schematic diagram of the air cushion surge chamber in the hydropower station.

The compressibility of the water and wall within the air chamber was neglected. H_a is the absolute head equal to the gage plus barometric pressure heads

$$H_a=H_P-Z_s+H_{atm}-\left(R_S+\frac{1}{2gA_S^2}\right)|Q_S|Q_S \tag{4}$$

where H_P is the piezometric head at the bottom of air chamber; Z_s is the elevation of the air–water interface in air chamber; H_{atm} is the absolute barometric pressure head; Q_s is the inflow rate to the air chamber; R_s is the head loss coefficient of the impedance hole of the air chamber; A_s is the cross-section area of the air chamber.

The air volume was allowed to vary for inflow to and outflow from air chamber. The integrated continuity equation $dV_a/dt = -Q_s$, can be written as

$$Q_t = Q - Q_S = Q - A_S \frac{dZ_s}{dt} \tag{5}$$

where Q and Q_t are the flow rates at the inlet and outlet pipe of the bottom of the air chamber.

The piezometric head at the bottom of the air chamber, H_P, is associated with the inflow rate to the air chamber Q_s, which can be expressed as

$$\left.\begin{array}{c} H_P = C_2 - C_1 Q_s \\ C_1 = \left(\frac{1}{\left(\frac{1}{B_{P1}}\right)+\left(\frac{1}{B_{M2}}\right)}\right); C_2 = C_1\left(\left(\frac{C_{P1}}{B_{P1}}\right) + \left(\frac{C_{M2}}{B_{M2}}\right)\right) \end{array}\right\} \tag{6}$$

where $C_{P1}, B_{P1}, C_{M2}, B_{M2}$ are boundary parameters that can be calculated from pressure heads and flow rates at the upstream and downstream pipe of the air chamber, and are discussed in the section concerning boundary treatment.

After combining Equations (3)–(6), the pressure head, flow rate, and water level at the air chamber can be obtained.

2.3. Control Equations of Hydraulic Turbine

The unit characteristic curve of the hydraulic turbine consists of the flow characteristic curve and the moment characteristic curve. Using modified Suter transformation [11], the flow function and torque function of the hydraulic turbine are as follows:

$$WH(x,y) = \frac{1}{\left(\frac{Q_{11}}{Q_{11r}} + c\right)^2 + \left(\frac{N_{11}}{N_{11r}}\right)^2} = \frac{h}{\left(q + c\sqrt{h}\right)^2 + n^2} \tag{7}$$

$$WB(x,y) = \frac{M_{11}}{M_{11r}} = \frac{m}{h} \tag{8}$$

$$\left\{\begin{array}{ll} x = arctan\left[\left(q + c\sqrt{h}\right)/n\right], & n \geq 0; \\ x = arctan\left[\left(q - c\sqrt{h}\right)/n\right] + \pi, & n < 0; \end{array}\right. \tag{9}$$

where x is the relative flow angle; y is the relative guide vane opening; $WH(x,y)$ represents the flow functions; $WB(x,y)$ is the torque function; Q_{11} is the unit flow rate; Q_{11r} is the unit flow rate at rated operating conditions; $q = Q_{11}/Q_{11r}$ is the relative unit flow; N_{11} is the unit speed; N_{11r} is the unit speed at rated operating conditions; $n = N_{11}/N_{11r}$ is the relative unit speed;H is the water head pressure; H_r is the head pressure at rated operating conditions; $h = H/H_r$ is the relative head; M_{11} is the unit torque; M_{11r} is the unit torque at rated operating conditions; $m = M_{11}/M_{11r}$ is the relative unit moment; and the subscripts 11 and $11r$ indicate the unit value and the rated value, respectively.

When load rejection occurs, the unit speed equation is as follows:

$$n = n_0 + \frac{\Delta t}{T_a}(1.5m_0 - m_{00}) \tag{10}$$

where T_a is the unit inertia time constant: $T_a = \frac{GD^2 N_r^2}{365 P_r}$; GD^2 is the unit rotational moment of inertia; P_r is rated power output; N_r is the rated speed; and the subscripts "0" and "00" indicate the first one time step and the first two time steps of the calculation time step, respectively.

The head balance equation [11,12] is given by

$$h = [C_{P1} - C_{M2} - (B_{P1} + B_{M2})Q_r q + C_3|q|q]/H_r \tag{11}$$

where coefficient $C_3 = Q_r^2\left(\frac{1}{2gA_1^2} - \frac{1}{2gA_2^2}\right)$; A_1 is the inlet cross-sectional area of the worm shell; and A_2 is the outlet cross-sectional area of the tailpipe. Combining Equations (7), (8), (10) and (11), the head, flow rate, speed, and torque of the unit can be calculated.

3. Numerical Solution by Using the Second-Order Finite Volume Method

The matrix form of the water hammer equations (Equations (1) and (2)) can be expressed as follows:

$$\frac{\partial \boldsymbol{U}}{\partial t} + \boldsymbol{A}\frac{\partial \boldsymbol{U}}{\partial x} = \boldsymbol{S} \tag{12}$$

where $\boldsymbol{U} = \binom{H}{V}$, $\boldsymbol{A} = \begin{pmatrix} V & a^2/g \\ g & V \end{pmatrix}$, $\boldsymbol{S} = \binom{0}{gS_0 - J}$.

For the pipe water hammer problem, the Mach number is small, so the effect of the convective term can be neglected. The classical water hammer equation can be obtained by solving Equation (3) with the Riemann problem solution method.

$$\frac{\partial \boldsymbol{U}}{\partial t} + \frac{\partial \boldsymbol{F}}{\partial x} = \boldsymbol{S} \tag{13}$$

where $\boldsymbol{F} = \bar{\boldsymbol{A}}\boldsymbol{U}, \bar{\boldsymbol{A}} = \begin{pmatrix} 0 & a^2/g \\ g & 0 \end{pmatrix}$.

The finite volume method was used to discretize the computational domain in the x-axis and t-axis, as shown in Figure 2, to form multiple computational control volumes with the fixed-grid length Δx and then compute the control volumes.

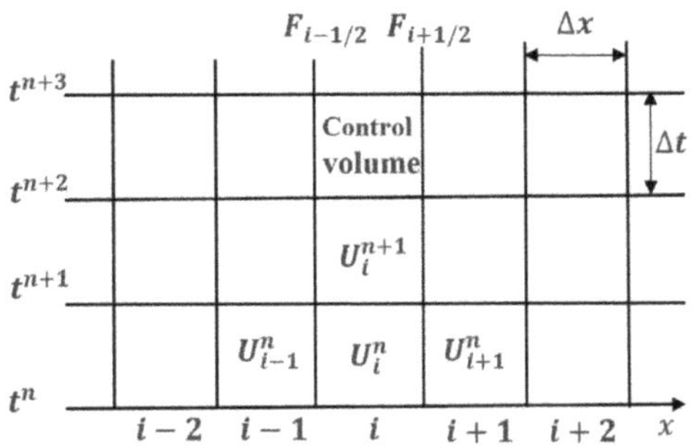

Figure 2. Grid system in the FVM Godunov scheme.

For the ith control volume, the integration of Equation (13) between control interfaces $i - 1/2$ and $i + 1/2$ yields

$$\mathbf{U}_i^{n+1} = \mathbf{U}_i^n - \frac{\Delta t}{\Delta x}\left(\mathbf{f}_{i+1/2}^n - \mathbf{f}_{i-1/2}^n\right) + \frac{\Delta t}{\Delta x}\int_{i-1/2}^{i+1/2} \mathbf{s}dx \tag{14}$$

where $\mathbf{U}_i$ is the average value of $\mathbf{u}$ within $[i - 1/2, i + 1/2]$; the superscripts n and $n + 1$ indicate the t and $t + \Delta t$ time levels, respectively; and $\mathbf{f}_{i+1/2}$ and $\mathbf{f}_{i-1/2}$ are the mass and momentum fluxes at the control interfaces $i - 1/2$ and $i + 1/2$, which are determined by solving a local Riemann problem at each cell interface [2,3].

3.1. Computation of Flux Term

In the Godunov approach, the numerical flux is determined by solving a local Riemann problem at each cell interface [4]. Applying Rankine–Hugoniot conditions $\Delta \mathbf{f} = \overline{A}\Delta \mathbf{u} = \overline{\lambda}_i \Delta \mathbf{u}$ where the eigenvalues $\overline{\lambda}_1 = -a$ and $\overline{\lambda}_2 = a$, the fluxes at $i + 1/2$ for all internal nodes and for $t \in [t^n, t^{n+1}]$ can be calculated by

$$\mathbf{f}_{i+1/2} = \overline{\mathbf{A}}_{i+1/2} u_{i+1/2} = \frac{1}{2}\overline{\mathbf{A}}_{i+1/2}\left\{ \begin{pmatrix} 1 & a/g \\ g/a & 1 \end{pmatrix} \mathbf{U}_L^n - \begin{pmatrix} -1 & a/g \\ g/a & -1 \end{pmatrix} \mathbf{U}_R^n \right\} \tag{15}$$

in which $\overline{\mathbf{A}}_{i+1/2} = \mathbf{A}$; $\mathbf{U}_L^n$= average value of $\mathbf{u}$ to the left of interface $i + 1/2$ at time n; and $\mathbf{U}_R^n$ = average value of $\mathbf{u}$ to the right of interface $i + 1/2$ at time n.

The estimation approach of $\mathbf{U}_L^n$ and $\mathbf{U}_R^n$ determines the accuracy order of the numerical scheme. In the first-order accuracy, $\mathbf{U}_L^n = \mathbf{U}_i^n$ and $\mathbf{U}_R^n = \mathbf{U}_{i+1}^n$. Herein, the MUSCL–Hancock method is used to achieve second-order accuracy in space and time, while the MINMOD limiter is suggested to avoid the spurious oscillations. The details of the MUSCL-Hancock method and the MINMOD limiter can also be found in a reference book (Toro 2009) [12].

The MUSCL–Hancock approach achieves a second-order extension of the Godunov scheme if the intercell flux $\mathbf{f}_{i+1/2}$ is computed according to the following steps [6]:

Step (1): Data Reconstruction. The data cell average values $\mathbf{U}_i^n$ are locally replaced by piece-wise linear functions in each cell $[x_{i-1/2}, x_{i+1/2}]$, and $\mathbf{U}_i^n$ at the extreme points are,

$$\mathbf{U}_i^L = \mathbf{U}_i^n - \frac{\Delta x}{2}\Delta_i,\ \mathbf{U}_i^R = \mathbf{U}_i^n + \frac{\Delta x}{2}\Delta_i \tag{16}$$

where Δ_i is a suitably chosen slope vector. The MINMOD limiter was used here to increase the order of accuracy of a scheme while avoiding spurious oscillations. Namely,

$$\Delta_i = \mathrm{MINMOD}\left(\sigma_i^n, \sigma_{i-1}^n\right) = \begin{cases} \sigma_i^n, if, \left|\sigma_i^n\right| < \left|\sigma_{i-1}^n\right|, and, \sigma_i^n \sigma_{i-1}^n > 0 \\ \sigma_{i-1}^n, if, \left|\sigma_i^n\right| > \left|\sigma_{i-1}^n\right|, and, \sigma_i^n \sigma_{i-1}^n > 0 \\ 0, if, \sigma_i^n \sigma_{i-1}^n < 0 \end{cases} \tag{17}$$

where $\sigma_i^n = \left(\mathbf{U}_{i+1}^n - \mathbf{U}_i^n\right)/\Delta x$ and $\sigma_{i-1}^n = \left(\mathbf{U}_i^n - \mathbf{U}_{i-1}^n\right)/\Delta x$.

Step (2): Evolution. For each cell $[x_{i-1/2}, x_{i+1/2}]$, the boundary extrapolated values $\mathbf{U}_i^L$, $\mathbf{U}_i^R$ in Equation (16) are evolved by a time $0.5\Delta t$ according to

$$\begin{aligned} \overline{\mathbf{U}}_i^L &= \mathbf{U}_i^L + \tfrac{1}{2}\tfrac{\Delta t}{\Delta x}\left[\mathbf{f}\left(\mathbf{U}_i^L\right) - \mathbf{f}\left(\mathbf{U}_i^R\right)\right], \\ \overline{\mathbf{U}}_i^R &= \mathbf{U}_i^R + \tfrac{1}{2}\tfrac{\Delta t}{\Delta x}\left[\mathbf{f}\left(\mathbf{U}_i^L\right) - \mathbf{f}\left(\mathbf{U}_i^R\right)\right] \end{aligned} \tag{18}$$

Step (3): The Riemann Problem. To compute intercell flux $\mathbf{f}_{i+1/2}$, the conventional Riemann problem with data can be calculated by

$$\mathbf{U}_L^n \equiv \overline{\mathbf{U}}_i^R, \mathbf{U}_R^n \equiv \overline{\mathbf{U}}_{i+1}^L \tag{19}$$

Insert Equation (19) into Equation (15) and a second-order scheme for flux terms at $i + 1/2$ for all internal cell and for $t = [t^n, t^{n+1}]$ is obtained.

3.2. Incorporation of Source Term

When considering the pipe friction resistance, the second-order Runge–Kutta solution is used to obtain the second-order calculation accuracy explicit results, and the calculation process is as follows.

First step:

$$\overline{U}_i^{n+1} = U_i^n - \frac{\Delta t}{\Delta x}\left(\mathbf{f}_{i+(1/2)}^n - \mathbf{f}_{i-(1/2)}^n\right) \tag{20}$$

Second step:

$$\overline{\overline{U}}_i^{n+1} = \overline{U}_i^{n+1} + \frac{\Delta t}{2} S\left(\overline{U}_i^{n+1}\right) \tag{21}$$

Last step:

$$U_i^{n+1} = \overline{U}_i^{n+1} + \Delta t S\left(\overline{\overline{U}}_i^{n+1}\right) \tag{22}$$

The time step should satisfy the CFL convergence condition [1,11], i.e.,

$$Cr = \frac{a\Delta t}{\Delta x} \leq 1 \tag{23}$$

$$N = \frac{L}{\Delta x} \tag{24}$$

where Cr is the Courant number; N is the number of pipe grids; and L is the pipe length.

For the water hammer problem, the Courant number actually refers to the relative relationship between the time step Δt and the space step Δx [1,12]. When Cr is greater than 1, the calculation result is unstable; when Cr is less than 1, and the closer to 0, the more serious the numerical dissipation is, and the accuracy of the calculation result is worse. Therefore, the range of the Courant number is Cr less than or equal to 1, and preferably equal to 1 or close to 1.

3.3. Virtual-Boundary Strategy

Boundary conditions including the interior device boundary in the hydraulic system of the hydropower station contain the upstream head-constant reservoir, air chamber, turbine, and the downstream reservoir. As discussed above, in the second-order Godunov scheme, the head and flow rate of the ith control volume at time $t + \Delta t$ are calculated by combining the parameters of the upstream two (($i-2$)th, ($i-1$)th) and the downstream two (($i+1$)th, ($i+2$)th) control volumes at time t. Therefore, numerically, boundary conditions are expected to provide numerical fluxes $\mathbf{f}_{1/2}$, $\mathbf{f}_{N+1/2}$, which are required in order to update the extreme cells I_1 and I_N to the next time level $n + 1$.

In this paper, as shown in Figure 3, virtual control volumes I_{-1} and I_0 adjacent to I_1 and virtual control volumes I_{N+1} and I_{N+2} adjacent to I_N were proposed to realize second-order Godunov scheme at the upstream and downstream control volumes of the computational domain, respectively. The corresponding fluxes $\mathbf{f}_{1/2}$ and $\mathbf{f}_{N+1/2}$ were computed in the same method as the interior control volumes.

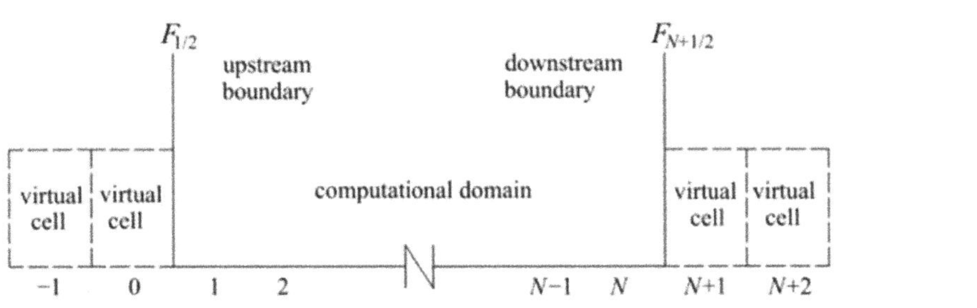

Figure 3. Grid diagram introducing virtual cells at the boundaries of reservoir and hydropower unit.

The head and flow rate in the virtual control volumes were assumed to be identical with those at boundaries, namely

$$\mathbf{U}_{-1}^{n+1} = \mathbf{U}_0^{n+1} = \mathbf{U}_{1/2} = \begin{pmatrix} H_{1/2} \\ V_{1/2} \end{pmatrix} \tag{25}$$

$$\mathbf{U}_{N+1}^{n+1} = \mathbf{U}_{N+2}^{n+1} = \mathbf{U}_{N+1/2} = \begin{pmatrix} H_{N+1/2} \\ V_{N+1/2} \end{pmatrix} \tag{26}$$

In the Godunov scheme, the Rankine–Hugoniot condition across each wave of speed $\bar{\lambda}_i$ gives the following relations,

$$\frac{a}{g}(V_{i+1/2} - V_R) - (H_{i+1/2} - H_R) = 0 \tag{27}$$

$$\frac{a}{g}(V_{i+1/2} - V_L) + (H_{i+1/2} - H_L) = 0 \tag{28}$$

At the upstream reservoir boundary, from the Riemann invariance equation (Equation (27)), it follows that

$$H_{1/2} - \frac{a}{g}V_{1/2} = H_1^n - \frac{a}{g}V_1^n \tag{29}$$

At the downstream reservoir boundary, from the Riemann invariance equation (Equation (28)), it follows that:

$$H_N^n + \frac{a}{g}V_N^n = H_{N+1/2} + \frac{a}{g}V_{N+1/2} \tag{30}$$

where V_1^n and H_1^n are the velocity and pressure head of the first control volume adjacent to the upstream reservoir; V_N^n and H_N^n are the velocity and pressure head of the last control volume adjacent to the downstream reservoir; and $H_{1/2}$ and $H_{N+1/2}$ are the head pressures of the upstream and downstream reservoirs, respectively.

For the upstream and downstream boundaries of the hydraulic turbine, from the turbine control equations, only the physical variable values of the virtual control volumes at the worm housing and at the tail pipe are required to derive the physical variable values at the turbine. Therefore, combining the Riemann invariance equations (Equations (27) and (28)), it is obtained

$$C_{P1} = H_N^n + \frac{a}{g}V_N^n \tag{31}$$

$$B_{P1} = \frac{a}{gA_1} \tag{32}$$

$$C_{M2} = H_1^n + \frac{a}{g}V_1^n \tag{33}$$

$$B_{M2} = \frac{a}{gA_2} \tag{34}$$

where V_N^n and H_N^n are the flow rate and head of the last control volume of the upstream pipe at the snail shell; and V_1^n and H_1^n are the flow rate and head of the first control volume of the downstream pipe at the right end of the tail pipe, respectively. The obtained Equations (31) to (34) are brought into Equation (11) to solve the head balance equation under virtual boundary conditions.

Similarly, for the upstream and downstream boundaries of air chamber, from the control equations of the air chamber, combining the Riemann invariance equations (Equations (27) and (28)), Equations (31)–(34) can be obtained and brought into Equation (6) to solve the head balance equation under virtual boundary conditions.

4. Numerical Solution by Using Method of Characteristics

The momentum and continuity Equations (1) and (2) are transformed into four ordinary differential equations by the MOC [1].

$$C^+ : \begin{cases} \frac{g}{a}\frac{dH}{dt} + \frac{dV}{dt} + \frac{fV|V|}{2D} = 0 \\ \frac{dx}{dt} = +a \end{cases} \tag{35}$$

$$C^- : \begin{cases} -\frac{g}{a}\frac{dH}{dt} + \frac{dV}{dt} + \frac{fV|V|}{2D} = 0 \\ \frac{dx}{dt} = -a \end{cases} \tag{36}$$

where f is the Darcy–Weisbach friction factor; D is pipe diameter.

As shown in Figure 4, integration of C^+ along characteristic lines from interior (fixed grid) point A to point P, and integration of C^- along characteristic lines from interior (fixed grid) point B to point P, can be written as

$$C^+ : H_i^{n+1} = C_P - B_P Q_i^{n+1} \tag{37}$$

$$C^- : H_i^{n+1} = C_M + B_M Q_i^{n+1} \tag{38}$$

in which Q is the flow rate; and the coefficients C_P, B_P, C_M, and B_M are known constants when the equations are applied.

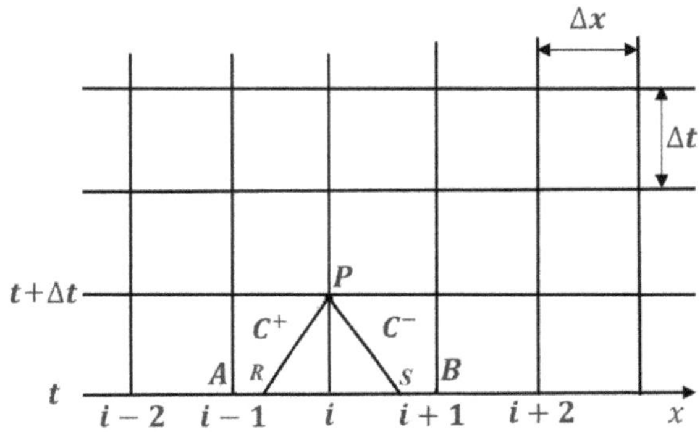

Figure 4. Definition of the sketch in the x–t plane with a fixed MOC grid for the water hammer problem.

When the Courant number $Cr \leq 1$ ($Cr = \frac{a\Delta t}{\Delta x}$), the space–line interpolation fixed-grid MOC scheme can give the values of the coefficients C_P, B_P, C_M, and B_M along the C^+ and C^- characteristic lines as follows:

$$C_P = H_{PR} + B \cdot Q_{PR} \tag{39}$$

$$B_P = B + R \cdot Cr \cdot |Q_{PR}| \tag{40}$$

$$C_M = H_{PS} - B \cdot Q_{PS} \tag{41}$$

$$B_M = B + R \cdot Cr \cdot |Q_{PS}| \tag{42}$$

in which, B is a function of the physical properties of the fluid and the pipeline, often called the pipeline characteristic impedance, and $B = a/gA$, A is the cross-section area; R is the pipeline resistance coefficient $R = f\Delta x/(2gDA^2)$; as shown in Figure 4, Q_{PR} and H_{PR} are the flow rate and pressure head at R node; Q_{PS} and H_{PS} are the flow rate and pressure head at S node; their values can be calculated by interpolation,

$$Q_{PR} = Q_i^n - Cr \cdot (Q_i^n - Q_{i-1}^n) \tag{43}$$

$$Q_{PS} = Q_i^n - Cr \cdot (Q_i^n - Q_{i+1}^n) \tag{44}$$

$$H_{PR} = H_i^n - Cr \cdot (H_i^n - H_{i-1}^n) \tag{45}$$

$$H_{PS} = H_i^n - Cr \cdot (H_i^n - H_{i+1}^n) \tag{46}$$

Combining Equations (39) and (40), H_i^{n+1} and Q_i^{n+1} at the interior node can be obtained. Similarly, the pressure head and flow rate of the boundary nodes adjacent to the hydropower unit can be calculated by combining Equation (39), the control equations of the hydraulic turbine, and Equation (40).

5. Results and Discussion

The main purposes of this section are (1) to investigate the accuracy, stability, and efficiency of second-order FVM GTS and MOC in a simple reservoir–pipe–valve system; (2) to validate the proposed second-order FVM model by comparing the calculated and measured data of load rejection in a hydropower plant with a complicated pipe system; (3) to explore the possible computation error caused by the MOC scheme in a complex pipe system of the hydropower plant; and (4) to study the effect of air chamber parameters on the error of MOC scheme simulating the hydraulic events in the hydropower plant.

5.1. Water Hammer Problem in a Simple Reservoir–Pipe–Valve System

The classical "reservoir–pipe–valve" system is used to verify the accuracy of the proposed. The upstream is a reservoir, and the downstream is a valve connected by a single pipe. The pipe is 800 m long, which is divided into 16 control volumes. The water hammer wave velocity is 1000 m/s, and the upstream reservoir head is 20 m. The initial velocity of the pipe is 0.15 m/s. The water hammer problem is caused by the instantaneous valve closure. It is assumed that the pipe wall is smooth, which means any dissipation is caused by the numerical calculation.

Figures 5 and 6 show the water hammer solutions for the simple system using MOC and second-order FVM, respectively, to investigate the effect of different Courant numbers *Cr* (1.0, 0.7, 0.5, 0.3, 0.1) on the calculation results of the two solution schemes. The accuracy and efficiency of FVM and MOC water hammer calculations were analyzed.

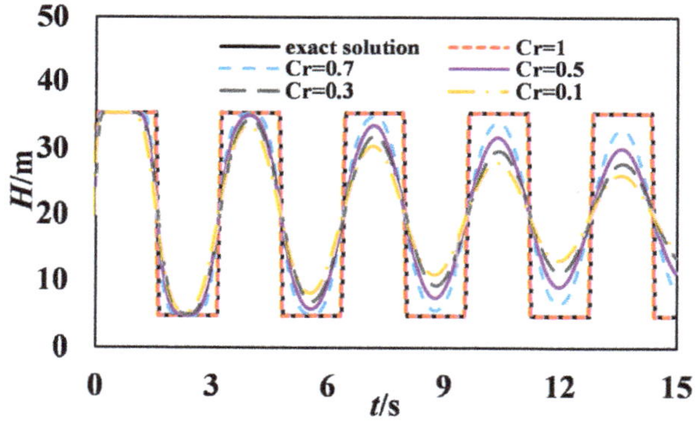

Figure 5. Pressure head calculated by the MOC scheme with different Courant numbers.

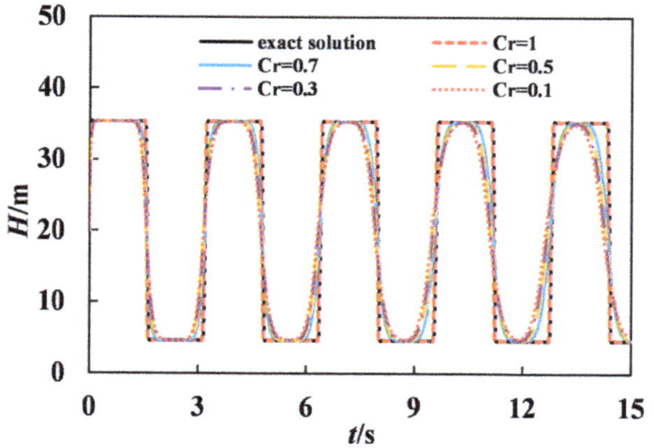

Figure 6. Pressure head calculated by second-order FVM with different Courant numbers.

The results in Figures 5 and 6 indicate that when $Cr = 1.0$, the results of both FVM and MOC calculations were identical with the exact solution (i.e., the analytical solution is obtained by the analytical method when the Courant number was equal to 1). When $Cr < 1.0$, both computational results had numerical dissipation. For the same Cr, the numerical dissipation of MOC was more severe, e.g., for $Cr = 0.1$, the initial energy (peak pressure) of MOC was dissipated by 26% in 15 s, while the FVM in the second-order Godunov scheme was only dissipated by 1.06%.

Figure 7 shows that when $Cr < 1.0$, the second-order FVM was more stable and less dissipative than the MOC scheme for the same number of control volumes ($N_S = 32$). At $Cr = 0.3$, in order to reach the same numerical accuracy, MOC needed $N_S = 256$, while only $N_S = 32$ was used in the second-order FVM scheme. Table 1 displays that the for the same computation accuracy, the computation time in the MOC scheme (0.19 s with $N_S = 256$) was about 5 times that in the second-order FVM scheme (0.037 s, $N_S = 32$). Therefore, when $Cr < 1.0$, the second-order FVM scheme is more efficient than the MOC scheme for the same computation accuracy.

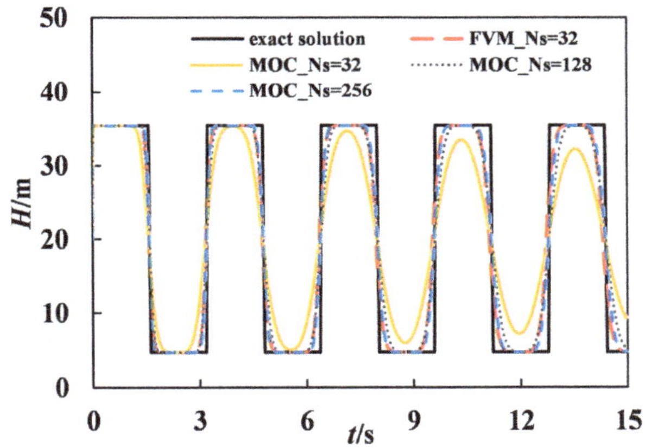

Figure 7. Comparison of pressure heads calculated by second-order FVM and MOC with different grid numbers at $Cr = 0.3$.

Table 1. Computation time of MOC and second-order FVM with different numbers of grids.

Number of Grids	MOC Calculation Time/s	FVM Calculation Time/s
32	0.012	0.037
128	0.069	0.555
256	0.19	1.849

Overall, for water hammers in a simple pipe, the second-order FVM model is accurate, efficient, and stable even for Courant numbers less than one. For the given Courant number and the same accuracy, the proposed model is far more efficient than the MOC model.

5.2. Hydraulic Transients in Actual Hydropower Plant

5.2.1. Project Overview

One real hydropower station has two turbine units through branch pipes, one air chamber, and pressurized pipes between the upstream and downstream reservoirs. The layout of the water transmission system is shown in Figure 8. The pipe parameters are shown in Table 2, while the parameters of the turbine units are shown in Table 3.

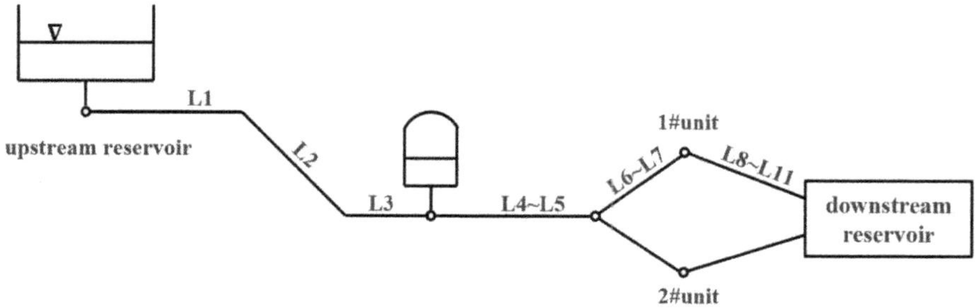

Figure 8. Schematic diagram of a hydropower plant layout with an air cushion chamber.

Table 2. Parameters of the water pipe system in the numerical simulations.

Pipe Number	Pipe Length/m	Wave Speed/(m·s^{-1})	Roughness
L1	15.39	976.4	0.014
L2	169.26	976.4	0.014
L3	20.77	976.4	0.014
L4	56.4	976.4	0.015
L5	26.6	976.4	0.014
L6	100.33	1202.3	0.013
L7	5.4	1210.8	0.013
L8	14	1045.1	0.014
L9	70.94	1045.1	0.014
L10	25.52	1152.75	0.014
L11	13.6	1152.75	0.014

Table 3. Unit parameters of the water turbine in the numerical simulations.

Unit Parameters	Numerical Value
Single machine capacity (MW)	150
Rated head (m)	105.8
Rated flow rate (m^3·s^{-1})	148.8
Rated speed (r·min^{-1})	200
Power Rating (kW)	139,000
Rotational inertia (t·m^2)	10,920

In this section, two field experiment cases on load rejection are introduced to investigate the accuracy of the numerical models.

Field Experiment Case A of Load Rejection: The water levels of upstream and downstream reservoirs are 412.4 m and 290.97 m, respectively. The guide vane closing law adopts "two stages": initial guide vane opening 73.8%, first closing time 3.62 s, second closing time 32.53 s (closing to 10% of no load), inflection point guide vane opening 60%, guide vane inactivity time 0.27, and total closing time 43.04 s. The turbine operating parameters are based on the rated parameters of the unit.

Field Experiment Case B of Load Rejection: The water levels of upstream and downstream reservoirs are 406.08 m and 290.6 m, respectively. The law of guide vane closing adopts "two-stage": initial guide vane opening 74.3%, the first stage closing time is 5.64 s, the second stage closing time is 30.53 s, and the inflection point guide vane opening is 61.13%. The total closing time is 36.17 s. The rated parameters of the turbine are used for the operating parameters.

5.2.2. Comparison to Field Experimental Data

When MOC is used to model the complicated pipe system, there are often two treatment methods, including (1) MOC (Scheme 1) being used to adjust the wave speed so that $Cr = 1$, and (2) MOC (Scheme 2) being used to keep the wave speed invariant and increase the number of pipe grids so that Cr is as close to 1 as possible. According to the above two schemes, the adjusted wave speed a, number of grids N, and Courant number Cr are shown in Table 4.

Table 4. Wave speed, grid number, and Courant number in pipe sections in the MOC simulations.

Pipe Number	MOC (Scheme 1)			MOC (Scheme 2)		
	$a/(\mathrm{m\cdot s^{-1}})$	N	Cr	$a/(\mathrm{m\cdot s^{-1}})$	N	Cr
L1	961.875	4	1.000	976.400	31	0.983
L2	984.070	43	1.000	976.400	346	0.998
L3	1038.500	5	1.000	976.400	42	0.987
L4	1007.143	14	1.000	976.400	115	0.995
L5	950.000	7	1.000	976.400	54	0.991
L6	1194.405	21	1.000	1202.300	166	0.995
L7	1350.000	1	1.000	1210.800	8	0.897
L8	1166.667	3	1.000	1045.100	26	0.970
L9	1043.235	17	1.000	1045.100	133	0.980
L10	1063.333	6	1.000	1152.750	44	0.994
L11	1133.333	3	1.000	1152.750	23	0.975

As discussed in Section 5.1, it is clear that the second-order FVM still maintains high computational accuracy even when Cr is less than 1. Therefore, for complex pipe components and devices of the hydropower plant, the characteristics of each pipe section (pipe length, wave speed, etc.) keep invariance in the second-order FVM simulation, and only the Courant number and the number of pipe section grids need to be adjusted. The number of pipe section grids is determined as follows.

In the calculation of the second-order FVM, to ensure the stability of the calculation, the pipe grid must satisfy the Courant condition, taking the ith pipe as an example, i.e.,

$$C_{r_i} = \frac{a_i \Delta t}{\Delta x_i} \leq 1 \tag{47}$$

where in order to ensure the synchronization of the calculation at all pipe sections, the calculation time step Δt for each pipe section is the same; Δx_i is the grid length of the ith pipe section, m; a_i is the wave speed of the ith pipe section, $\mathrm{m\cdot s^{-1}}$; Cr_i is the Courant number of the ith pipe section.

In the FVM calculation, the grid number N_i is calculated by the following equation.

$$N_i = \frac{L_i}{\Delta x_i} \tag{48}$$

where N_i is the grid number of the ith pipe; L_i is the length of the ith pipe, m.

Substitute Equation (36) into Equation (35) to obtain the following equation

$$N_i = C_{r_i} \frac{L_i}{a_i \Delta t} \tag{49}$$

From Equation (37), it can be seen that in order to ensure that N_i is an integer, it is necessary to adjust the Courant number Cr of the ith pipe, and the principle of adjustment is as follows: the range of the Courant number is less than or equal to 1, and preferably equal to 1 or close to 1. Using the abovementioned method, the wave speed a, grid number N, and Cr of each segment in the second-order FVM calculation can be obtained, as shown in Table 5.

Table 5. Wave speed, grid number, and Courant number in pipe sections in the FVM simulations.

Pipe Number	a/(m·s^{-1})	FVM N	Cr
L1	976.400	3	0.761
L2	976.400	43	0.992
L3	976.400	5	0.940
L4	976.400	14	0.969
L5	976.400	6	0.881
L6	1202.300	20	0.959
L7	1210.800	1	0.897
L8	1045.100	3	0.896
L9	1045.100	16	0.943
L10	1152.750	5	0.903
L11	1152.750	2	0.678

The maximum rotational speed during load rejection is an important index of hydraulic transient control. For the two field experiment cases on load rejection, the simulation results of MOC (Scheme 1), MOC (Scheme 2), and second-order FVM are given to compare the experimental rotational speed in Table 6 and Figures 9a and 10a, and the corresponding transient pressures at the worm gear are displayed in Figures 9b and 10b.

Table 6. Maximum rotational speed during 100% load rejection in two field experiment cases and the calculation results of MOC and FVM.

Experiment Case	Experimental Rotational Speed (r·min^{-1})	FVM (r·min^{-1})	MOC (Scheme 1) (r·min^{-1})	MOC (Scheme 2) (r·min^{-1})
A	283.84	282.416	281.455	280.946
B	279.2	279.076	278.081	277.672

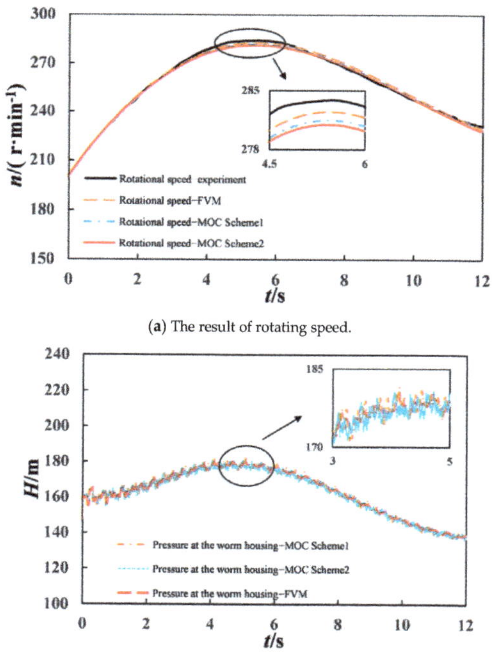

(**a**) The result of rotating speed.

(**b**) The result of pressure head at the volute.

Figure 9. Comparison of the calculation results (FVM and MOC) and the measured data in Field Experiment Case A of Load Rejection.

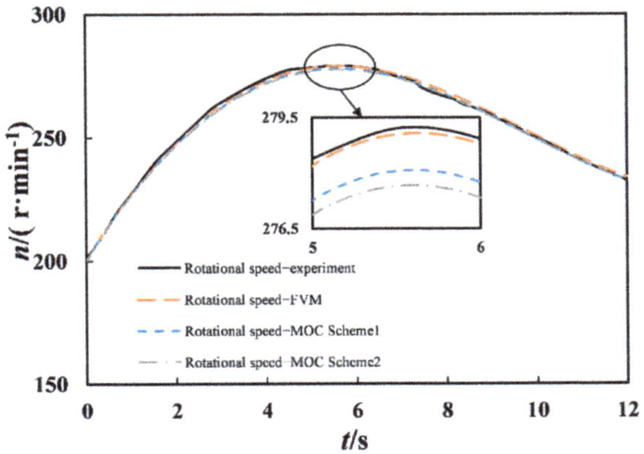

(**a**) The result of rotating speed.

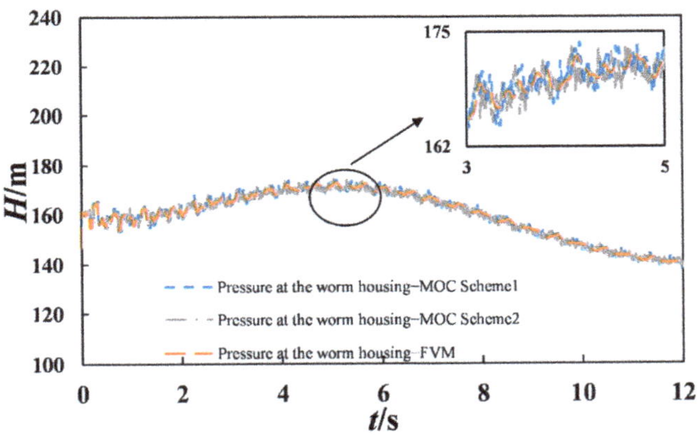

(**b**) The result of pressure head at the volute.

Figure 10. Comparison of the calculation results (FVM and MOC) and the measured data in Field Experiment Case B of Load Rejection.

Figures 9a and 10a show that the MOC (Scheme 1), MOC (Scheme 2), and FVM simulation results of unit rotation speed during 100% load rejection basically matched with the experimental results. However, the second-order FVM model better reproduced the experimental data, and the simulation results were more accurate than those of MOC. The reason for the larger calculation error caused by MOC is that MOC adjusts the water hammer wave speed or reduces the Courant number. Compared with MOC (Scheme 2), the simulation results of MOC (Scheme 1) were slightly better. The reason for this result is that the Courant number Cr in MOC (Scheme 2) in each pipe section was less than one, which led to more serious dissipation of the MOC (Scheme 2) than that of the MOC (Scheme 1) with slight wave speed adjustment. However, FVM does not need to adjust the wave velocity of the pipe, and only needs to reduce the Cr condition appropriately. Compared with MOC, FVM not only simplified the simulation process, but also had better calculation accuracy.

As shown in Figures 9b and 10b, for the simulation of the pressure at the worm gear, the FVM simulation results were more stable with less fluctuation than those of both MOC (Scheme 1) and MOC (Scheme 2).

5.2.3. Effect of Air Chamber Parameters on the Error of MOC Scheme

The abovementioned results demonstrated that the second-order FVM method proposed in this paper could perfectly reproduce the exact solution and the field experimental data in the simple pipe system and the real complex pipe system. In this section, the second-order Godunov FVM simulation results are taken as the benchmark to study the effect of air chamber parameters on the error of MOC (Scheme 1).

A. The effect of static water depth in design condition

Here, the air chamber control constant method CT_0 is used, keeping $P_0 \cdot V_0$ constant. When the cross-section area of air chamber remains unchanged, the $P_0 \cdot l_0$ value can be treated constant, in which l_0 is the air length of air chamber. So, the static water depth L_{s0} under design condition can be derived from the total height of the air chamber and air length. In this paper, five static water depths L_{s0} were selected to study their effects on 100% load rejection hydraulic transients of the power station. The MOC and FVM simulation results are shown in Table 7 and Figure 11.

Table 7. Comparison of the maximum rotational speeds calculated by FVM and MOC with differently designed air chamber water depths.

Design Water Depth(m)	Maximum Rotational Speed (FVM) ($r \cdot min^{-1}$)	Maximum Rotational Speed (MOC) ($r \cdot min^{-1}$)
4.8	278.068	277.068
5.4	278.55	277.555
6	279.076	278.081
6.6	279.626	278.629
7.2	280.154	279.148

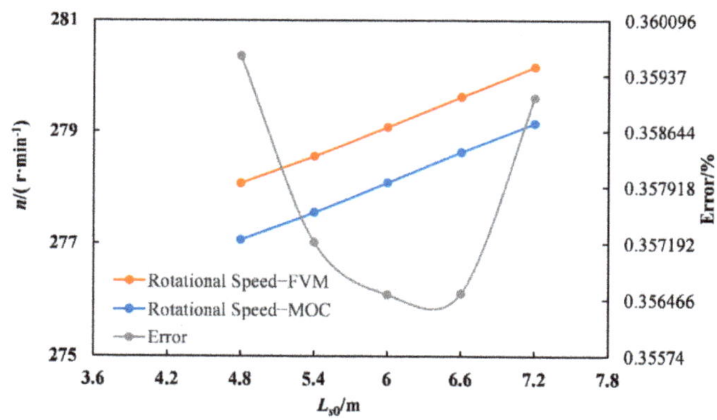

Figure 11. The maximum rotational speeds calculated by FVM and MOC with different design air-chamber water depths and the MOC computation error.

As shown in Table 7 and Figure 11, with the increase of L_{s0}, the maximum rotational speed gradually increased. The error of MOC calculation had a trend of decreasing and then increasing, with a slight change between 0.35% and 0.36%.

B. The effect of air cushion height

Under the given values of the design air pressure and design water depth, as the roof elevation of the air chamber increased, the air cushion height (air length) l_0 increased. Five air cushion height (air length) l_0 were selected to study their effects on 100% load rejection hydraulic transients of the power station. The MOC and FVM simulation results are shown in Table 8 and Figure 12.

Table 8. Comparison of the maximum rotational speeds calculated by FVM and MOC with different air cushion heights.

Air Cushion Height (m)	Maximum Rotational Speed (FVM) ($r \cdot min^{-1}$)	Maximum Rotational Speed (MOC) ($r \cdot min^{-1}$)
4	279.076	278.081
5	278.202	277.203
6	277.459	276.451
7	279.626	275.836
8	276.363	275.336

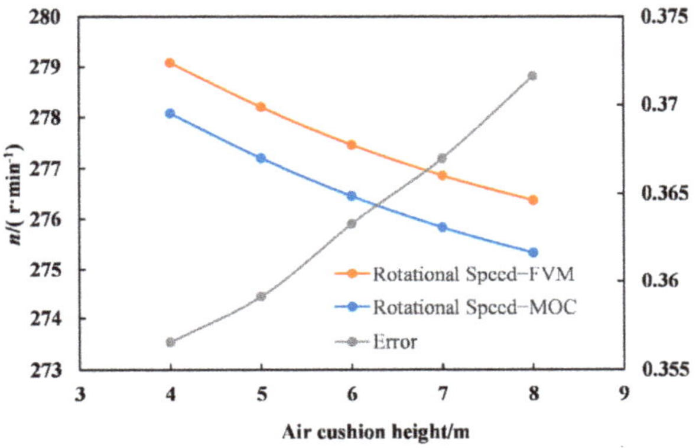

Figure 12. The maximum rotational speeds calculated by FVM and MOC with different air cushion heights and the MOC computation error.

As shown in Table 8 and Figure 12, with the increase of air cushion height (air length) l_0, the maximum rotational speed gradually decreased. The error of MOC calculation had a trend of decreasing, with change between 0.355% and 0.375%. It indicates that the simulation effect of MOC became worse with the increase of air cushion height. For high head hydropower plants, when the air cushion height is large, it is advisable to use FVM for simulation in order to ensure the calculation accuracy.

C. Effect of polytropic exponent k

The thermodynamic process of the closed air chamber was between isothermal and isentropic, and polytropic exponent k ranged from 1.0 to 1.4, which is the range commonly recognized and adopted at present [1]. Five polytropic exponents k were selected to study their effects on 100% load rejection hydraulic transients of the power station. The MOC and FVM simulation results are shown in Table 9 and Figure 13.

Table 9. Comparison of the maximum rotational speeds calculated by FVM and MOC with different polytropic exponents.

Polytropic Exponent	Maximum Rotational Speed (FVM) ($r \cdot min^{-1}$)	Maximum Rotational Speed (MOC) ($r \cdot min^{-1}$)
1	279.077	278.082
1.1	279.077	278.081
1.2	279.076	278.084
1.3	279.076	278.082
1.4	276.076	278.081

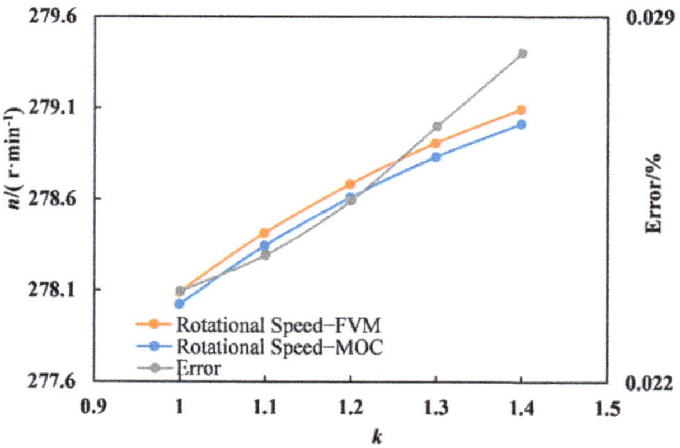

Figure 13. The maximum rotational speeds calculated by FVM and MOC with different polytropic exponents and the MOC computation error.

As shown in Table 9 and Figure 13, with the increase of polytropic exponent k, the maximum rotational speed gradually increased. The error of MOC calculation had a trend of increasing, with slight change between 0.022% and 0.029%.

6. Conclusions

In this paper, the second-order FVM Godunov scheme model was developed to simulate hydraulic transients and load rejection in a hydropower plant with an air chamber. The virtual boundary strategy was proposed to simply and effectively handle the complicated boundary problems. The results of the proposed model were compared with MOC results, the exact solution, and the measured data. The main conclusions are as follows.

(1) The second-order FVM Godunov scheme model can more accurately, stably, and efficiently simulate the water hammer problem in pipe systems. When the Courant number was Cr = 1, both calculated results of FVM and MOC were consistent with the exact solution. When the Courant number was $Cr < 1$, both computational results had numerical dissipation. As the Courant number gradually decreased, the second-order FVM simulation results were more stable. For the given Courant number, the numerical dissipation of MOC was more serious. The second-order FVM is more efficient for the same accuracy.

(2) For the load-rejection process of hydropower units containing an air chamber, the results calculated by the proposed FVM model were basically consistent with the measured rotational speed variation, which verifies that the second-order FVM model can be accurate for the simulation analysis of load-rejection process of hydropower units containing complex pipe systems.

(3) For complex pipe systems, the second-order FVM model better reproduced the experimental data, and the simulation results were more accurate than those of MOC. The reason for the larger calculation error caused by MOC is that MOC adjusts the water hammer wave speed or reduces the Courant number. The second-order FVM does not need to adjust the wave velocity of the pipe, and only needs to reduce the Cr condition appropriately. Compared with MOC, FVM not only simplifies the simulation process, but also has better calculation accuracy.

(4) The error of MOC calculation is associated with the air chamber parameters. For the current case of hydropower plant, with the increase of static water depth in the design condition, the error of MOC calculation had a trend of decreasing and then increasing, with a slight change between 0.35% and 0.36%. With the increase of air cushion height (air length), the error of MOC calculation had a trend of decreasing, with a change between 0.355% and 0.375%. With the increase of polytropic exponent k, the error of MOC calculation had a trend of increasing, with a slight change between 0.022% and 0.029%.

Overall, the second-order FVM model was robust in simulating the water hammer problems in a simple or complex pipe system. Considering the higher accuracy, stability, and efficiency, the high-order FVM is feasible and suggested for water hammer simulation in real hydraulic systems with more complicated pipe components and devices.

Author Contributions: Writing—original draft preparation, J.L., G.W. and J.W.; resources, writing, review and editing, L.Z. All authors have read and agreed to the published version of the manuscript.

Funding: This research was funded by National Natural Science Foundation of China, grant numbers 51839008 and 51679066.

Data Availability Statement: Some or all data, models, or code that support the findings of this study are available from the corresponding author upon reasonable request.

Conflicts of Interest: The authors declare no conflict of interest.

References

1. Wylie, E.B.; Streeter, V.L.; Suo, L. *Fluid Transients in Systems*; Prentice Hall: Englewood Cliffs, NJ, USA, 1993.
2. Zhou, L.; Li, Y.; Zhao, Y.; Ou, C.; Zhao, Y. An accurate and efficient scheme involving unsteady friction for transient pipe flow. *J. Hydroinform.* **2021**, *23*, 879–896.
3. Zhou, L.; Li, Y.; Karney, B.; Cheng, Y.; Liu, D. Godunov-type solutions for transient pipe flow implicitly incorporating Brunone unsteady friction. *J. Hydraul. Eng.* **2021**, *147*, 04021021.
4. Guinot, V. Riemann solvers for water hammer simulations by Godunov method. *Int. J. Numer. Meth. Eng.* **2000**, *49*, 851–870. [CrossRef]
5. Zhao, M.; Ghidaoui, M.S. Godunov-type solutions for water hammer flows. *J. Hydraul. Eng.* **2004**, *130*, 341–348.
6. Zhou, L.; Wang, H.; Liu, D.Y.; Ma, J.J.; Wang, P.; Xia, L. A second-order finite volume method for pipe flow with water column separation. *J. Hydro-Environ.* **2017**, *17*, 47–55.
7. Zhou, L.; Wang, H.; Bergant, A.; Tijsseling, A.S.; Liu, D.; Guo, S. Godunov-type solutions with discrete gas cavity model for transient cavitating pipe flow. *J. Hydraul. Eng.* **2018**, *144*, 04018017. [CrossRef]
8. Elong, A.J.; Zhou, L.; Karney, B.; Fang, H.; Cao, Y.; Assam, S.L.Z. Flood Prediction with Two-Dimensional Shallow Water Equations: A Case Study of Tongo-Bassa Watershed in Cameroon. *Appl. Sci.* **2022**, *12*, 11622. [CrossRef]
9. Zhou, L.; Li, W.J.; Liu, D.Y.; Ou, C. Second-order Godunov mathematical model of water flow impinging trapped air mass in pipeline. *Int. J. Hydrogen Energy* **2021**, *39*, 95–99.
10. Xue, Z.; Zhou, L.; Karney, B.; Liu, D.; Wang, P. Primitive Form Godunov-Type Scheme for Two-Phase Homogeneous Water Hammer Flows. *J. Hydraul. Eng.* **2020**, *146*, 04020018. [CrossRef]
11. Yang, K.L *Hydraulic Transient and Regulation in Power Plant and Pumping Station*; China Water & Power Press: Beijing, China, 2000; pp. 116–121.
12. Toro, E.F. *Riemann Solvers and Numerical Methods for Fluid Dynamics*; Springer: Berlin/Heidelberg, Germany, 2009; pp. 115–125.

Disclaimer/Publisher's Note: The statements, opinions and data contained in all publications are solely those of the individual author(s) and contributor(s) and not of MDPI and/or the editor(s). MDPI and/or the editor(s) disclaim responsibility for any injury to people or property resulting from any ideas, methods, instructions or products referred to in the content.

water

MDPI

Article

Rapid Filling Analysis with an Entrapped Air Pocket in Water Pipelines Using a 3D CFD Model

Duban A. Paternina-Verona [1], Oscar E. Coronado-Hernández [1,*], Hector G. Espinoza-Román [2], Vicente S. Fuertes-Miquel [3] and Helena M. Ramos [4]

1 Facultad de Ingeniería, Universidad Tecnológica de Bolívar, Cartagena 131001, Colombia
2 Grupo INMEDIT S.A.S., Facultad de Ingeniería, Universidad de Cartagena, Cartagena 130001, Colombia
3 Department of Hydraulic and Environmental Engineering, Universitat Politècnica de València, 46022 Valencia, Spain
4 Department of Civil Engineering, Architecture and Georesources, CERIS, Instituto Superior Técnico, University of Lisbon, 1049-001 Lisbon, Portugal
* Correspondence: ocoronado@utb.edu.co

Abstract: A filling operation generates continuous changes over the shape of an air–water interface, which can be captured using a 3D CFD model. This research analyses the influence of different hydro-pneumatic tank pressures and air pocket sizes as initial conditions for studying rapid filling operations in a 7.6 m long PVC pipeline with an irregular profile, using the OpenFOAM software. The analysed scenarios were validated using experimental measurements, where the 3D CFD model was suitable for simulating them. In addition, a mesh sensitivity analysis was performed. Air pocket pressure patterns, water velocity oscillations, and the different shapes of the air–water interface were analysed.

Keywords: filling events; entrapped air pocket; thermodynamic behaviour; CFD; air-water interface

Citation: Paternina-Verona, D.A.; Coronado-Hernández, O.E.; Espinoza-Román, H.G.; Fuertes-Miquel, V.S.; Ramos, H.M. Rapid Filling Analysis with an Entrapped Air Pocket in Water Pipelines Using a 3D CFD Model. *Water* **2023**, *15*, 834. https://doi.org/10.3390/w15050834

Academic Editor: Yakun Guo

Received: 1 February 2023
Revised: 18 February 2023
Accepted: 19 February 2023
Published: 21 February 2023

Copyright: © 2023 by the authors. Licensee MDPI, Basel, Switzerland. This article is an open access article distributed under the terms and conditions of the Creative Commons Attribution (CC BY) license (https://creativecommons.org/licenses/by/4.0/).

1. Introduction

Filling events are frequently performed for water utilities companies in order to operate drinking water distribution systems to supply the required water demands [1–4]. To analyse the performing of filling operations is crucial for the understanding of the pressurised liquid phase (water) and the compression of the gas phase (entrapped air pocket). Conduits in storm drainage networks can also be pressurised when a return period occurrence is higher compared to the designed one [5–7]. The air–water interface exhibits a dynamic behaviour in pressurised installations while the water phase replaces part of the portion occupied by the entrapped air pocket. The phenomenon can generate pressure surges and temperature gradients [2]. Air valves are used as protection devices to avoid risky conditions with regard to over-pressures [8–12], where several studies have been developed using experimental measurements and analytical formulations [8–10,13]. Water filling processes have been analysing using elastic [14] and rigid [1,11,12,15] column models, which can provide information regarding the water filling velocity, air pocket pressure, air pocket temperature, air density, and position of the air–water interface.

In addition, filling operations have been modelled using computational fluid dynamics (CFD) in order to represent the hydraulic and thermodynamic evolution of the water and air phases, respectively [16,17]. A 3D CFD analysis can provide detailed information in each cell of an entire water installation compared to analytical models [16]. Besharat et al. [18] developed a 2D CFD model for studying the dynamic behaviour during the compression of an entrapped air pocket, which showed a good agreement in regards to experimental measurements. Martins et al. [19] studied the over-pressure peaks in rapid filling events, where the influence of an air pocket size was evaluated. Zhou et al. [4] studied different hydraulic and thermodynamic phenomena of an entrapped air pocket in a vertical pipeline during rapid filling events. Fang et al. [20] studied the motion of entrapped air pockets in a storm-water drainage system. Aguirre-Mendoza et al. [17,21] developed a 2D CFD model to analyse the effects of orifice sizes in the control of filling processes of water pipelines. Many researches have used the SST k-ω model to analyse the dynamic interaction between water and air [17,22–24]. CFD models for simulating a two-phase transient flow in pipelines have rarely been used to analyse their numerical accuracy, to verify the behaviour of the air–water interface, even to verify the pressure patterns of the entrapped air pocket. Air–water mixing accumulates multiple hydraulic–thermodynamic phenomena that have been studied to a limited extent in the literature. The aim of this research is to analyse the thermodynamic behaviour of an entrapped air pocket and the hydraulic transients during a rapid filling event with the upstream end closed in a water pipeline with an irregular profile using a 3D CFD model (OpenFOAM software v8). The experimental tests were performed at the hydraulic lab at the Instituto Superior Técnico (University of Lisbon, Lisbon, Portugal). A 7.6 m long PVC pipe with nominal diameter DN63 was used, considering several initial hydro-pneumatic tank pressures and air pocket sizes. Experimental measurements were used to validate the 3D CFD model.

2. Experimental Facility

The experimental facility corresponds to a water pipeline with a total length of 7.6 m and nominal diameter DN63 (internal diameter of 51.4 mm), which has two branch pipes (left and right). Figure 1 presents the experimental scheme. $L_{left,d}$ and $L_{right,d}$ correspond to inclined pipe for the left and the right branch pipes, respectively, with lengths of 1.5 m each. Horizontal branch pipes $L_{left,h}$ and $L_{right,h}$ have lengths of 2.05 and 1.85 m, respectively. The left branch is connected to an electro-pneumatic ball valve. A hydro-pneumatic tank is used to supply different initial absolute pressure. The ball valve located at downstream end at the right branch pipe was closed for all experimental tests. A pressure transducer was located at the highest point of the installation to measure the air pocket pressure.

An initial air pocket size was defined with a length L_{iap} located at the left branch pipe. The filling event starts with the opening of the electro-pneumatic ball valve. After that, the entrapped air pocket is compressed causing the air pocket pressure to increase its tendency. The water column located inside of the branch pipe remains at rest, acting as a boundary condition. A total of six (6) experimental measurements of pressure oscillations of the air pocket trapped in the branched pipe were performed. Their initial characteristics are shown in Table 1.

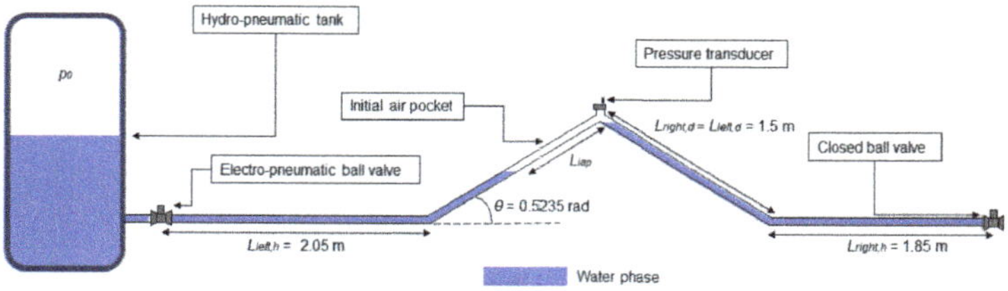

Figure 1. Experimental setup.

Table 1. Initial conditions of the experimental tests.

Test	p_0 (Pa)	L_{iap} (m)
1	121,325	0.46
2	176,325	0.46
3	151,325	0.46
4	121,325	0.96
5	176,325	0.96
6	151,325	0.96

3. Three-Dimensional Model

CFD models have been used for analysing the air–water interface during different hydraulic events [24–28]. Two-dimensional CFD models have been implemented for studying filling events [4,17,18,21]. In this research, a 3D CFD model was used to simulate a rapid filling process using the computational software OpenFOAM [29], developed by OpenFOAM Foundation. It allows to simulate multiphase compressible flows, considering an interface capture of fluid. A geometric domain in multiple blocks with small finite volumes should be defined in the software. The air–water interface is represented under user-defined conditions. Computational fundamentals are described as follows.

3.1. Governing Equations

The CFD model involves the solution of a set of mathematical equations involving different physical principles, such as mass conservation, momentum, transport, turbulence, and thermodynamic principles.

3.1.1. Partial Vof Equations

The volume of fluid equations are partially evaluated using a partial volume of fluid (PVoF) model, where the combination of fluids (air and water) is solved numerically using the phase fraction (α_w) for the calculation of density and dynamic viscosity (ρ and μ) of each cell of the geometrical domain [30,31]. The equations of the density and dynamic viscosity are

$$\rho = \alpha_w \rho_w + (1 - \alpha_w)\rho_a \tag{1}$$

$$\mu = \alpha_w \mu_w + (1 - \alpha_w)\mu_a \tag{2}$$

where ρ_a = air density, ρ_w = water density. μ_a and μ_w correspond to dynamic viscosity of air and water, respectively. Transport model associated to PVoF model uses a rate change of volume fraction, convective transport terms, and compressibility of the fluid combination. The equation of the PVoF transport model is

$$\frac{\partial \alpha_w}{\partial t} + \nabla \cdot (\alpha_w \vec{u}) + \nabla \cdot ((1 - \alpha_w)\alpha_w u_r) = 0 \tag{3}$$

where $\vec{u}$ = flow velocity vector in a cell, and u_r = velocity field.

3.1.2. Navier–Stokes Equations

The behaviour of multiphase flows can be analysed by the Navier–Stokes equations for compressible flows using the interface capture of the phase fraction of the PVoF model. The continuity formulation is

$$\frac{\partial \rho}{\partial t} + \nabla \cdot (\rho \vec{u}) = 0 \tag{4}$$

where ρ = density, and $\vec{u}$ = flow velocity vector. The momentum equation is represented by the following expression:

$$\frac{\partial (\rho \vec{u})}{\partial t} + \nabla \cdot (\rho \vec{u} \vec{u}) = -\nabla p + \nabla \cdot (\mu \nabla \vec{u}) + \rho \vec{g} - F_s \tag{5}$$

where ∇p = pressure gradient term, $\vec{g}$ = gravitational acceleration vector, and F_s = surface tension forces.

3.1.3. Thermodynamic Equations

The principle of conservation of energy is based on the sum of the rates of change in thermodynamic energy and mechanical energy during fluid motion [29]. Equation (6) shows the principle of the conservation of energy.

$$\frac{\partial (\rho C_p T)}{\partial t} + \nabla \cdot (\rho \vec{u} C_p T) = \nabla \cdot \vec{q} + S_T \tag{6}$$

where C_p = specific heat at constant pressure, T = temperature, $\vec{q}$ = heat flux, and S_T = thermal energy source. The heat flux is calculated using the Fourier conduction law, which depends on the product of the thermal diffusivity (a_{eff}) and the temperature gradient (∇T), as shown in Equation (7):

$$\vec{q} = -a_{\text{eff}} \nabla T \tag{7}$$

The thermodynamic behaviour of air density is evaluated through the equation of state, which relates air pressure (p_a), the universal gas constant (R), and air phase temperature (T_a). Equation (8) represents the thermodynamic ratio of the above-mentioned parameters of the air phase.

$$\rho_a = \frac{p_a}{RT_a} \tag{8}$$

3.1.4. SST k-ω Turbulence Model

The SST k-ω turbulence model was used to represent the turbulent phenomenon occurring in the analysed pipeline with entrapped air because aerodynamic flows are facing adverse pressure gradients [32]. This model was implemented for a suitable resolution of the viscous laminar sub-layer and the effects of vorticity in areas far from the walls [33]. The fundamental equations of the SST k-ω turbulence model are

$$\frac{D(\rho k)}{Dt} = S_k - \beta^* \rho k \omega + \nabla \cdot (\rho D_k \nabla k) - \frac{2}{3}\rho k(\nabla \cdot \vec{u}) + \rho G \tag{9}$$

$$\frac{D(\rho \omega)}{Dt} = \nabla \cdot (\rho D_\omega \nabla \omega) + \frac{\rho \gamma G}{\nu_t} - \frac{2}{3}\rho \gamma \omega (\nabla \cdot \vec{u}) - \rho \beta \omega^2 + S_\omega + \rho(1 - F_1) CD_{k\omega} \tag{10}$$

where k = turbulent kinetic energy, ω = specific turbulence dissipation rate. D_k and D_ω are diffusive terms for k and ω, respectively. G = generation of turbulence kinetic energy, ν_t = turbulent kinematic viscosity, F_1 = blending function, and $CD_{k\omega}$ = closure coefficient of k and ω. Coefficients such as S_k, S_ω, β, β^*, and γ depend on turbulence model. In addition, wall functions were used to calculate the logarithmic region between viscous sub-layer and far-wall zone.

3.2. Numerical Approach

The Pressure-Implicit with Splitting of Operators (PISO) algorithm was used for the numerical resolution of the governing equations, using the coupling system as a function of pressure and flow velocity in finite volumes. Pressure equation and momentum were corrected twice. On the other hand, non-orthogonal and outer correction were applied to solve the pressure term associated to transient flows. The numerical approach of the time and divergence schemes of the governing equations was performed using first-order and implicit Gaussian approximations to guarantee a stable numerical approximation. Second-order linear schemes were applied for the resolution of Laplacian terms and gradients.

3.3. Mesh Computational and Boundary Conditions

Figure 2 shows the spatial distribution of geometric domain associated to the pipeline of irregular profile for representing the filling processes. Geometric domain was distributed by unstructured cells, which are adequate for complex geometries [34]. The opening of the electro-pneumatic ball valve was simulated using a dynamic mesh function. The geometric domain was distributed with a total of 151,539 cells, where 150,847 cells are tetrahedral cells and 692 cells are polyhedral cells. The 3D CFD model presents a non-orthogonality of 23.0, maximum skewness of 1.96, maximum aspect ratio of 117.3. Mesh sensitivity analysis was performed through tests with structured and unstructured meshes, which were performed in the contribution of Besharat et al. [18] for an irregular pipeline of similar dimensions.

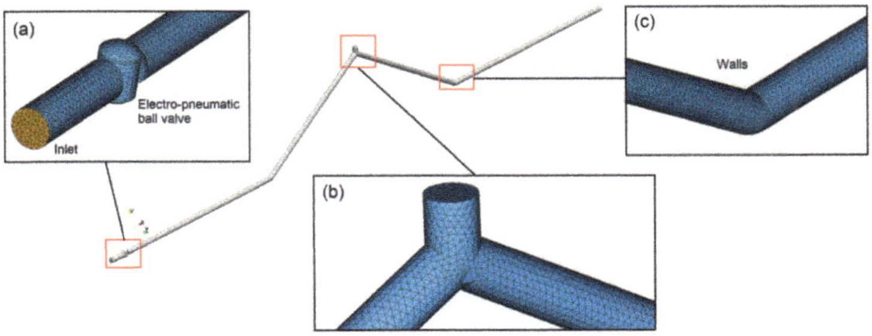

Figure 2. Geometric domain with detail in distribution of cells and boundary conditions: (**a**) inlet and electro-pneumatic ball valve, (**b**) upstream end, and (**c**) detail of walls in right pipe branch.

Different boundaries were defined in 3D CFD model, which are described as shown in Table 2.

Table 2. Description of the existing boundaries in geometrical domain.

Boundary	Characteristics
Inlet	Corresponds to the boundary where water inflow is generated during filling events.
Walls	This boundary corresponds to the walls of the existing pipeline and accessories.
VSI (Valve–Sliding Interface)	Corresponds to the sliding interface that ensures continuity of flow over the electro-pneumatic ball valve.

A no-slip boundary condition was defined in all boundaries with regard to walls. Initial pressure (inlet boundary) corresponds to absolute pressure (p_0), which was supplied by the hydro-pneumatic tank. The system is at rest with an initial air temperature of 293 K. The air pocket is found to atmospheric pressure (p = 101,325 Pa).

4. Results and Discussion

A 3D CFD model was developed to analyse the thermodynamic and hydraulic variables during rapid filling processes in water pipelines. A detailed analysis was performed using Test No. 1 and Test No. 2, which correspond to hydraulic scenarios with L_{iap} = 0.46 m and initial gauge pressures of 20 and 75 kPa, respectively. The CFD model was validated using the experimental measurements of the air pocket pressure gauge oscillations, which were conducted at the hydraulic lab at the Instituto Superior Técnico of the University of Lisbon, Portugal. The comparison between the numerical results and experimental measurements was performed. Figure 3a shows the behaviour of the air pocket gauge pressure pulses for Test No. 1, where a maximum value of 46 kPa was reached at t = 0.44 s. The gauge pressure patterns decreased to a value of 4 kPa (at t = 0.77 s). The 3D CFD model is capable of representing the air pocket gauge pressure oscillations with regard to the experimental measurements with a good accuracy. Figure 3b shows the air pocket gauge pressure oscillations for Test No. 2, where a maximum value of 214 kPa was reached, while the CFD model shows a maximum gauge pressure of 190 kPa at t = 0.37s and t = 0.38 s, respectively. Subsequently, the minimum pressure gauge of the first oscillation of the CFD model reaches a value of 22 kPa at t = 0.59 s. The gauge undergoes four oscillations, where the last one has peak maximum and minimum values of 156 kPa and 40 kPa at t = 1.78 s and t = 2.0 s, respectively.

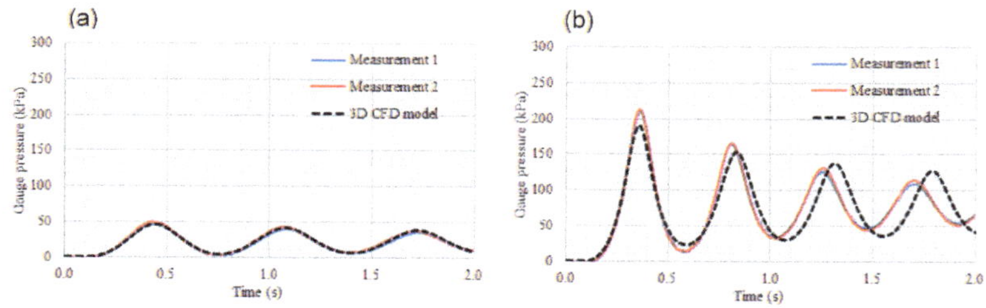

Figure 3. Analysis of air pocket pressure pulses: experimental measurements versus 3D CFD model. (**a**) Test No. 1, and (**b**) Test No. 2.

4.1. Air–Water Interaction

The dynamic behaviour of the air–water interaction was explored using a 3D CFD model. The model shows how the air–water interface is not totally perpendicular to the main direction of the pipe. A complex air–water interface is presented from 1.0 to 2.0 s, which cannot be predicted using 1D models. The air–water interface exhibits a compression process, reaching the peak of the air pocket gauge pressure. Figure 4 shows the behaviour of the air–water interface for Tests No. 1 and No. 2. In addition, the streamlines and contours data are presented for the gas phase (entrapped air pocket) in order to observe the variation in the air volume in the geometric domain of the pipeline with an irregular profile (see Figure 5).

The entrapped air pocket shows dynamic behaviour for Test No. 1, considering an inlet gauge pressure supplied by the hydro-pneumatic tank of 20 kPa (see Figure 4a), where the air phase is compressing, generating a vorticity on the vertices of the left and right branch pipes (L_{left} and L_{right}) at t = 0.25 s. The filling event generates an initial compression of the entrapped air pocket from the initial fraction V/V_p = 7.8% to 6.2% (see the first peak of the air pocket pressure in Figure 5) at t = 0.44 s. Here, V/V_p corresponds to the volume of fluid over the total volume of the analysed pipeline. After that, the entrapped air pocket expands because the effect of the accumulated compression energy generates a rupture of the air–water interface above the water column, as shown at t = 0.50 s (see Figure 4a), and at the same time, the streamlines of the air velocity show values varying between 0.4 and 1.2 m/s. The entrapped air pocket reached maximum values between 0.4 and 0.8 m/s over the air–water interface at t = 1.0 s. The water and air are mixed from t = 1.5 to 2.0 s (see Figure 4a). The volume of the entrapped air pocket oscillates between 6.2% and 7.8% for the analysed tests (Figure 5).

Figure 4b shows the air–water interface for the filling event for Test No. 2. There is an increasing trend in the air velocity, which is occurring for the compression of the gas phase. The entrapped air pocket begins to move with values greater than 1.2 m/s, as shown from t = 0.25 to 0.50 s. Small bubbles are formed because the transient event is rapidly occurring for the initial gauge pressure of 75 kPa. Test No. 2 shows that the air pocket is compressed, where an initial volume of 7.8% V/V_p reaches a value of 3.2% V/V_p at t = 0.37 s (see Figure 5). Afterwards, the air pocket volume gradually expands from 3.2% to 6.4% (at t = 0.6 s). The second oscillation of the air pocket volume shows peak values of 3.4% V/V_p and 6.3% V/V_p. During the last pulse, values of 3.6% V/V_p and 5.8% V/V_p are reached.

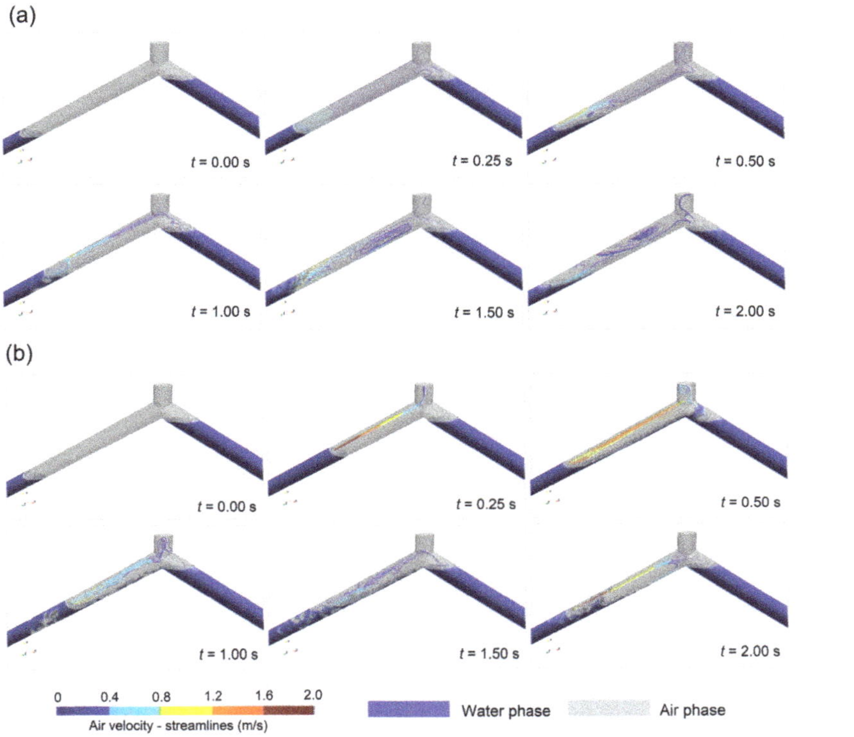

Figure 4. Dynamic behaviour of the entrapped air pocket during the analysed filling process. (**a**) Test No. 1, and (**b**) Test No. 2.

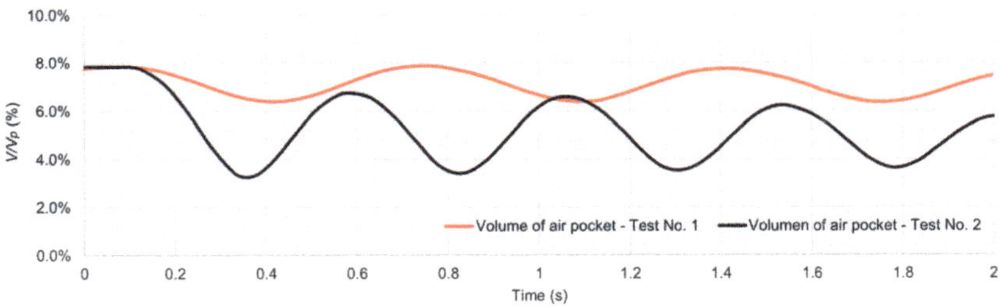

Figure 5. Analysis of air pocket volume.

4.2. Water Velocity

The analysed transient event produces permanent changes in the water velocity patterns. The 3D CFD model provides the numerical information on these variations. Figure 6 shows the water velocity oscillations for Tests No. 1 and No. 2 along the entire pipeline.

The water column is capable of generating a rapid compression of the entrapped air pocket, where a maximum water velocity of 0.56 m/s (at t = 0.20 s) is reached for Test No. 1 (see Figure 6). After that, a decreasing trend is detected, reaching a null value at t = 0.36 s. Negative water velocities are found in the left pipe branch. For the analysed time periods, there are three peaks and drops of the water flow pulses. The lowest value was −0.5 m/s at t = 0.5 s. Figure 7a presents the vector field of the water velocities for time values of 0.1 s,

0.2 s, 0.8 s, and 1.6 s for both tests. Negative water velocities are occurring at t = 0.4 and 0.6 s. The results of Test No. 2 are more more critical compared to Test No. 1, because a maximum value of 2.1 m/s (at t = 0.18 s) is attained, and a minimum value of −1.6 m/s is found at t = 0.37 s. The water velocity shows a similar trend compared to Test No. 1 (see Figure 7b).

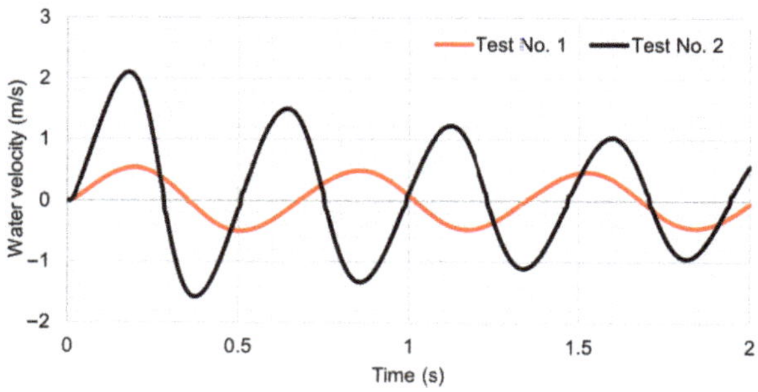

Figure 6. Water velocity patterns during filling events.

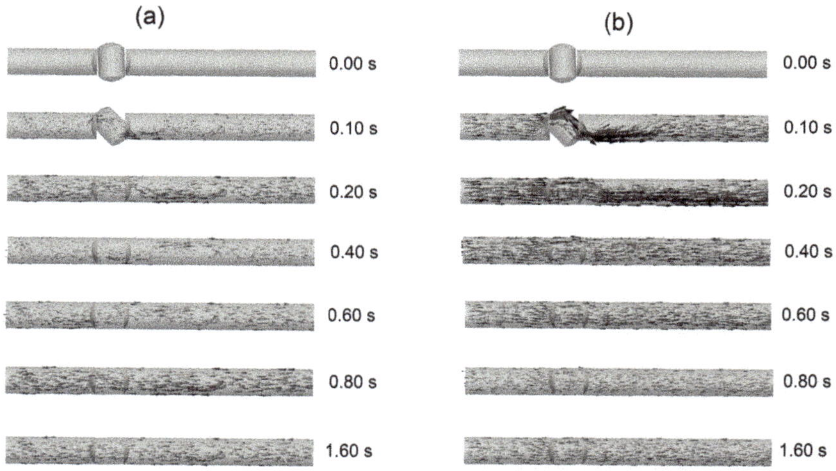

Figure 7. Vector field of water velocity patterns: (**a**) Test No. 1, and (**b**) Test No. 2.

4.3. Air Pocket Behaviour

The rapid changes produced by the filling event rapidly cause a compression of the entrapped air pocket. Figure 8 shows how the temperature changes are occurring in the entrapped air pocket, where polytropic behaviour is presented.

Figure 8a shows the temperature behaviour of the entrapped air pocket for Test No. 1. It shows a maximum value of 42 °C (at t = 0.40 s). After that, the temperature pattern decreases to a value of 32 °C at t = 0.5 s. When the air pocket is being compressed for the second oscillation, the temperature pattern rises up to 40 °C at t = 1.0 s, where a heterogeneous temperature gradient is shown. Figure 8b shows the thermodynamic evolution for Test No. 2. A maximum temperature of 120 °C was reached at t = 0.35 s according to the 3D CFD model. The temperature pulse tends to decrease to a value varying from 40 °C to 60 °C. Finally, the air pocket temperature reaches values lower than 30 °C after t = 1.0 s. The air temperature changes are strongly included by the initial gauge pressure supplied by the hydro-pneumatic tank. In all the tests, the temperature gradients are non-uniform over the air phase, due to the heat transfer between the fluids, showing that the temperature in the air–water interface zone is not higher than 30 °C.

The analysis of the permissible overheating in water pipelines should be conducted for design purposes using mathematical models that involve the analysis of the air–water interface, because thermoplastic materials (e.g., PVC) should not be exposed to values higher than 60 °C to avoid undesirable deformations.

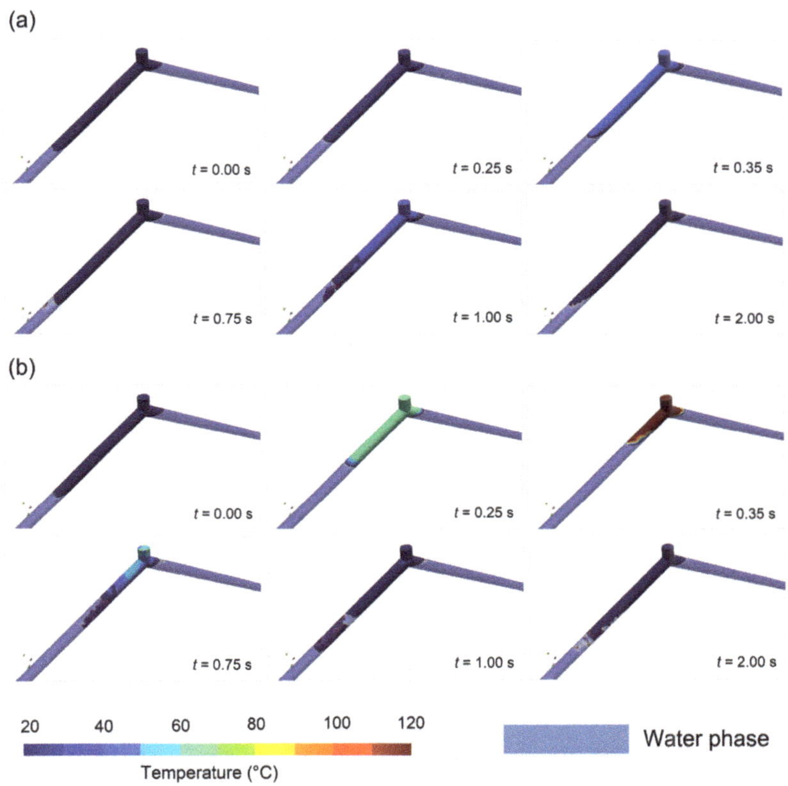

Figure 8. Temperature gradients in CFD models: (**a**) Test No. 1, and (**b**) Test No. 2.

5. Conclusions

The study of transient flows during different filling events through 3D CFD models allowed to analyse the different behaviours presented that are related to the gauge pressure oscillations, collapse of the air–water interface, water flow velocities, and thermodynamic phenomena in the air phase. This study highlights the good numerical accuracy presented by the 3D CFD model in predicting the gauge pressure oscillations of the air pocket measured in the experimental facility. The pressure damping of the entrapped air pocket leads to the generation of water flow velocity transitions by changing the displacement of the water flow at different time instants over the hydraulic pipeline and mixing level of the air and water, which becomes more intense when the inlet gauge pressure is higher. The main focus of this research was to obtain the relevant information related to the hydraulic phenomena associated with water and the thermodynamic phenomena associated with air, which help the user to identify the magnitude of these phenomena and relate them to transient events that may occur in hydraulic installations of different scales. From the hydraulic–thermodynamic phenomena analysed, the following statements can be defined:

- The entrapped air pocket exhibits constant volume changes over time. The water column compresses the air pocket to a limit state and then it tends to expand to release the compression energy that was accumulated during the filling event. This compression–expansion phenomenon of the air pocket is cyclic over time. The dynamic behaviour of the trapped air during the volume changes can be visualised in detail through streamlines and velocity contours, where such events manifest their volume reduction and expansion through vortices located at the air–water interface zones that can physically reach maximum velocities of 1.2 m/s (in Test No. 2).
- Transient flows can result in backflows towards the pumping source, which can lead to a loss of the hydraulic efficiency during filling events. During the first few seconds, the velocity transitions are abrupt, and then they gradually dissipate over time due to the damping pressure of the entrapped air pocket. These backflows are more critical in scenarios with higher inlet gauge pressures. In addition, the CFD model predicts in detail the vector field of the water flow by showing the changes in the water trajectory at different time instants.
- Temperature changes are inevitable during filling processes, considering that the compression of the trapped air causes a change in the thermodynamic conditions of the air phase. In such phenomena, the trapped air pocket undergoes adiabatic behaviour, which has been reproduced in previous research on filling processes. The 3D CFD model in the analysed tests shows a non-uniform temperature distribution, showing that away from the air–water interface the highest temperatures occur (up to 120 °C in Test No. 2), whereas in the air pocket near the air–water interface, temperatures between 20 °C and 30 °C occur.

The development of the three-dimensional model allowed for representing an integral behaviour of the hydraulic–thermodynamic phenomena from a physical approach. The application of an unstructured mesh with the number of cells defined in this research favoured the convergence of the numerical solution and an adequate simulation with a Courant number less than 1.0. Moreover, the unstructured mesh favoured the detailed visualisation of the contours of the different variables analysed. The application of 3D CFD models to simulate hydraulic and thermodynamic phenomena in pipes with entrapped air represents a high computational cost considering the rapidity of the hydraulic event; however, it guarantees a more detailed and complete level of information compared to other computational models.

It is important to evaluate the selection of the pipe class in large-scale water installations because the compression of entrapped air pockets can produce absolute pressure values higher than those values reached when considering a monophasic fluid (water). The current regulations around the world do not consider a complete model of a two-phase flow when designing these systems. In this sense, the utilisation of a 3D CFD can be used to compute the extreme values of the air pocket oscillations to increase the reliability during the filling operation.

For future research, the three-dimensional behaviour of hydraulic events with air expulsion orifices, which has not been studied in detail in the literature using 3D CFD models, should be studied. Additionally, it is important to test the simulation of large-scale hydraulic scenarios, considering the presence and absence of air expulsion orifices.

Author Contributions: Conceptualisation, D.A.P.-V. and O.E.C.-H.; methodology, D.A.P.-V. and H.G.E.-R.; software, D.A.P.-V. and H.G.E.-R.; validation, O.E.C.-H. and D.A.P.-V.; formal analysis, D.A.P.-V.; investigation, D.A.P.-V. and O.E.C.-H.; resources, H.M.R. and O.E.C.-H.; writing—original draft preparation, D.A.P.-V.; writing—review and editing, D.A.P.-V. and O.E.C.-H.; visualisation, H.M.R. and V.S.F.-M.; supervision, H.M.R. and V.S.F.-M. All authors have read and agreed to the published version of the manuscript.

Funding: This research project was funded through Grant No. INV03CI2214 of the Universidad Tecnológica de Bolívar.

Institutional Review Board Statement: Not applicable.

Informed Consent Statement: Not applicable.

Data Availability Statement: Not applicable.

Acknowledgments: The authors would like to thank Dirección de Investigación, Innovación y Emprendimiento of the Universidad Tecnológica de Bolívar for funding this research project through Grant No. INV03CI2214.

Conflicts of Interest: The authors declare no conflicts of interest.

Notation

The following notations are used in this manuscript:

a_{eff}	thermal conductivity (W/(m·K))
C_p	specific heat (J/(kg K))
D_k	diffusion term for k (m^2/s)
D_ω	diffusion term for ω (m^2/s)
F_1	blending function (–)
F_s	surface tension (kg m/s^2)
$\vec{g}$	gravitational acceleration vector (m/s^2)
G	turbulent kinetic energy generation (m^2/s^3)
k	turbulent kinetic energy (m^2/s^2)
L_{iap}	initial air pocket (m)
p	static pressure (N/m^2)
p_0	initial absolute pressure (N/m^2)
$\vec{q}$	heat flux vector (W/m^2)
R	universal gas constant (J/(K·mol))
T	temperature (°C)
t	time (s)
$\vec{u}$	velocity vector (m/s)
u_r	velocity field (m/s)
y^+	distance function (–)
α_w	phase of fraction (water) (–)
μ	dynamic viscosity (Ns/m^2)
ν_t	turbulent kinematic viscosity (m^2/s)
ρ	density (kg/m^3)
ω	dissipation frequency (1/s)
Subscripts	
a	refers to air (e.g., air density)
w	refers to water (e.g., water density)
m	refers to the mixture between air and water (e.g., mixed density)
t	refers to a turbulent condition (e.g., turbulent dynamic viscosity)

References

1. Fuertes, V. Hydraulic Transients with Entrapped Air Pockets. Ph.D. Thesis. Department of Hydraulic Engineering, Polytechnic University of Valencia, Editorial Universitat Politècnica de València, Valencia, Spain, 2001.
2. Fuertes-Miquel, V.S.; Coronado-Hernández, O.E.; Mora-Meliá, D.; Iglesias-Rey, P.L. Hydraulic modeling during filling and emptying processes in pressurized pipelines: A literature review. *Urban Water J.* **2019**, *16*, 299–311. [CrossRef]
3. Zhou, L.; Liu, D.; Karney, B. Investigation of hydraulic transients of two entrapped air pockets in a water pipeline. *J. Hydraul. Eng.* **2013** *139*, 949–959. [CrossRef]
4. Zhou, L.; Wang, H.; Karney, B.; Liu, D.; Wang, P.; Guo, S. Dynamic behavior of entrapped air pocket in a water filling pipeline. *J. Hydraul. Eng.* **2018**, *144*, 04018045. [CrossRef]
5. Vasconcelos, J.G. Dynamic Approach to the Description of Flow Regime Transition in Stormwater Systems. Ph.D. Thesis, University of Michigan Library, Ann Arbor, MI, USA, 2005.
6. Vasconcelos, J.G.; Wright, S.J. Experimental investigation of surges in a stormwater storage tunnel. *J. Hydraul. Eng.* **2005**, *131*, 853–861. [CrossRef]
7. Chosie, C D.; Hatcher, T.M.; Vasconcelos, J.G. Experimental and numerical investigation on the motion of discrete air pockets in pressurized water flows. *J. Hydraul. Eng.* **2014**, *140*, 04014038. [CrossRef]
8. Zhou, F.; Hicks, F.; Steffler, P. Transient flow in a rapidly filling horizontal pipe containing trapped air. *J. Hydraul. Eng.* **2002**, *128*, 625–634. [CrossRef]

9. Zhou, F.; Hicks, F.; Steffler, P. Analysis of effects of air pocket on hydraulic failure of urban drainage infrastructure. *Can. J. Civ. Eng.* **2004**, *31*, 86–94. [CrossRef]
10. De Martino, G.; Fontana, N.; Giugni, M. Transient flow caused by air expulsion through an orifice. *J. Hydraul. Eng.* **2008**, *134*, 1395–1399. [CrossRef]
11. Fuertes-Miquel, V.S.; López-Jiménez, P.A.; Martínez-Solano, F.J.; López-Patiño, G. Numerical modelling of pipelines with air pockets and air valves. *Can. J. Civ. Eng.* **2016**, *43*, 1052–1061. [CrossRef]
12. Coronado-Hernández, Ó.E.; Besharat, M.; Fuertes-Miquel, V.S.; Ramos, H.M. Effect of a commercial air valve on the rapid filling of a single pipeline: A numerical and experimental analysis. *Water* **2019**, *11*, 1814. [CrossRef]
13. Martin, C.; Lee, N.H. Rapid expulsion of entrapped air through an orifice. In *BHR Group Conference Series Publication*; St. Edmunds, B., Ed.; Professional Engineering Publishing: London, UK, 1998; Volume 39, pp. 125–132.
14. Zhou, L.; Pan, T.; Wang, H.; Liu, D.; Wang, P. Rapid air expulsion through an orifice in a vertical water pipe. *J. Hydraul. Res.* **2019**, *57*, 307–317. [CrossRef]
15. Romero, G.; Fuertes-Miquel, V.S.; Coronado-Hernández, Ó.E.; Ponz-Carcelén, R.; Biel-Sanchis, F. Transient phenomena generated in emptying operations in large-scale hydraulic pipelines. *Water* **2020**, *12*, 2313. [CrossRef]
16. Zhou, L.; Liu, D.Y.; Ou, C.Q. Simulation of flow transients in a water filling pipe containing entrapped air pocket with VOF model. *Eng. Appl. Comput. Fluid Mech.* **2011**, *5*, 127–140. [CrossRef]
17. Aguirre-Mendoza, A.M.; Oyuela, S.; Espinoza-Román, H.G.; Coronado-Hernández, O.E.; Fuertes-Miquel, V.S.; Paternina-Verona, D.A. 2D CFD Modeling of Rapid Water Filling with Air Valves Using OpenFOAM. *Water* **2021**, *13*, 3104. [CrossRef]
18. Besharat, M.; Tarinejad, R.; Aalami, M.T.; Ramos, H.M. Study of a compressed air vessel for controlling the pressure surge in water networks: CFD and experimental analysis. *Water Resour. Manag.* **2016**, *30*, 2687–2702. [CrossRef]
19. Martins, N.M.; Delgado, J.N.; Ramos, H.M.; Covas, D.I. Maximum transient pressures in a rapidly filling pipeline with entrapped air using a CFD model. *J. Hydraul. Res.* **2017**, *55*, 506–519. [CrossRef]
20. Fang, H.; Zhou, L.; Cao, Y.; Cai, F.; Liu, D. 3D CFD simulations of air-water interaction in T-junction pipes of urban stormwater drainage system. *Urban Water J.* **2021**, *19*, 74–86. [CrossRef]
21. Aguirre-Mendoza, A.M.; Paternina-Verona, D.A.; Oyuela, S.; Coronado-Hernández, O.E.; Besharat, M.; Fuertes-Miquel, V.S.; Iglesias-Rey, P.L.; Ramos, H.M. Effects of Orifice Sizes for Uncontrolled Filling Processes in Water Pipelines. *Water* **2022**, *14*, 888. [CrossRef]
22. Wang, H.; Zhou, L.; Liu, D.; Karney, B.; Wang, P.; Xia, L.; Ma, J.; Xu, C. CFD approach for column separation in water pipelines. *J. Hydraul. Eng.* **2016**, *142*, 04016036. [CrossRef]
23. Wu, G.; Duan, X.; Zhu, J.; Li, X.; Tang, X.; Gao, H. Investigations of hydraulic transient flows in pressurized pipeline based on 1D traditional and 3D weakly compressible models. *J. Hydroinform.* **2021**, *23*, 231–248. [CrossRef]
24. Paternina-Verona, D.A.; Coronado-Hernández, O.E.; Fuertes-Miquel, V.S. Numerical modelling for analysing drainage in irregular profile pipes using OpenFOAM. *Urban Water J.* **2022**, *19*, 569–578. [CrossRef]
25. Besharat, M.; Coronado-Hernández, O.E.; Fuertes-Miquel, V.S.; Viseu, M.T.; Ramos, H.M. Backflow air and pressure analysis in emptying a pipeline containing an entrapped air pocket. *Urban Water J.* **2018**, *15*, 769–779. [CrossRef]
26. Besharat, M.; Coronado-Hernández, O.E.; Fuertes-Miquel, V.S.; Viseu, M.T.; Ramos, H.M. Computational fluid dynamics for sub-atmospheric pressure analysis in pipe drainage. *J. Hydraul. Res.* **2019**, *58*, 553–565. [CrossRef]
27. Hurtado-Misal, A.D.; Hernández-Sanjuan, D.; Coronado-Hernández, O.E.; Espinoza-Román, H.; Fuertes-Miquel, V.S. Analysis of Sub-Atmospheric Pressures during Emptying of an Irregular Pipeline without an Air Valve Using a 2D CFD Model. *Water* **2021**, *13*, 2526. [CrossRef]
28. Paternina-Verona, D.A.; Coronado-Hernández, O.E.; Espinoza-Román, H.G.; Besharat, M.; Fuertes-Miquel, V.S.; Ramos, H.M. Three-Dimensional Analysis of Air-Admission Orifices in Pipelines during Hydraulic Drainage Events. *Sustainability* **2022**, *14*, 14600. [CrossRef]
29. Greenshields, C.; Weller, H. *Notes on Computational Fluid Dynamics: General Principles*; CFD Direct Ltd.: Reading, UK, 2022.
30. Hirt, C.W.; Nichols, B.D. Volume of fluid (VOF) method for the dynamics of free boundaries. *J. Comput. Phys.* **1981**, *39*, 201–225. [CrossRef]
31. Bombardelli, F.A.; Hirt, C.; García, M.H.; Matthews, B.; Fletcher, C.; Partridge, A.; Vasquez, S. Computations of curved free surface water flow on spiral concentrators. *J. Hydraul. Eng.* **2001**, *127*, 629–631. [CrossRef]
32. Menter, F.R. Two-equation eddy-viscosity turbulence models for engineering applications. *AIAA J.* **1994**, *32*, 1598–1605. [CrossRef]
33. Menter, F.R. Review of the shear-stress transport turbulence model experience from an industrial perspective. *Int. J. Comput. Fluid Dyn.* **2009**, *23*, 305–316. [CrossRef]
34. Blazek, J. *Computational Fluid Dynamics: Principles and Applications*; Butterworth-Heinemann: Oxford, UK, 2015.

Disclaimer/Publisher's Note: The statements, opinions and data contained in all publications are solely those of the individual author(s) and contributor(s) and not of MDPI and/or the editor(s). MDPI and/or the editor(s) disclaim responsibility for any injury to people or property resulting from any ideas, methods, instructions or products referred to in the content.

water

MDPI

Article

Energy Analysis of a Quasi-Two-Dimensional Friction Model for Simulation of Transient Flows in Viscoelastic Pipes

Kai Wu [1], Yujie Feng [1,*], Ying Xu [2], Huan Liang [3] and Guohong Liu [1]

[1] State Key Laboratory of Urban Water Resource and Environment, School of Environment, Harbin Institute of Technology, Huanghe Road No. 73, Harbin 150090, China
[2] School of Energy and Architecture Engineering, Harbin University of Commerce, Harbin 150028, China
[3] Industrial Control Energy Saving Business Division, Beijing Huada Zhibao Electronic System Co., Ltd., Beijing 100020, China
* Correspondence: yujief@hit.edu.cn

Abstract: Quasi-two-dimensional (quasi-2D) friction models have been widely investigated in transient pipe flows. In the case of viscoelastic pipes, however, the effect of different values of the Reynolds number (Re) on pressure fluctuations (which can lead to water hammer) have not been considered in detail. This study establishes a quasi-2D friction model employing an integral total energy method and investigates the work due to frictional and viscoelastic terms at different Re values. The results show that viscoelastic work (W_P) and frictional work (D_f) increase with an increase in Re. However, when the initial Re values are high, the D_f values are much larger than the W_P values. In addition, for Re < 3×10^5, the 1D model underestimated the viscoelastic terms. There was no significant difference between the two models for Re > 3×10^5. In the case of different initial Re values, the two models produced almost the same values for W_P. This study provides a theoretical basis for investigating transient flow from the perspective of energy analysis.

Keywords: transient flow; viscoelastic pipe; quasi-2D model; energy analysis; Reynolds numbers

Citation: Wu, K.; Feng, Y.; Xu, Y.; Liang, H.; Liu, G. Energy Analysis of a Quasi-Two-Dimensional Friction Model for Simulation of Transient Flows in Viscoelastic Pipes. *Water* **2022**, *14*, 3258. https://doi.org/10.3390/w14203258

Academic Editors: Kamil Urbanowicz and Helena M. Ramos

Received: 23 August 2022
Accepted: 10 October 2022
Published: 15 October 2022

Publisher's Note: MDPI stays neutral with regard to jurisdictional claims in published maps and institutional affiliations.

cc BY

Copyright: © 2022 by the authors. Licensee MDPI, Basel, Switzerland. This article is an open access article distributed under the terms and conditions of the Creative Commons Attribution (CC BY) license (https://creativecommons.org/licenses/by/4.0/).

1. Introduction

In recent years, viscoelastic pipes have been increasingly used in urban water-supply systems [1]. The transient-flow pressure fluctuation generated by such pipes is different from the pressure fluctuation of traditional elastic pipes because of their different viscoelastic properties; when pumps suddenly stop or valves close quickly, a water-hammer accident result. The peak values of pressure fluctuation are relatively small; however, pressure-fluctuation attenuation and phase delay are large. Therefore, a correct understanding of the role of viscoelastic characteristics in the transient-flow pressure-fluctuation process is necessary.

Accurate numerical calculation of transient flow in viscoelastic pipes must consider the effect of not only unsteady friction (UF) of the pipe wall but also wall viscoelasticity. Some researchers have studied the pressure fluctuation of transient flows in viscoelastic pipes. Rieutord and Blanchard [2] pointed out that the acceleration of pressure attenuation is due to a time offset between the pressure and the retarded strain of the pipe wall. Covas et al. [3] proposed that, owing to retarded deformation, the maximum pressure of the transient flow of viscoelastic pipes is smaller than that of elastic pipes, and the attenuation of the pressure wave is faster. In addition, Covas et al. [4] found that the viscoelastic effect is more evident in a transient flow when the retarded time of the viscoelastic pipes is less than the propagation time of the pressure wave along the pipes.

Numerous studies on the influence of UF on the transient flow of elastic pipes have been conducted. The results show that a one-dimensional (1D) quasi-steady friction model can accurately simulate the maximum value of pressure fluctuation in elastic pipelines, but in most cases, it cannot accurately describe the peak-pressure damping [1]. One-dimensional UF model [5] and two-dimensional (2D) friction models [6] can accurately simulate not only the maximum peak of pressure fluctuations but also the attenuation of pressure fluctuation. Pezzinga [6] compared the error values of 1D- and 2D-model-simulated maximum and minimum pressure heads with experimental results and confirmed the accuracy of the 2D-model-simulated shear stress. In a study of different friction models of the transient flow through a simulated viscoelastic pipeline, Firkowski et al. [7] examined the effects of UF and experimentally obtained creep functions. Urbanowicz et al. [8] investigated quasi-steady and unsteady frictions in viscoelastic pipes and presented a simplified effective numerical-convolution integral to describe retarded strain. Sun et al. [9] concluded that a 1D quasi-steady friction model could accurately simulate the pressure-fluctuation attenuation of viscoelastic pipes at a certain water temperature.

In previous studies, the viscoelastic parameter was mistakenly thought to be energy dissipation, but an increasing number of studies have shown that there is energy transfer between the fluid and the pipe wall [10]. Karney [11] divided the energy involved in transient flow into four parts: the kinetic energy of the fluid, the internal energy associated with fluid compressibility and pipeline-elasticity effects, the energy dissipated by friction, and the work done at the ends of the pipeline. Through integral processing, the basic momentum and continuity equations of transient flow can be used to analyse the energy of the entire pipeline. Duan et al. [10] used Fourier methods and energy analysis to show that energy transfer was related to the ratio of the pressure wave period to the retarded time (T/τ). Duan et al. [12] used local transient analysis (LTA) and integral total energy (ITE) methods to study the influence of friction on transient flow in a 1D UF model and 2D k-ε turbulence model. The results showed that the friction dissipation calculated by the 1D model was smaller than that calculated by the 2D model after the first pressure-fluctuation period. Meniconi et al. [13] introduced nondimensional parameters to analyse the energy dissipation of the transient flow in viscoelastic pipes, finding numerically that the energy dissipation was related to the Reynolds number (Re) and valve opening. Riasi et al. [14] employed a 2D k-ε turbulence model to infer nondimensional parameters related to Re and the viscoelasticity of the pipe. The influence of these parameters on the pressure fluctuation, wall-shear stress, velocity profiles, turbulence production, and dissipation was investigated.

As mentioned, Re is an important parameter in transient flow. Hence, an understanding of how it influences viscoelastic and frictional effects on the energy dissipation of transient flow is critical for the design and operation of a pipeline system. In this study, transient flow with different initial Re values was investigated for both 1D and 2D models. First, the energy equation of the 2D model was derived using the ITE method. The contributions of the viscoelastic and friction terms to the work done in both friction models were analysed and discussed, and the change rule of the work done under different Re values were clarified. The trends of work done and energy dissipation at different Re values were examined from the perspective of energy analysis, and the laws governing fluctuation of friction and viscoelastic terms at different stages of transient flow were studied. Our study established a quasi-2D model based on energy analysis to analyse the friction and viscoelastic effects of transient flow at different Reynolds number.

2. Materials and Methods

2.1. Governing Equations

The continuity and momentum equations of the quasi-2D transient flow model are expressed as [4,5]:

$$\frac{g}{a^2}\frac{\partial H}{\partial t}+\frac{\partial u}{\partial x}+\frac{1}{r}\frac{\partial (rv)}{\partial r}+2\frac{\partial \varepsilon_r}{\partial t}=0, \tag{1}$$

$$\frac{\partial u}{\partial t}+g\frac{\partial H}{\partial x}-\frac{1}{\rho r}\frac{\partial (r\tau)}{\partial r}=0,. \tag{2}$$

where H is the pressure head, Q is the discharge, a is the wave speed, r is the radial distance from the pipe centre, x is the axial coordinate along the pipe, t is the time, u is the longitudinal velocity, v is the radial velocity, τ is the shear stress, ε_r is the retarded strain, g is the gravitational acceleration, and ρ is the density.

By integrating Equations (1) and (2) over the cross-sectional area of the pipeline, the continuity and momentum equations for the 1D transient flow in viscoelastic pipes can be expressed as [4]:

$$\frac{\partial H}{\partial t} + \frac{a^2}{gA}\frac{\partial Q}{\partial x} + \frac{2a^2}{g}\frac{\partial \varepsilon_r}{\partial t} = 0, \tag{3}$$

$$\frac{\partial Q}{\partial t} + gA\frac{\partial Q}{\partial x} + \frac{\pi D}{\rho}\tau_w = 0, \tag{4}$$

where A is the cross-sectional area of the pipeline, and τ_w is the pipe-wall shear stress evaluated using the Darcy–Weisbach formula.

2.2. Kelvin-Voigt Model

The behaviour of the viscoelastic pipes was simulated using the Kelvin-Voight (K-V) model [15,16]. Each of the K-V components consists of a spring and sticky pot, representing the instantaneous strain and retarded strain components, respectively. The total creep compliance is given by:

$$J(t) = J_0 + \sum_{k=1}^{N} J_k\left(1 - e^{-\frac{t}{\tau_k}}\right), \tag{5}$$

where $J_0 = 1/E_0$, E_0 is the elastic modulus, $J_k = 1/E_k$ is the creep compliance of the k-th element, τ_k is the retarded time of the k-th element, and E_k is the elasticity modulus of the k-th element.

For viscoelastic pipes, the total strain ε is given by the sum of the instantaneous strain (ε_e) and retarded component (ε_r):

$$\varepsilon = \varepsilon_e + \varepsilon_r. \tag{6}$$

Using the K-V model with N elements, the retarded strain is the sum of the single-element deformations [3]:

$$\partial\varepsilon_r(t) = \sum_{k=1\cdots N} \partial\varepsilon_{rk}(t). \tag{7}$$

The retarded strain is [4]:

$$\varepsilon_r(x,t) = \sum_{k=1\cdots N} \varepsilon_{rk}(x,t) = \sum_{k=1\cdots N} \frac{\gamma\alpha D}{2e}\int_0^t \left[H(x,t-t') - H_0(x)\right]\frac{J_k}{\tau_k}e^{\frac{-t'}{\tau_k}}\,dt', \tag{8}$$

where t' is the time variation.

The retarded strain rate is algebraically expressed as [17,18]:

$$\frac{\partial\varepsilon_r(t)}{\partial t} = \sum_{k=1\cdots N}\frac{\partial\varepsilon_{rk}(t)}{\partial t} = \sum_{k=1}^{N}(AH - V_E), \tag{9}$$

with

$$V_E = BH_0 + (A-B)H(x,t-\Delta t) + C\varepsilon_{rk}(x,t-\Delta t), \tag{10}$$

$$A = \frac{\alpha D}{2e}\gamma\frac{J_k}{\Delta t}\left(1 - e^{-\frac{\Delta t}{\tau_k}}\right), B = \frac{\alpha D}{2e}\gamma\frac{J_k}{\Delta t}e^{-\frac{\Delta t}{\tau_k}}, C = \frac{e^{-\frac{\Delta t}{\tau_k}}}{T_k}, \tag{11}$$

where γ is the bulk weight, α is the constraint coefficient, D is the pipe diameter, e is the wall thickness, H is the pressure head, and $\Delta t = L/(a \times N_x)$. Here, N_x is the number of pipe meshes.

2.3. Numerical Scheme of 1D and Quasi-2D Models

The solutions for transient flows in viscoelastic pipes were calculated by quasi-2D and corresponding 1D models using the method of characteristics (MOC). The characteristic forms of the quasi-2D and corresponding 1D transient-flow equations are:

$$\frac{dH}{dt} \pm \frac{a}{g}\frac{du}{dt} + \frac{a^2}{gr}\frac{\partial(rv)}{\partial r} \mp \frac{a}{g}\frac{1}{\rho r}\frac{\partial(r\tau)}{\partial r} + \frac{2a^2}{g}\sum_{k=1}^{N}(AH - V_E) = 0, \tag{12}$$

$$\frac{dH}{dt} \pm \frac{a}{g}\frac{dQ}{dt} \pm \frac{a^2}{g}\frac{4\tau_v}{\rho D} + \frac{2a^2}{g}\sum_{k=1}^{N}(AH - V_E) = 0. \tag{13}$$

Because of the steep radial velocity gradient, a gradually encrypted radial grid was applied from the pipe centre to the pipe wall. This is shown in Figure 1, where j is the subscript representing the radial grid number, r_j is the radial distance between the outer surface of cylinder j and the pipe centre, and r_{cj} is the radial distance between the centre of cylinder j and the pipe centre.

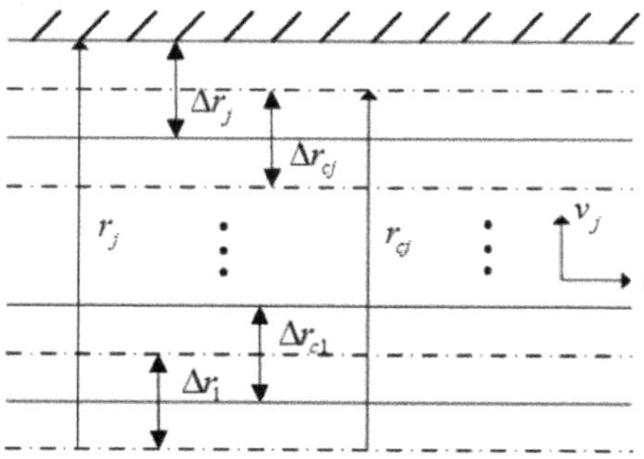

Figure 1. Radial difference grid.

Integrating Equation (11) on the characteristic lines between times $n\Delta t$ and $(n+1)\Delta t$ (Figure 1) results in the following discretised expressions:

$$(1+F)H_i^{n+1} - \theta C_{q,j} q_{i,j-1}^{n+1} + \theta C_{q,j} q_{i,j}^{n+1} - \omega C_{u1,j} u_{i,j-1}^{n+1} + \left(\frac{a}{g} + \omega C_{u2,j}\right) u_{i,j}^{n+1} - \omega C_{u3,j} u_{i,j+1}^{n+1} = K_{pi,j}, \tag{14}$$

$$(1+F)H_i^{n+1} - \theta C_{q,j} q_{i,j-1}^{n+1} + \theta C_{q,j} q_{i,j}^{n+1} + \omega C_{u1,j} u_{i,j-1}^{n+1} - \left(\frac{a}{g} + \omega C_{u2,j}\right) u_{i,j}^{n+1} + \omega C_{u3,j} u_{i,j+1}^{n+1} = K_{ni,j}, \tag{15}$$

with

$$F = \frac{2a^2\Delta t}{g}\sum_{k=1}^{N}\frac{\alpha D}{2e}\gamma\frac{J_k}{\Delta t}e^{-\Delta t/\tau_k}, \tag{16}$$

$$K_{pi,j} = H_{i-1}^n + (1-\theta)C_{qj}\left(q_{i-1,j-1}^n - q_{i-1,j}^n\right) + (1-\varepsilon)C_{u1,j}u_{i-1,j-1}^n$$
$$+ \left[a/g - (1-\varepsilon)C_{u2,j}\right]u_{i-1,j}^n + (1-\varepsilon)C_{u3,j}u_{i-1,j+1}^n + \frac{2a^2\Delta t}{g}\sum_{k=1}^{N}V_E, \tag{17}$$

$$K_{ni,j} = H_{i+1}^n + (1-\theta)C_{qj}\left(q_{i+1,j-1}^n - q_{i+1,j}^n\right) - (1-\varepsilon)C_{u1,j}u_{i+1,j-1}^n$$
$$- \left[\frac{a}{g} - (1-\varepsilon)C_{u2,j}\right]u_{i+1,j}^n - (1-\varepsilon)C_{u3,j}u_{i+1,j+1}^n + \frac{2a^2\Delta t}{g}\sum_{k=1}^{N}V_E, \tag{18}$$

$$C_{qj} = \frac{a\Delta t^2}{g r_{cj}(r_j - r_{j-1})},\ C_{u1,j} = \frac{a\Delta t \nu_{Tj-1} r_{j-1}}{g r_{cj}(r_{cj} - r_{cj-1})(r_j - r_{j-1})},$$

$$C_{u3,j} = \frac{a\Delta t v_{Tj} r_j}{g r_{cj}\left(r_{cj+1} - r_{cj}\right)\left(r_j - r_{j-1}\right)},\ C_{u2,j} = C_{u1,j} + C_{u3,j}, \tag{19}$$

where ε and θ are the weighting coefficients, $q(=rv)$ is the radial flux, and the source terms $K_{pi,j}$, and $K_{ni,j}$ are known values, the elements of which depend on H, u, and q at the previous time level.

When Equation (14) is subtracted from Equation (15), we obtain:

$$\omega C_{u1,j} u_{i,j-1}^{n+1} - \left(\frac{a}{g} + \omega C_{u2,j}\right) u_{i,j}^{n+1} + \omega C_{u3,j} u_{i,j+1}^{n+1} = 0.5\left(K_{ni,j} - K_{pi,j}\right). \tag{20}$$

Instead, if they are added, we obtain:

$$(1+F)H_i^{n+1} - \theta C_{q,j} q_{i,j-1}^{n+1} + \theta C_{q,j} q_{i,j}^{n+1} = 0.5\left(K_{ni,j} + K_{pi,j}\right). \tag{21}$$

These equations enable us to calculate the axial velocity, nodal pressure head, and radial flux.

Integrating Equation (12) on the characteristic lines between times $n\Delta t$ and $(n + 1)\Delta t$, the discretised forms of the 1D characteristic equations are expressed as:

$$(1+F)H_i^{n+1} + BQ_i^{n+1} + \frac{4\varepsilon a\Delta t}{\rho g D}\tau_{w,i}^{n+1} = C_p, \tag{22}$$

$$(1+F)H_i^{n+1} - BQ_i^{n+1} - \frac{4\varepsilon a\Delta t}{\rho g D}\tau_{w,i}^{n+1} = C_m\ , \tag{23}$$

with

$$\tau_{w,i}^{n+1} = \rho v_T \frac{u_{i,N_{R+1}}^{n+1}}{\Delta r_{cN_{r+1}}} = \rho v_T \frac{u_{i,N_{R+1}}^{n+1}}{\left(D/2 - r_{cN_{r+1}}\right)} \tag{24}$$

$$C_p = H_{i-1}^n + \frac{a}{gA}Q_{i-1}^n - \frac{4a\Delta t}{\rho g D}\tau_{w,i-1}^n + \frac{2a^2\Delta t}{g}\sum_{k=1}^{N} V_E, \tag{25}$$

$$C_m = H_{i+1}^n - \frac{a}{gA}Q_{i+1}^n + \frac{4a\Delta t}{\rho g D}\tau_{w,i+1}^n + \frac{2a^2\Delta t}{g}\sum_{k=1}^{N} V_E\ . \tag{26}$$

A five-region turbulent flow model can be introduced to calculate the turbulent eddy viscosity (v_T). Combining Equations (21) and (22), the pressure head and discharge may be obtained as:

$$H_i^{n+1} = \frac{C_p + C_m}{2(1+F)}, \tag{27}$$

$$Q_i^{n+1} = \frac{C_p - C_m}{2B} - \frac{4\varepsilon a\Delta t}{\rho g D B}\tau_{wi}^{n+1}. \tag{28}$$

2.4. ITE Method Based on Quasi-2D Friction Model

The energy equation corresponding to the quasi-2D model is obtained by integrating the basic differential equation. However, in addition to axial flow, radial flow is considered in the quasi-2D model; therefore, radial integration should be considered when the equation associated with the flow rate is integrated [10].

$$\int_0^L \int_0^{D/2} \left[Eq.\ (3) \times \rho^2 g H\right] dr dx;\quad \int_0^L \int_0^{D/2} [Eq.\ (4) \times u] dr dx \tag{29}$$

The energy equation of the quasi-2D model is as follows (a detailed derivation is provided in Appendix A):

$$\frac{dU}{dt} + \frac{dT}{dt} + D_f + W_E + W_P = 0, \tag{30}$$

where U is the total internal energy of the system, T is the total kinetic energy of the system, D_f is the total rate of frictional dissipation, W_E is the total rate of work from the ends of the pipe, and W_P is the total rate of work from the pipe wall, expressed as follows:

$$U = \frac{\rho g^2 A}{2a^2} \int_0^L H^2(x,t)dx, \tag{31}$$

$$T(t) = \frac{\pi \rho D}{4} \int_0^L \int_0^{D/2} u^2 dr dx, \tag{32}$$

$$D_f(t) = -\frac{\pi D}{2} \int_0^L \int_0^{D/2} \frac{u}{r} \frac{\partial (r\tau)}{\partial r} dr dx, \tag{33}$$

$$W_E = \frac{\pi \rho D g}{2} \int_0^{D/2} [u(L,t)H(L,t) - u(0,t)H(0,t)]dr, \tag{34}$$

$$W_P = 2\rho A g \int_0^L H \frac{\partial \varepsilon}{\partial t} dx. \tag{35}$$

3. Results: Experimental Setup and Validation

The experimental data in this study were obtained from the Water Engineering Laboratory at the University of Perugia, Italy [19]. High-density polyethylene (HDPE) pipe was used in the experiment. A pressure tank was used to maintain the pressure upstream of the pipes, and a fast-closing pneumatic valve was used to generate excitation of the transient flow downstream of the pipes. The inner diameter (D) of the pipe was 93.3 mm, and the wall thickness (e) was 8.1 mm. The length of the experimental pipes (L) used in this study was 128.6 m.

Five groups of experimental data, each with a different Re, initial flow rate, and initial pressure, were analysed in this study. The specific experimental parameters are listed in Table 1. The limiting coefficient of the pipe was 0.97, and the relative roughness of the pipe was 8.57×10^{-4}. The K-V parameters were calibrated according to literature [19], and their values were as follows: 1D model (E_0 = 1556 N/mm^2, E_1 = 7820 N/mm^2, τ_1 = 582.6 ms, E_2 = 18,370 N/mm^2, τ_2 = 59.76 ms, E_3 = 6842 N/mm^2, and τ_3 = 21,570 ms); 2D model (E_0 = 1563 N/mm^2, E_1 = 9596 N/mm^2, τ_1 = 562.3 ms, E_2 = 19,490 N/mm^2, τ_2 = 52.18 ms, E_3 = 5834 N/mm^2, and τ_3 = 19,680 ms).

Table 1. Experimental settings [19].

Case	Q (L/s)	Re	H_0 (m)	T_c (s)
1	1	13,380	21.63	0.0875
2	2.04	28,320	21.13	0.0752
3	2.95	40,950	20.74	0.1188
4	4.03	55,940	20.34	0.1575
5	5.02	69,680	19.82	0.1533

The numerical results of the 1D and quasi-2D models at Re = 55,940 are compared to the experimental results in Figure 2a. The figure shows that the simulation results of the quasi-2D model are very close to the experimental results in terms of peak value, valley value, and phase, especially in the second half-circle of transient flow (30–50 L/a); the same is true for the 1D model. Figure 2b shows the simulation results for the pressure head at different Re values, in which the peak pressure increases with Re. The pressure head decreases with the increase of Reynolds number, whereas the phase is unchanged at different Reynolds numbers.

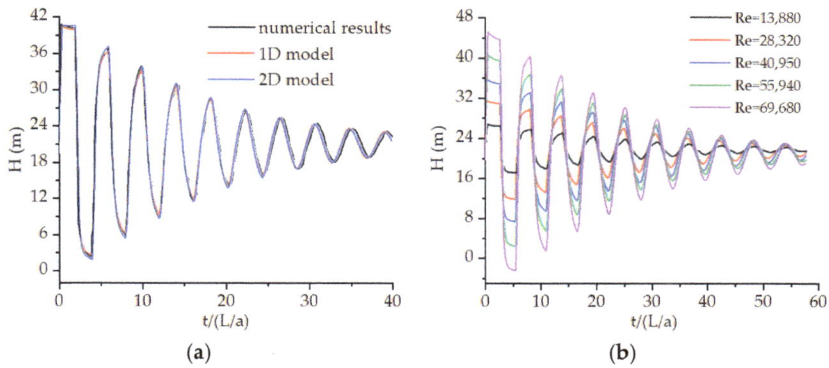

Figure 2. Variation in pressure head with time: (**a**) results of experiment and two simulations at Re = 55,940; (**b**) simulation results at different Re values.

Figure 3 shows the variation trend of the total, retarded, and instantaneous strain in the two cases. In the critical region of transient flow, the retarded strain accounted for a much smaller proportion of the total strain than the instantaneous strain. Although the retarded strain was relatively small, it can be seen from Figure 3 that, with an increase in L/a, the retarded strain decreased further and the instantaneous strain also showed a decreasing trend. At the later stage of transient flow (30 L/a), the total and instantaneous strain overlapped substantially.

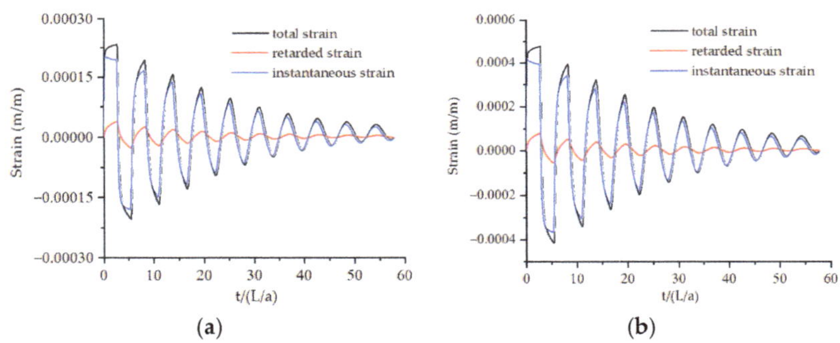

Figure 3. Results of strain at different Re values: (**a**) Re = 13,880; (**b**) Re = 28,320.

From Figure 3, it can also be seen that the instantaneous strain reached its maximum value at the initial moment of the transient flow, whereas the retarded strain reached its maximum value in the third phase of transient flow (3 L/a); thus, the total strain reached its maximum value at 3 L/a. This suggests that the pressure fluctuation in viscoelastic pipes exhibits a delay compared with that in elastic pipes.

Figure 4 depicts the changes in the instantaneous and retarded strains at five Re values. Both tended to increase with Re. At the same time, there was a little effect on the peak value of the retarded strain at different Re values. The maximum peak values of the instantaneous and retarded strains changed almost linearly with Re. At the same Re value, the maximum instantaneous strain was approximately five times the maximum retarded strain.

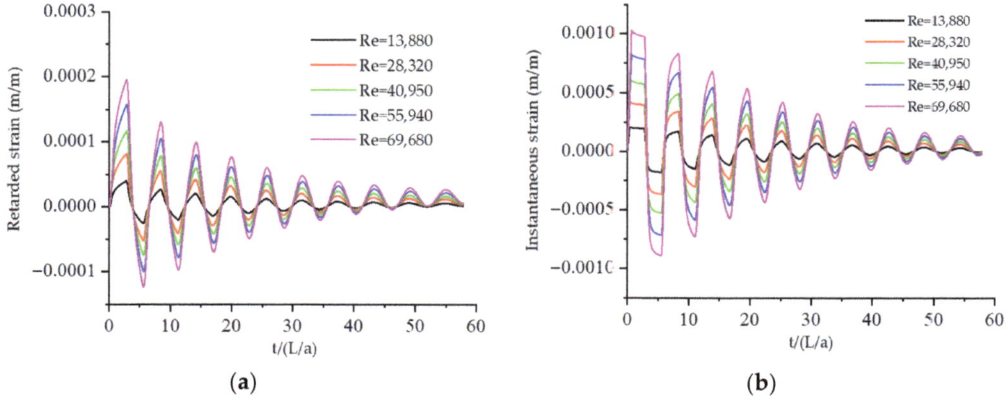

Figure 4. Variation of (**a**) retarded and (**b**) instantaneous strain at different Re values.

Figure 5 illustrates the time difference (Δt) between the extreme value (peak no.) of retarded strain and instantaneous strain (the time of retarded strain/the time of instantaneous strain). According to the line graph, the time difference is the biggest at the first few peak numbers of the Re values, whereas it is closer at the end peak number. This means that different Reynolds numbers have greater influence on the first few cycles of strain, that is, viscoelasticity is obvious in the early phase of transient flow. During the whole period of transient flow, the time difference between retarded strain and instantaneous strain not only shows an upward trend with the peak number monotonic increasing but increases as Re values increase. It would be expected that the retarded strain effect would become smaller with time because the velocity decreases with time; and the difference of Re values also affects viscoelasticity during the transient flow, though not greatly.

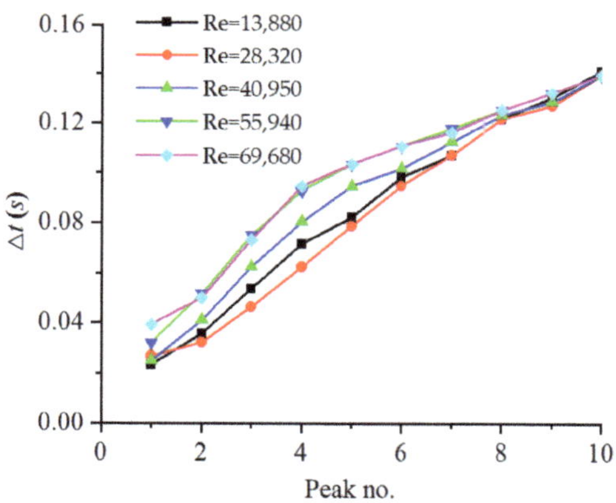

Figure 5. The time difference between the extreme value of retarded strain and instantaneous strain.

To further illustrate the viscoelastic effect, the ratio of retarded strain to total strain ($\varepsilon_r/\varepsilon$) is shown in Figure 6. As can be seen in the line graph, the ratio at the valley decreases from 0.12 in the first valley to 0.05 in the last valley, whereas there is a minor fluctuation in this peak numbers (from 0.175 in the first peak to 0.125 in the last peak, but the smallest value is 0.11 in the five peak). The ratios are consistent at different Re values, which means that the viscoelastic term has no significant effect on transient flow at different Reynolds numbers.

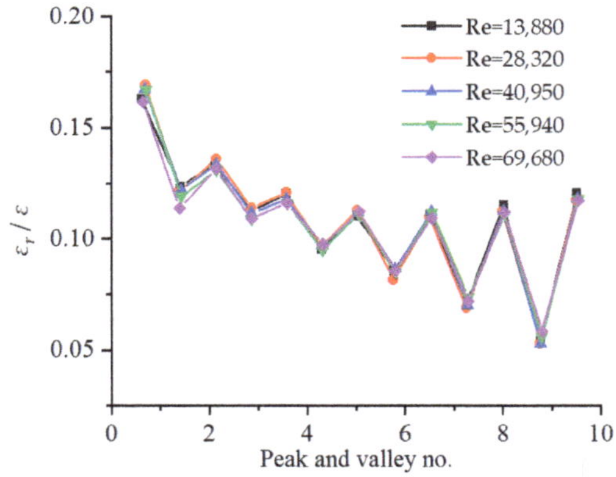

Figure 6. Ratio of retarded strain to total strain.

4. Discussion: Energy Analysis at Different Reynolds Numbers

To understand the influence of different Re values on the energy variations of viscoelastic pipes, distinctions between the viscoelasticity and friction terms were analysed.

4.1. D_f in 1D and 2D Models

The energy changes in the viscoelastic and friction terms of the 1D and 2D friction models under different initial Re values were further analysed based on the ITE method (i.e., Equation (29)). Combined with Equation (33), the simulation results of the work done by the friction term D_f are shown in Figure 7. The results show that larger D_f values correspond to greater values of the cumulative sum of the wall-shear stress of the entire pipeline. This indicates that the work done by the friction term becomes larger as Re values increase.

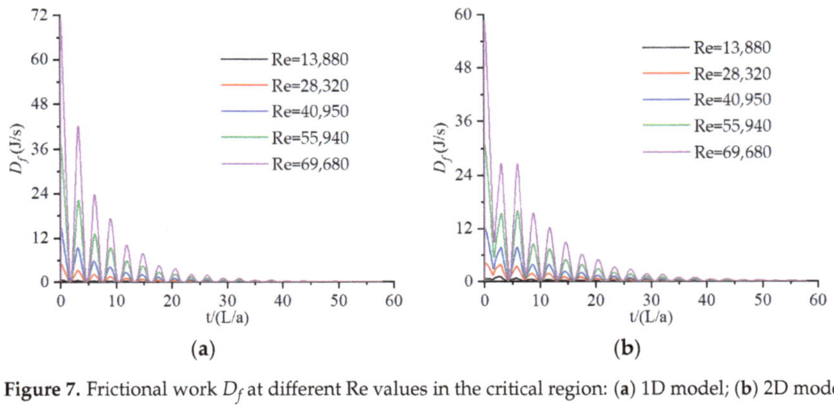

Figure 7. Frictional work D_f at different Re values in the critical region: (**a**) 1D model; (**b**) 2D model.

In Figure 7a,b, the D_f values calculated using the 2D model are larger than those calculated using the 1D model. It can be seen that the 1D model underestimated the instantaneous wall shear stress under flow conditions with initial Re values in the range of 1.0×10^5–3.0×10^5.

As shown in Figure 7, with an increase in the initial Re, D_f differed only slightly between the 1D and 2D models. This means that the instantaneous wall-shear stress calculated by the 1D model was close to that calculated by the 2D model, with an initial Re value of 4.0×10^5–7.0×10^5.

For all Re values, D_f was maximal near the initial time and gradually decayed with time in both the 2D and 1D models.

Furthermore, the D_f values were positive for both models in the critical region of turbulence. Thus, the D_f values show the dissipation of the friction energy in the transient flow. This is consistent with the description of transient flow energy in pipelines [10]. The D_f values also increased with an increasing initial Re.

4.2. W_P in 1D and 2D Models

The simulation results for W_P (the work done by the viscoelastic term of the pipe wall per unit time) are shown in Figure 8. The W_P values gradually increased with the initial Re values.

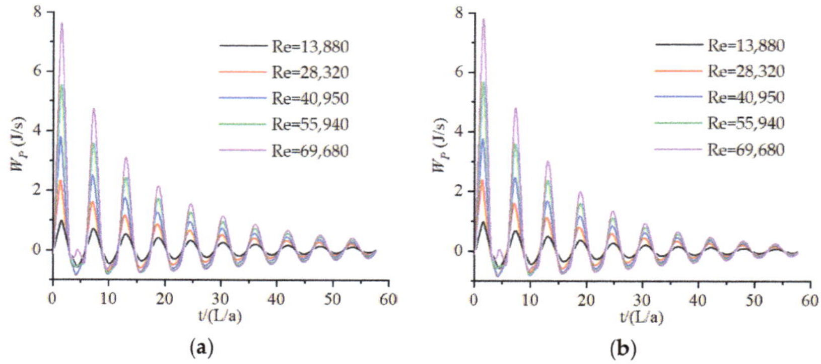

Figure 8. Viscoelastic work, W_P, at different Re values in the critical region: (**a**) 1D model; (**b**) 2D model.

The W_P values from the 1D model were similar at all Re values to those from the 2D model because the governing equations of both models were consistent in calculating the retarded strain. The work due to the viscoelastic term of the pipe wall in the two friction models was essentially the same under different initial Re values. In particular, the maximum value of W_P calculated by both friction models appeared at approximately 2 L/a for all Re values, and the W_P values gradually decayed over time. In the critical region of turbulence, W_P had both positive and negative values. This indicates that the interaction between the fluid and the pipe wall during transient flow involves both energy transfer and energy dissipation, which agrees with the description of the energy variation of transient flows in pipelines in the literature [10].

By comparing Figures 7 and 8, it can be seen that the work done by the viscoelastic and friction terms was similar at low Re. With an increase in the initial Re, D_f became significantly greater than W_P. This shows that the work done by the friction term had a significant influence on the energy dissipation when the initial Re values were relatively high [13].

To further illustrate the dependence of the viscoelastic effect at different flow conditions, the W_P values of viscoelastic term were compared with the 1D model at different Reynolds number (Figure 9). From these figures, the difference of W_P values between 1D model considering both friction and viscoelastic effects, and only considering the viscoelastic effect, increase with Re values, but compared with the increase of D_f values at different Reynolds number, there was a slight increase in W_P values in orders of magnitude. This is because the viscoelastic effect does not dominate under different initial Reynolds numbers [10,20].

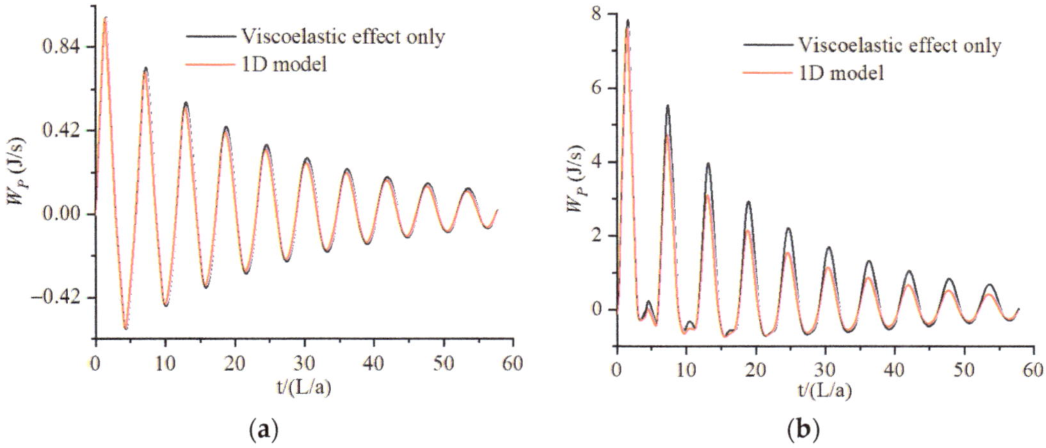

Figure 9. Work done of W_P by 1D model including only viscoelastic effect: (**a**) Re = 13,880; (**b**) Re = 69,680.

4.3. Energy integral in 2D Models

Figure 10 shows the cumulative energies of the viscoelastic and friction terms, along with their sum (the total dissipational energy change) for the entire pipeline under different initial Re values.

The cumulative work done by the friction term increased with time, most steeply between 0 and 5 L/a; after 20 L/a, the energy dissipation gradually stabilised. This indicates that energy dissipation was the dominant mode of energy change.

The positive energy change of the viscoelastic term was the largest between 0 and 2.5 L/a, and the negative energy change was largest between 2.5 and 5 L/a, as shown in Figure 10. The energy change in the viscoelastic term had a sinusoidal fluctuation and an overall upward trend over time. This indicates that the energy change between the viscoelastic pipes and the fluid in the pipe not only led to energy conversion but also energy dissipation in the initial stage of the transient flow in the pipes. The energy changes became predominantly dissipative as time elapsed.

It can also be seen from Figure 10 that the energy changes of both the viscoelastic and friction terms increased with Re. At different initial Re values, the energy variation trend of the friction term was essentially the same. However, for the energy change of the viscoelastic term, the fluctuation range was larger and the duration longer when the initial Re values were small.

In particular, the proportion of energy dissipation generated by the friction term in the total energy change increased continuously with an increase in the initial Re values. Accordingly, the proportion of energy change generated by the viscoelastic term in the total energy change decreased. More specifically, the proportions of the viscoelastic terms in the total energy variation were 22.1% (Re = 13,880), 18.2% (Re = 28,320), 15.1% (Re = 40,950), 11.9% (Re = 55,940), and 10.1% (Re = 69,680) when the transient flow time was approximately 60 L/a.

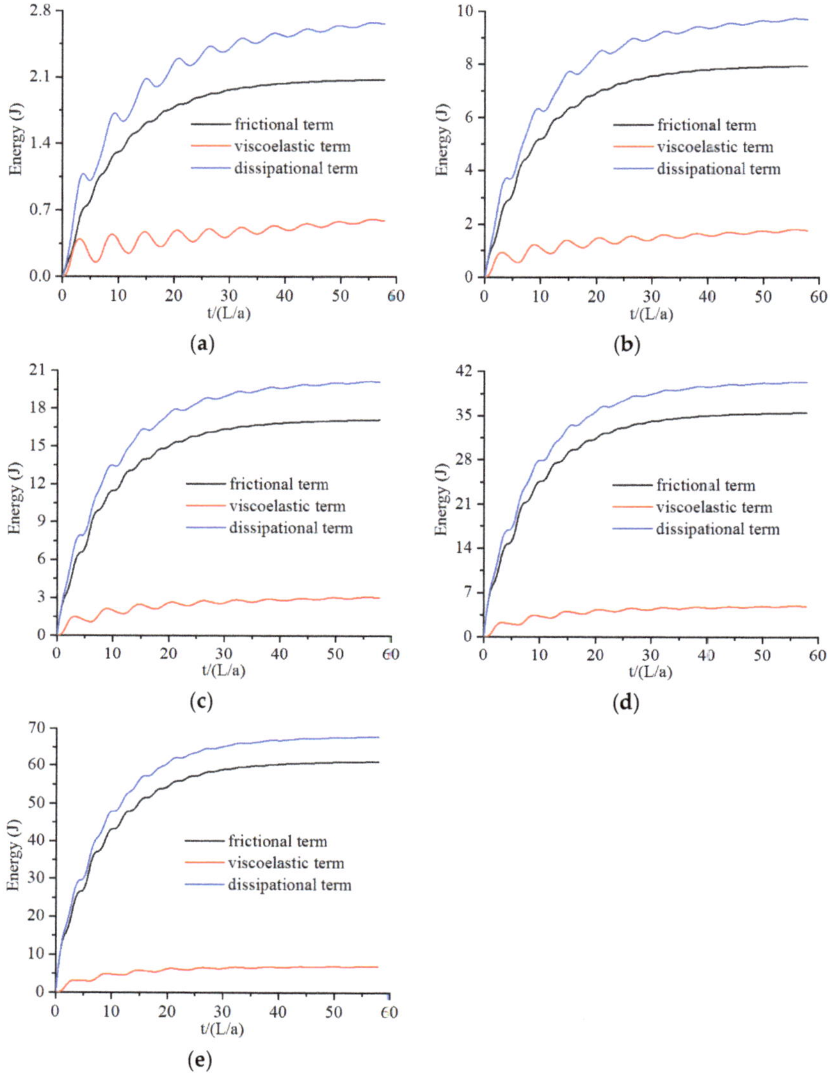

Figure 10. Energy variation of 2D model in critical region at different Re values: (**a**) Re = 13,880; (**b**) Re = 28,320; (**c**) Re = 40,950; (**d**) Re = 55,940; (**e**) Re = 69,680.

5. Conclusions

In this study, the ITE method was used to derive the energy equation of a quasi-2D model of viscoelastic pipes, and the work changes of the friction term and viscoelastic term in 1D and 2D models under different initial Re values were compared. The results are as follows:

(1) When the initial Re was $< 3.0 \times 10^5$, the 1D (but not the 2D) model underestimated the work due to the friction term D_f. For Re $> 3.0 \times 10^5$, this error was reduced.

(2) The work done by the viscoelastic term of the pipe wall remained approximately constant under different initial Re values for both the 1D and 2D models.

(3) With an increase in Re values, both the viscoelastic and frictional work in the 1D and 2D models increased with time. However, $D_f > W_P$ for a large initial Re.

(4) The energy dissipation of the friction term increased significantly initially then slowed down, and finally approached a constant value.

(5) The energy change of the viscoelastic term exhibited sinusoidal fluctuation during the early stage of the transient flow. At a smaller initial Re, the fluctuation in energy lasted for a long time, presented an overall upward trend, and finally approached a constant value.

(6) With an increase in initial Re values, the proportion of the energy dissipation generated by the friction term in the total energy change increased continuously, whereas the proportion of the energy change generated by the viscoelastic term in the total energy change decreased correspondingly.

Author Contributions: Conceptualization, K.W. and Y.F.; methodology, Y.X.; software, H.L.; validation, G.L., Y.F. and Y.X.; formal analysis, K.W.; investigation, G.L.; resources, Y.F.; data curation, H.L.; writing—original draft preparation, K.W.; writing—review and editing, Y.X.; visualization, Y.F.; supervision, Y.F.; project administration, Y.X.; funding acquisition, Y.X. All authors have read and agreed to the published version of the manuscript.

Funding: This research was funded by the National Natural Science Foundation of China (Grant No. 51978202), the Science Foundation of Harbin University of Commerce (2020CX07), and the Natural Science Fund of Heilongjiang Province (LH2019E111 and LH2020E028).

Data Availability Statement: All relevant data are included in the paper. The data presented in this study are available on request from the corresponding author.

Conflicts of Interest: The authors declare no conflict of interest.

Nomenclature

The following symbols are used in this paper:

A	cross-sectional area of the pipeline
a	wave speed
D	pipe diameter
D_f	total rate of frictional dissipation
E_k	elasticity modulus of the *k*-th element
e	wall thickness
g	gravitational acceleration
H	pressure head
J	creep compliance of the *k*-th element
j	subscript representing the radial grid number
N_r	number of segments along the radius
N_{r0}	number of cylinders along the radius
Q	discharge
q	radial flux
r	radial distance from the pipe centre
r_{cj}	radial distance between the centre of the cylinder *j* cross-section and the pipe centre
r_j	radial distance between the outer surface of the cylinder j cross-section and the pipe centre
T	total kinetic energy of the system
t	time
U	total internal energy of the system
u	longitudinal velocity
v	radial velocity
W_E	total rate of work from the ends of the pipe
W_P	total rate of work from the pipe wall
x	axial coordinate along the pipe

Greek Symbols	
α	constraint coefficient
ε, θ	weighting coefficients
ε_r	retarded strain
ρ	density
τ	shear stress
τ_k	retarded time of the k-th element
τ_w	pipe-wall shear stress
γ	bulk weight
Abbreviations	
quasi-2D	quasi-two-dimensional
1D	one-dimensional
Re	Reynolds number
K-V	Kelvin-Voight
HDPE	high-density polyethylene pipe
MOC	method of characteristics

Appendix A. Derivation of Equation (29)

We multiply Equation (3) by Hdx and integrate along the pipe length L:

$$\frac{g}{2a^2}\frac{d}{dt}\int_0^L H^2dx + \int_0^L H\frac{\partial u}{\partial x}dx + 2\int_0^L H\frac{\partial \varepsilon}{\partial t}dx = 0 \tag{A1}$$

Then, we integrate along the pipe diameter D:

$$\frac{g}{2a^2}\frac{d}{dt}\int_0^{D/2}\int_0^L H^2dxdr + \int_0^{D/2}\int_0^L H\frac{\partial u}{\partial x}dxdr + 2\int_0^{D/2}\int_0^L H\frac{\partial \varepsilon}{\partial t}dxdr = 0 \tag{A2}$$

In the quasi-2D model, the pressure changes are only along the axial direction of the pipeline, and slight changes in the radial direction are ignored. Therefore, the first and third terms in Equation (A2) can be treated as constants in the pipeline radial direction:

$$\frac{g}{2a^2}\cdot\frac{D}{2}\frac{d}{dt}\int_0^L H^2dx + \int_0^L\int_0^{D/2} H\frac{\partial u}{\partial x}drdx + 2\cdot\frac{D}{2}\int_0^L H\frac{\partial \varepsilon}{\partial t}dx = 0 \tag{A3}$$

Multiplying by $2\rho Ag/D$, we have

$$\frac{\rho Ag^2}{2a^2}\frac{d}{dt}\int_0^L H^2dx + \frac{\pi\rho Dg}{2}\int_0^L\int_0^{D/2} H\frac{\partial u}{\partial x}drdx + 2\rho Ag\int_0^L H\frac{\partial \varepsilon}{\partial t}dx = 0 \tag{A4}$$

Next, we multiply Equation (2) by u and integrate along the pipe diameter D and the pipe length L simultaneously:

$$\frac{1}{2}\frac{d}{dt}\int_0^L\int_0^{D/2} u^2drdx + g\int_0^L\int_0^{D/2} u\frac{\partial H}{\partial x}drdx - \frac{1}{\rho}\int_0^L\int_0^{D/2}\frac{u}{r}\frac{\partial(r\tau)}{\partial r}drdx = 0 \tag{A5}$$

Calculating the derivatives of the second term of Equation (A5), we have

$$\int_0^{D/2}\int_0^L u\frac{\partial H}{\partial x}dxdr = \int_0^{D/2}\left[-\int_0^L H\frac{\partial u}{\partial x}dx + u(L,t)H(L,t) - u(0,t)H(0,t)\right]dr \tag{A6}$$

Solving Equations (A4) and (A6) simultaneously yields

$$\begin{aligned}&\tfrac{\rho Ag^2}{2a^2}\tfrac{d}{dt}\textstyle\int_0^L H^2dx - \tfrac{\pi\rho Dg}{2}\int_0^{D/2}\int_0^L u\tfrac{\partial H}{\partial x}dxdr + 2\rho Ag\int_0^L H\tfrac{\partial \varepsilon}{\partial t}dx\\ &+\tfrac{\pi\rho Dg}{2}\textstyle\int_0^{D/2}[u(L,t)H(L,t) - u(0,t)H(0,t)]dr = 0\end{aligned} \tag{A7}$$

Finally, solving Equations (A5) and (A7) simultaneously yields:

$$\begin{aligned} &\frac{\rho A g^2}{2a^2}\frac{d}{dt}\int_0^L H^2 dx + \frac{\pi\rho D}{4}\frac{d}{dt}\int_0^L\int_0^{D/2} u^2 dr dx - \frac{\pi D}{2}\int_0^L\int_0^{D/2}\frac{u}{r}\frac{\partial(r\tau)}{\partial r} dr dx \\ &+2\rho A g\int_0^L H\frac{\partial\varepsilon}{\partial t}dx + \frac{\pi\rho D g}{2}\int_0^{D/2}[u(L,t)H(L,t) - u(0,t)H(0,t)]dr = 0 \end{aligned} \quad \text{(A8)}$$

References

1. Duan, H.F.; Pan, B.; Wang, M.L.; Chen, L.; Zheng, F.; Zhang, Y. State-of-the-art review on the transient flow modeling and utilization for urban water supply system (UWSS) management. *J. Water Supply Res. Technol.* **2020**, *69*, 858–893. [CrossRef]
2. Rieutord, E.; Blanchard, A. Pulsating viscoelastic pipe flow–water-hammer. *J. Hydraul. Res.* **1979**, *17*, 217–229. [CrossRef]
3. Covas, D.; Stoianov, I.; Mano, J.F.; Ramos, H.; Graham, N.; Maksimovic, C. The dynamic effect of pipe-wall visco-elasticity in hydraulic transients. Part I: Experimental analysis and creep characterization. *J. Hydraul. Res.* **2004**, *42*, 516–530. [CrossRef]
4. Covas, D.; Stoianov, I.; Mano, J.F.; Ramos, H.; Graham, N.; Maksimovic, C. The dynamic effect of pipe-wall viscoelasticity in hydraulic transients. Part II: Model development, calibration and verification. *J. Hydraul. Res.* **2005**, *43*, 56–70. [CrossRef]
5. Urbanowicz, K. Analytical expressions for effective weighting functions used during simulations of water hammer. *J. Theor. Appl. Mech.* **2017**, *55*, 1029–1040. [CrossRef]
6. Pezzinga, G. Quasi-2D model for unsteady flow in pipe networks. *J. Hydraul. Eng.* **1999**, *125*, 676–685. [CrossRef]
7. Firkowski, M.; Urbanowicz, K.; Duan, H.F. Simulation of unsteady flow in viscoelastic pipes. *SN Appl. Sci.* **2019**, *1*, 519. [CrossRef]
8. Urbanowicz, K.; Duan, H.F.; Bergant, A. Transient liquid flow in plastic pipes. *J. Mech. Eng.* **2020**, *66*, 77–90. [CrossRef]
9. Sun, Q.; Zhang, Z.; Wu, Y.; Xu, Y.; Liang, H. Numerical analysis of transient pressure damping in viscoelastic pipes at different water temperatures. *Materials* **2022**, *15*, 4904. [CrossRef]
10. Duan, H.F.; Ghidaoui, M.S.; Lee, P.J.; Tung, Y.-K. Energy analysis of viscoelasticity effects in pipe fluid transients. *J. Appl. Mech.* **2010**, *77*, 044503.1–044503.5. [CrossRef]
11. Karney, B.W. Energy relations in transient closed-conduit flow. *J. Hydraul. Eng.* **1990**, *116*, 1180–1196. [CrossRef]
12. Duan, H.F.; Meniconi, S.; Lee, P.J.; Brunone, B.; Ghidaoui, M.S. Local and integral energy-based evaluation for the unsteady friction relevance in transient pipe flows. *J. Hydraul. Eng.* **2017**, *143*, 04017015. [CrossRef]
13. Meniconi, S.; Brunone, B.; Ferrante, M.; Massari, C. Energy dissipation and pressure decay during transients in viscoelastic pipes with an in-line valve. *J. Fluids Struct.* **2014**, *45*, 235–249. [CrossRef]
14. Meniconi, S.; Duan, H.F.; Brunone, B.; Ghidaoui, M.S.; Lee, P.J.; Ferrante, M. Further developments in rapidly decelerating turbulent pipe flow modeling. *J. Hydraul. Eng.* **2014**, *140*, 04014028. [CrossRef]
15. Riasi, A.; Nourbakhsh, A.; Raisee, M. Energy dissipation in unsteady turbulent pipe flows caused by water hammer. *Comput. Fluids* **2013**, *73*, 124–133. [CrossRef]
16. Keramat, A.; Haghighi, A. Straightforward transient-based approach for the creep function determination in viscoelastic pipes. *J. Hydraul. Eng.* **2014**, *140*, 04014058. [CrossRef]
17. Gong, J.; Zecchin, A.C.; Lambert, M.F.; Simpson, A.R. Determination of the creep function of viscoelastic pipelines using system resonant frequencies with hydraulic transient analysis. *J. Hydraul. Eng.* **2016**, *142*, 04016023. [CrossRef]
18. Urbanowicz, K.; Firkowski, M.; Zarzycki, Z. Modelling water hammer in viscoelastic pipelines: Short brief. *J. Phys. Conf. Ser.* **2016**, *760*, 012037. [CrossRef]
19. Pezzinga, G.; Brunone, B.; Meniconi, S. Relevance of pipe period on Kelvin-Voigt viscoelastic parameters: 1D and 2-D inverse transient analysis. *J. Hydraul. Eng.* **2016**, *142*, 04016063. [CrossRef]
20. Pan, B.; Duan, H.F.; Meniconi, S.; Urbanowicz, K.; Che, T.C.; Brunone, B. Multistage Frequency-Domain Transient-Based Method for the Analysis of Viscoelastic Parameters of Plastic Pipes. *J. Hydraul. Eng.* **2020**, *146*, 04019068. [CrossRef]

water

MDPI

Article

Hydraulic Transient Simulation of Pipeline-Open Channel Coupling Systems and Its Applications in Hydropower Stations

Wei Zeng [1,2], Chao Wang [3,*] and Jiandong Yang [1]

1 State Key Laboratory of Water Resources and Hydropower Engineering Science, Wuhan University, Wuhan 430072, China
2 School of Civil, Environmental & Mining Engineering, The University of Adelaide, Adelaide, SA 5005, Australia
3 Power China Kunming Engineering Corporation Limited, Kunming 650051, China
* Correspondence: wc699621@whu.edu.cn

Abstract: Hydraulic systems may involve both pipelines and open channels, which challenges the hydraulic transient analysis. In this paper, a method of characteristics (MOC)-finite volume method (FVM) coupling method has been developed with the pipeline modelled using the MOC and the open channel modelled using the FVM. The coupling boundaries between these two simulation regions are developed based on Riemann invariants. The simulated parameters can be transmitted from the MOC region to the FVM region and in the reverse direction through the coupling boundaries. Validation of the method is conducted on a simple tank-pipe system by comparing the simulated result using 3D computational fluid dynamics (CFD) analysis. The new method is then applied to a prototype hydropower station with a sand basin located between the upstream reservoir and the turbines. The sand basin is modelled as an open channel coupled with the pipes in the system. The transient processes are also simulated by modelling the sand basin as a surge tank. The comparison with the results by the MOC-FVM coupling method shows the new coupling method can provide more reliable and accurate results. This is because the flow velocity in the horizontal direction in the sand basin is considered in the coupling method but neglected when the sand basin is modelled as a surge tank in the MOC.

Keywords: hydraulic transient; open channel flow; hydropower; method of characteristics; finite volume method

Citation: Zeng, W.; Wang, C.; Yang, J. Hydraulic Transient Simulation of Pipeline-Open Channel Coupling Systems and Its Applications in Hydropower Stations. *Water* **2022**, *14*, 2897. https://doi.org/10.3390/w14182897

Academic Editors: Helena M. Ramos and Kamil Urbanowicz

Received: 11 August 2022
Accepted: 7 September 2022
Published: 16 September 2022

Publisher's Note: MDPI stays neutral with regard to jurisdictional claims in published maps and institutional affiliations.

Copyright: © 2022 by the authors. Licensee MDPI, Basel, Switzerland. This article is an open access article distributed under the terms and conditions of the Creative Commons Attribution (CC BY) license (https://creativecommons.org/licenses/by/4.0/).

1. Introduction

Hydraulic transients occur regularly in hydraulic systems conveying fluids, such as water distribution systems, pumping stations [1], subsea pipelines [2] and hydropower stations [3]. They can cause large pressure surges which may damage the pipe assets and lead to operation instabilities when a control system is in operation together with the hydraulic system. Therefore, conducting hydraulic transient simulation is important to facilitate the design of hydraulic systems [4].

Traditional methods to simulate hydraulic transient events include the method of characteristics (MOC) [5,6], the finite-difference method, the finite volume method and frequency domain methods (e.g., transfer matrix method [7]). The MOC gradually becomes the most popular in real engineering practices. The transient simulation software programs to solve engineering problems are mostly based on MOC with some exceptions (e.g., [8]). One of the reasons for its popularity is its ability to consider complicated phenomena, such as the fluid-structure interaction [9], pipe wall viscoelasticity [10] and unsteady friction [11]. Another reason is its simplicity to be integrated with the modelling of other subsystems in a complicated system. For example, the transient simulation of a hydropower station involves the modelling of the turbines, pressure surge control elements (e.g., surge tanks), the control system (to control the operation of guide vanes) and sometimes the electrical system [3].

In engineering practice, some pressurized pipe flow in these hydraulic systems may be coupled with free-surface open channel flow. For example, a division canal may be used to deliver water to the inlet of a pump station. Another example used in some hydropower stations is the sand basin which can also suppress the pressure surge during the transient events. Traditional methods to simulate free surface open channel flow include the finite element method (FEM), the finite-difference method (FDM) and the finite volume method (FVM) [12]. Since the FVM possesses high efficiency and accuracy, modelling flexibility and clear physical meaning [13], the FVM is adopted in this study for open channel flow simulation.

One of the common measures in practice for simulating pipeline-open channel coupling systems is to neglect the effects of the open channel on the pressure response of the entire system. This is because the magnitude of the wave oscillation in the open channel is relatively small compared with the pressure surge in the pressurized pipe. Such measures are acceptable when the open channel is relatively short and only the pressure magnitude in the pipe is of interest. However, such approximation may reduce the accuracy of the simulation especially when the channel plays an important role in the transient response (e.g., the channel is located between two pressurized pipes). Another solution to such issues is the 1D-3D coupling simulation [14–17] with pipelines modelled using 1D MOC and other parts modelled with a 3D CFD method. For example, the ANSYS-fluent package can be used for 3D CFD simulation coupling with 1D MOC simulation in hydropower stations and pump stations. This method shows its advantages when the 3D region possesses a complicated geometry that is difficult to simplify to a 1D or 2D model, such as pump turbines in [16]. However, this method requires high computational resources, which conflicts with the requirement of fast computation and sometimes near-real time simulation for many engineering cases (e.g., hydropower stations, pump stations) involving open channels. The geometries of the open channels in these cases are relatively simple, and mostly can be simplified into a 1D model.

To address the engineering issue described above, this paper proposed a coupling method to simulate such coupling systems. The 1D MOC method is still used to model the pipes in the complicated system, the FVM is used to simulate the open channel flow in the system, and the coupling boundary is developed based on *Riemann* invariants. Such coupling method has been integrated into the comprehensive transient modelling software program TOPSYS which has been developed by the transient research group at Wuhan university [3,18] and validated through both model testing [19] and prototype testing [20]. This enables the simulation of complicated hydropower systems with considering the effects of the open channel components in the system.

The structure of the paper is as follows: Following the literature review in the introduction, the next section illustrates the methodologies to develop the coupling transient simulation. The 1D simulation method is then numerically validated with a 3D CFD case on a simple tank-pipe coupling system. The validated method is then applied to a prototype hydropower station to solve an engineering problem associated with a sand basin. The simulation results by the new method compared with the traditional transient simulation method (MOC) in this hydropower system illustrate the distinctive effects of the sand basin on both the magnitude and the period of the pressure surge wave as well as on the operation stability of the hydropower system.

2. Mathematical Model and Verifications

2.1. Transient Modelling of Pipeline Flow

The continuity and momentum equations of the pipeline transient flows are given as follows [21]:

$$\frac{\partial H}{\partial t} + \frac{a^2}{gA}\frac{\partial Q}{\partial x} = 0 \tag{1}$$

$$\frac{\partial H}{\partial x}+\frac{1}{gA}\frac{\partial Q}{\partial t}+f\frac{Q|Q|}{2gDA^2}=0 \tag{2}$$

in which H is the piezometric head; Q is the discharge; x is the distance; t is the time; a is the wave speed of the transient waves; g is the gravitational acceleration; f is the Darcy-Weisbach friction factor; D is the pipe diameter; and A is the cross-sectional area.

The method of characteristics is widely used to solve Equations (1) and (2) and is briefly introduced below. The partial differential equations can be transformed into ordinary differential equations by:

$$C^+:\ (Q_i-Q_{i-1})+\frac{gA}{a}(H_i-H_{i-1})+\frac{f}{2DA}Q_{i-1}|Q_{i-1}|\Delta t=0 \tag{3}$$

$$C^-:\ (Q_i-Q_{i+1})-\frac{gA}{a}(H_i-H_{i+1})+\frac{f}{2DA}Q_{i+1}|Q_{i+1}|\Delta t=0 \tag{4}$$

which are valid along with the positive (C^+) and negative (C^-) characteristic lines, as shown in Figure 1. The subscripts $i-1$, i and $i+1$ represent the space steps. By rearranging Equations (3) and (4), the following compatibility equations can be obtained:

$$C^+:\ Q_i=C_P-B{\cdot}H_i \tag{5}$$

$$C^-:\ Q_i=C_M+B{\cdot}H_i \tag{6}$$

where $C_P=Q_{i-1}+B{\cdot}H_{i-1}-R{\cdot}Q_{i-1}|Q_{i-1}|\Delta t$; $C_M=Q_{i+1}-B{\cdot}H_{i+1}-R{\cdot}Q_{i+1}|Q_{i+1}|\Delta t$; $B=gA/a$; $R=f/2DA$. Details of the method can be found in [21].

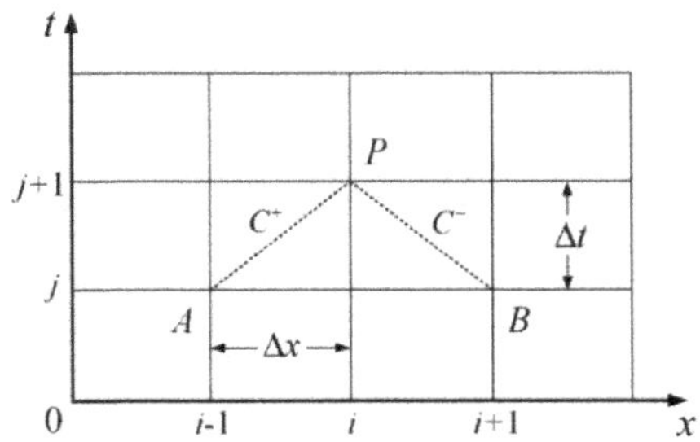

Figure 1. Characteristics lines in the x-t plane.

According to the space-time plane as shown in Figure 1, the parameters Q_i and H_i at time step $j+1$ can be calculated by using Equations (5) and (6) if Q_{i-1}, Q_{i+1}, H_{i-1} and H_{i+1} at time step j have already been obtained from the calculation in the time step $j-1$ or are already known from the steady state condition.

Apart from the MOC, the Preissmann four-point finite-difference method (FDM) can be also used to solve Equations (1) and (2). By using a forward scanning method [21], a linear relationship between the flow rate and the pressure head at each grid node can be obtained as:

$$Q_i=EE_i{\cdot}H_i+FF_i \tag{7}$$

where EE_i and FF_i are two terms that are related to the boundary conditions, the piezometric head and the discharge at node i and at the time step j. Details of the method can be found in [21].

It can be found from Equations (5)–(7) that a linear relationship between the flow rate and the pressure head can be obtained in both MOC and FDM. Both linear equations can be used to be coupled with the FVM which is used to solve the open channel flow. In the case studies in this paper, the MOC was used to simulate the pipeline flow. But it should be noted that the proposed coupling technique applies to FDM as well.

2.2. Modelling of Open Channel Flow

The continuity and momentum equations of the open channel flow are given as [22,23]

$$\frac{\partial \mathbf{U}}{\partial t} + \frac{\partial \mathbf{F}}{\partial x} = \mathbf{S} \tag{8}$$

with

$$\mathbf{U} = \begin{bmatrix} h \\ q \end{bmatrix} \ \mathbf{F} = \begin{bmatrix} q \\ uq + \frac{1}{2}gh^2 \end{bmatrix} \mathbf{S} = \begin{bmatrix} 0 \\ -\tau - gh\frac{\partial z}{\partial x} \end{bmatrix} \tag{9}$$

in which $\mathbf{U}$ is the vector of conserved variables, $\mathbf{F}$ is the vector of fluxes and each of its components is a function of the components of $\mathbf{U}$, $\mathbf{S}$ is the source term, u is the flow velocity in the x-direction which is along with the channel, h is the water depth, q is the discharge per unit width, z is the bottom elevation of the channel, τ is a parameter on the friction of the channel wall and can be expressed as $\tau = gn^2u|u|/h^{1/3}$ with n representing the Manning coefficient.

In this paper, this non-linear equation group is linearized locally and the fluxes at the grid interfaces are calculated using the *Roe*–average parameters. The Godunov-type finite volume method (FVM) [22,23] is applied to solve Equation (9). The grids in the FVM are shown in Figure 2, in which the grid node i + 1/2 is defined as the shared node of grid i and grid i + 1, and the subscripts L and R are the vectors on the left side and the right side of a grid node, respectively, and the total number of the grids is N.

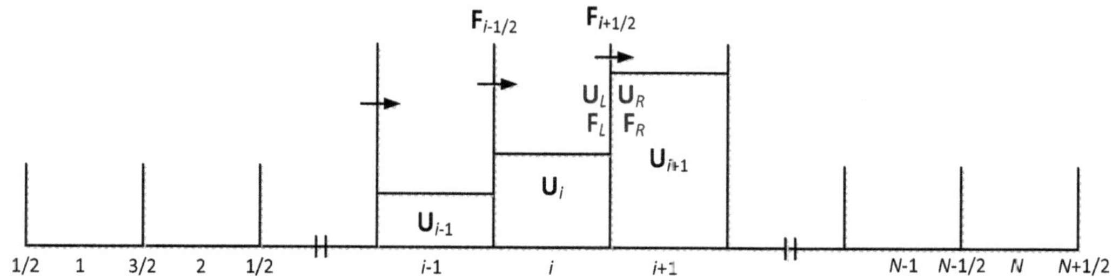

Figure 2. The schematic of the FVM grids.

Integrating the continuity equation and the momentum equation for the control volume i gives:

$$\frac{d}{dt}\int_{i-1/2}^{i+1/2} \mathbf{U}dx + \mathbf{F}_{i-1/2} - \mathbf{F}_{i+1/2} = \int_{i-1/2}^{i+1/2} \mathbf{S}dx \tag{10}$$

By setting $\mathbf{U}_i$ and $\mathbf{S}_i$ as the averaged values of the control volume i, Equation (10) can be then rewritten as:

$$\mathbf{U}_i^{k+1} = \mathbf{U}_i^k + \frac{\Delta t}{\Delta x}(\mathbf{F}_{i-1/2} - \mathbf{F}_{i+1/2}) + \Delta t\mathbf{S}_i \tag{11}$$

with the source term expressed as

$$\mathbf{S}_i = \begin{bmatrix} 0 \\ -gn^2u_i|u_i|/h_i^{1/3} - gh_i\frac{z_{i+1/2}-z_{i-1/2}}{\Delta x} \end{bmatrix} \tag{12}$$

in which the superscript k is the number of the time step, and the subscript i is the grid number. It should be noted that i represents the number of the grid node in the MOC and FDM. The symbols Δt and Δx are the duration of the time step and the size of the grid, respectively. The following processes show the procedures to calculate the vectors of fluxes $\mathbf{F}_{i-1/2}$ and $\mathbf{F}_{i+1/2}$ [22,23].

Equation (8) can be rewritten as:

$$\frac{\partial \mathbf{U}}{\partial t} + \mathbf{A}\frac{\partial \mathbf{U}}{\partial x} = 0 \quad (13)$$

in which **A** is the Jacob matrix of the flux function **F**(**U**) and can be expressed as:

$$\mathbf{A} = \frac{\partial \mathbf{F}(\mathbf{U})}{\partial \mathbf{U}} = \begin{vmatrix} \frac{\partial q}{\partial h} & \frac{\partial q}{\partial q} \\ \frac{\partial\left(uq+\frac{1}{2}gh^2\right)}{\partial h} & \frac{\partial\left(uq+\frac{1}{2}gh^2\right)}{\partial q} \end{vmatrix} = \begin{vmatrix} 0 & 1 \\ gh-u^2 & 2u \end{vmatrix} \quad (14)$$

The linearization of Equation (14) gives:

$$\frac{\partial \mathbf{U}}{\partial t} + \overline{\mathbf{A}}\frac{\partial \mathbf{U}}{\partial x} = 0 \quad (15)$$

in which $\overline{\mathbf{A}}$ is a constant matrix and can be calculated based on the known $\mathbf{U}_L$ and $\mathbf{U}_R$ by

$$\overline{\mathbf{A}} = \begin{bmatrix} 0 & 1 \\ g\overline{h}-\overline{u}^2 & 2\overline{u} \end{bmatrix} \quad (16)$$

in which $\overline{h} = \sqrt{h_L h_R}$. The eigenvalues of $\overline{\mathbf{A}}$ are:

$$\overset{\Delta}{\lambda}_1 = \overline{u} - \sqrt{g\overline{h}} = \overline{u} - \overline{c} \quad (17)$$

$$\overset{\Delta}{\lambda}_2 = \overline{u} + \sqrt{g\overline{h}} = \overline{u} + \overline{c} \quad (18)$$

and the corresponding right eigenvectors are:

$$\overline{\mathbf{K}}_1 = (1, \overline{u} - \overline{c})^T \quad (19)$$

$$\overline{\mathbf{K}}_2 = (1, \overline{u} + \overline{c})^T \quad (20)$$

The wave strength, $\overline{\alpha_i}$ can be obtained using the following equation

$$\Delta\mathbf{U} = \begin{bmatrix} \Delta u_1 \\ \Delta u_2 \end{bmatrix} = \begin{bmatrix} \Delta h \\ \Delta(hu) \end{bmatrix} = \begin{bmatrix} \Delta h \\ \overline{h}\Delta u + \overline{u}\Delta h \end{bmatrix} = \sum_{j=1}^{2} \overline{\alpha_j}\,\overline{\mathbf{K}}_j \quad (21)$$

with the solutions as:

$$\overline{\alpha}_1 = \frac{1}{2}\left(\Delta h - \frac{\overline{h}}{\overline{c}}\Delta u\right) \quad (22)$$

$$\overline{\alpha}_2 = \frac{1}{2}\left(\Delta h + \frac{\overline{h}}{\overline{c}}\Delta u\right) \quad (23)$$

in which $\Delta h = h_R - h_L$, $\Delta u = u_R - u_L$. The flux $\mathbf{F}_{i-1/2}$ and $\mathbf{F}_{i+1/2}$ can be then obtained using

$$\mathbf{F} = \frac{1}{2}(\mathbf{F}_L + \mathbf{F}_R) - \frac{1}{2}\sum_{j=1}^{2} \overline{\alpha}_j |\overline{\lambda}_j| \overline{\mathbf{K}}_j \quad (24)$$

2.3. Boundary Condition Based on Riemann Invariants

A general m-dimensional quasi-linear hyperbolic system can be given as [22,23]:

$$\frac{\partial \mathbf{W}}{\partial t} + \mathbf{A}(\mathbf{W})\frac{\partial \mathbf{W}}{\partial x} = 0 \quad (25)$$

in which $\mathbf{W} = [w_1, w_1, \ldots w_m]^T$ is the vector of dependent variables. The corresponding right eigenvector of the wave associated with i^{th} characteristic field with eigenvalue λ_i is:

$$\mathbf{K}_i = \left[k_i^1, k_i^2, \ldots k_i^m\right]^T \tag{26}$$

The generalized Riemann invariants are relations that hold for each wave, leading to m-1 ordinary differential equations as:

$$\frac{dw_1}{k_i^1} = \frac{dw_2}{k_i^2} = \frac{dw_3}{k_i^3} = \cdots = \frac{dw_m}{k_i^m} \tag{27}$$

For the open channel equations, the right eigenvectors are:

$$\mathbf{K}_1 = (1, u - c)^T \tag{28}$$

$$\mathbf{K}_2 = (1, u + c)^T \tag{29}$$

Thus, substituting Equation (28) into Equation (27) gives

$$dh = \frac{dq}{u - \sqrt{gh}} \tag{30}$$

By taking the relationship

$$q = hu \tag{31}$$

into Equation (30), it gives

$$dh = \frac{hdu + udh}{u - \sqrt{gh}} \tag{32}$$

which can be rearranged to

$$du + \sqrt{\frac{g}{h}}dh = 0 \tag{33}$$

The integration of Equation (33) gives

$$u + 2\sqrt{gh} = \text{constant} \tag{34}$$

According to Equation (34), the flow velocity and pressure head at the downstream boundary in the FVM region at the time step k+1 can be linked with those at the centre of the last FVM grid in the downstream side at the time step k. The expression is given as:

$$u_{N+1/2}^{k+1} + 2\sqrt{gh_{N+1/2}^{k+1}} = u_N^k + 2\sqrt{gh_N^k} \tag{35}$$

Similar processes can be conducted on the upstream coupling boundary using Equations (34) and (29). The result can be obtained as

$$u_{1/2}^{k+1} - 2\sqrt{gh_{1/2}^{k+1}} = u_1^k - 2\sqrt{gh_1^k} \tag{36}$$

It should be noted that the coupling boundary equations (Equations (35) and (36)) are only applicable to open channels with a subcritical flow which is the case for most engineering applications.

2.4. The Simulation Process

With all the equations listed above, this section gives the processes of the coupling simulation using the grid schematic as shown in Figure 3. To include the coupling boundaries that transmit data from both MOC to FVM and from FVM to MOC, a MOC-FVM-MOC coupling system with 10 grids is used to illustrate the simulation process. The horizontal and vertical coordinates represent the space step and the time step, respectively. The dots on the horizontal coordinate are the grid nodes. In the MOC regions, the solid diagonals represent the calculation process of the C+ (Equation (5)) and C− (Equation (6)) equations and the vertical lines which are superposed on the vertical coordinates mean the boundary conditions that are determined by the system. In the FVM region, the vertical dot-dashed lines represent the calculation process of Equation (11). The dash lines in the leftmost grids in the FVM region represent the coupling boundary of Equation (36) and those in the rightmost grids in the FVM region represent the coupling boundary of Equation (35). Starting from $k = 0$ with the initial parameters known, the simulation processes are represented by these lines in Figure 3.

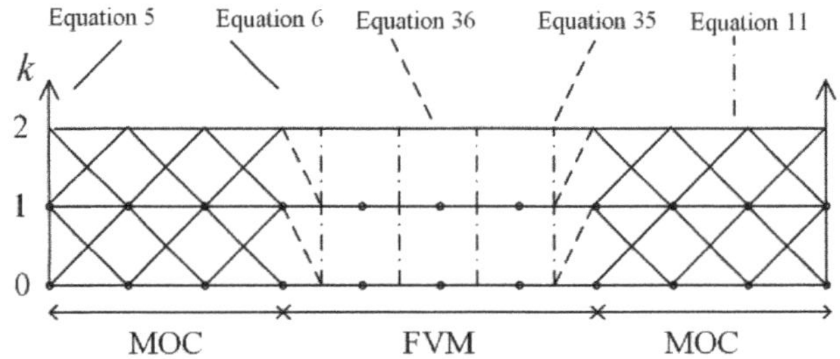

Figure 3. The calculation process of the coupling method represented by grids.

3. Numerical Validations

As shown in Figure 4, a tank-pipe system was used to validate the MOC-FVM coupling method. The tank at the upstream side of the pipe is 250 m in length and 5 m in width with an initial water depth of 2 m. The bottom of the tank is horizontal. The pipeline is 250 m in length with the area of the cross-section equal to 1 m^2. The wave speed of transient waves in the pipeline is 1000 m/s. The initial flow rate in the pipe is 0 m/s with the valve fully closed at the downstream end of the pipeline. Water is transmitted to the pipeline through the valve and the flow rate increases from 0 m^3/s to 2m^3/s in 5 s and keeps constant at 2 m^3/s after 5 s. The friction losses in the tank and pipeline are neglected in this validation case.

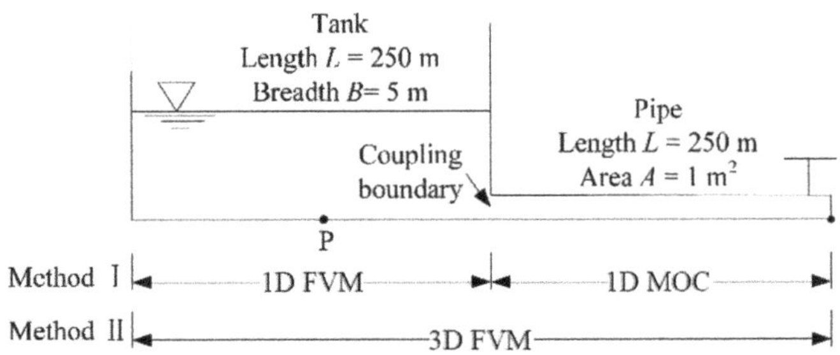

Figure 4. Schematic of the tank-pipe system.

Two methods were applied to simulate this tank-pipe case. The first method (Method I) is the proposed MOC-FVM coupling method. The time step in the simulation is 0.005 s and the spatial step is 5 m for the 1D meshes. The 3D FVM method (Method II) is used to validate the proposed coupling method. In Method II, the whole system including the tank and the pipe is modelled in three dimensions using ANSYS Fluent (3D FVM based simulation). The size of the meshes is 0.2 m and the fluid is assumed as inviscid.

The water depth at position P in Figure 4 is monitored in both simulations. The variations of the water depths from these two simulations are compared in Figure 5, which shows the simulated results are almost identical. The comparison illustrates the proposed

MOC-FVM coupling method is able to simulate the unsteady flow for a pipeline-open channel coupling system.

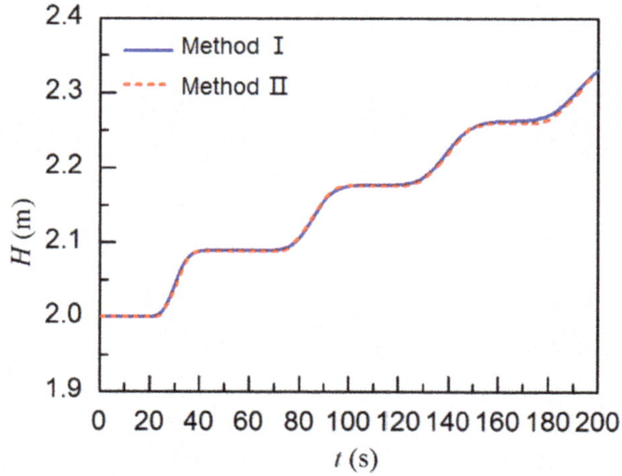

Figure 5. Comparison of the simulated results by the MOC-FVM coupling method and the 3D FVM.

4. Transient Simulation of a Hydropower Station with a Sand Basin

4.1. System Configuration and Modelling

The schematic of the hydropower station is shown in Figure 6. There are three Francis turbines in the station and they share one main penstock on the upstream side. A rectangle open channel is planned to build between the upstream reservoir and the turbines to serve as a sand basin. The length of the sand basin is set as L_s and the width is w. The initial water depth in the sand basin is h. the Manning coefficient in the sand basin is 0.014. The length of the tunnel between the upstream reservoir and the sand basin is L_1 and the length of the penstock between the sand basin and the bifurcation is L_2. The sum of L_1 and L_2 is 3370 m and the diameter of the tunnel and penstock are both 8.7 m. The wave speeds of the transient wave in the pipelines are calculated based on the theoretical equation [5] and are around 1100 m/s. The Darcy-Weisbach coefficient is 0.02. Some other basic information about the hydropower station is shown in Table 1.

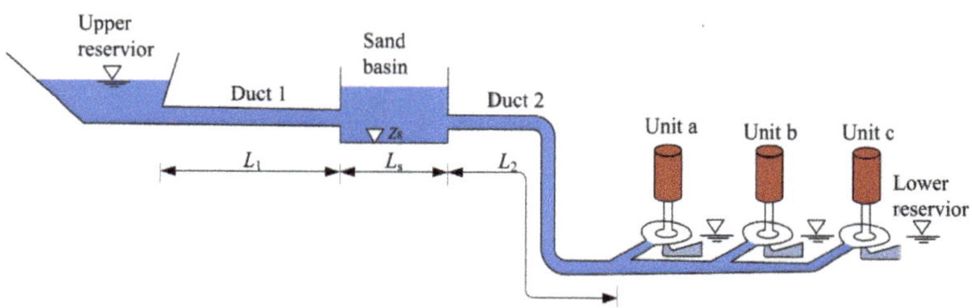

Figure 6. Schematic of the hydropower station with a sand basin.

Table 1. Basic information about the hydropower station.

Runner Inlet Diameter (m)	Guide Vane Height (m)	Upstream Water Level (m)	Downstream Water Level (m)	Rated Rotational Speed (rpm)	Rated Output (Mw)	Rated Flow Rate (m^3/s)	Rotational Inertia (t.m^2)
2.3	0.7	1073	829	429.6	10.54	29	726

The proposed MOC-FVM method has been integrated into the hydraulic transient simulation program–TOPSYS as shown in Figure 7. The pipe system in the hydropower station is simulated using the proposed MOC-FVM coupling method. The tunnel, penstock, upstream reservoir and branch pipes are simulated using the MOC with the sand basin simulated using the FVM. To simulate the hydraulic transient process of the hydropower station, the turbines and the controller of the servomotor that drives the guide vanes are also modelled with details of the modelling shown in [3]. The characteristic curves are adopted to model the turbines and a PID controller is used to control the movement of the guide vanes.

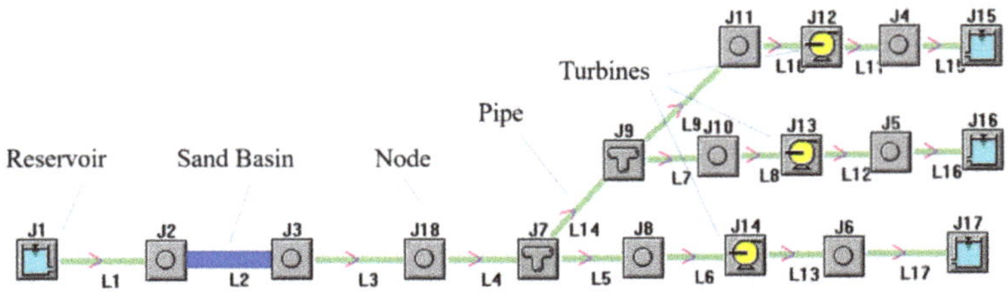

Figure 7. Modelling of the hydropower system in TOPSYS.

4.2. Full Load-Rejection

4.2.1. Simulation Results

In this case, all the three turbines reject their loads simultaneously. The guide vanes of the turbines are closed linearly within 10 s. The transient waves caused by the closure of the guide vanes will be partially reflected by the sand basin and the reflected waves will be superimposed with the incident waves. The wave reflection and superposition process will be affected by the position of the sand basin. Thus, three scenarios with L_1 = 1600 m, 2000 m and 2400 m, respectively, were simulated to illustrate the effects of the position of the sand basin on the transient performance of the hydropower station. The pressure heads at the inlet of the spiral case for these three scenarios were compared in Figure 8, which shows the maximum water hammer pressure decreases when the sand basin gets closer to the turbines. The wave fluctuations after 10 s are the water hammer waves and their periods are associated with the wave traveling time in the penstocks.

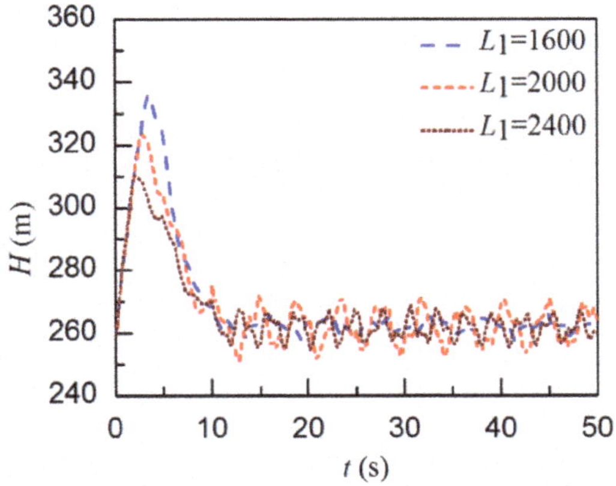

Figure 8. Comparison of the pressure heads at the inlet of the spiral case with different positions of the sand basin.

Another two scenarios with L_1 = 2400 m were conducted by changing the bottom elevation and the width of the sand basin, respectively. The comparison of the pressure heads at the inlet of the spiral case for these two scenarios with the original scenario is shown in Figure 9. The results show that the bottom elevation and width of the basin do not distinctively affect the water hammer pressure. This is because the pressure reaches its maximum shortly (within 5 s) after the load rejection, while the wave oscillation caused by the sand basin has a much large period (as shown in Figure 10).

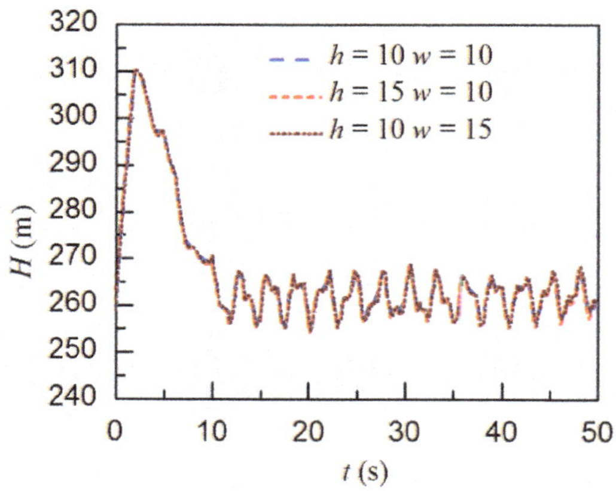

Figure 9. Comparison of the pressure heads at the inlet of the spiral case with different sizes of the sand basin.

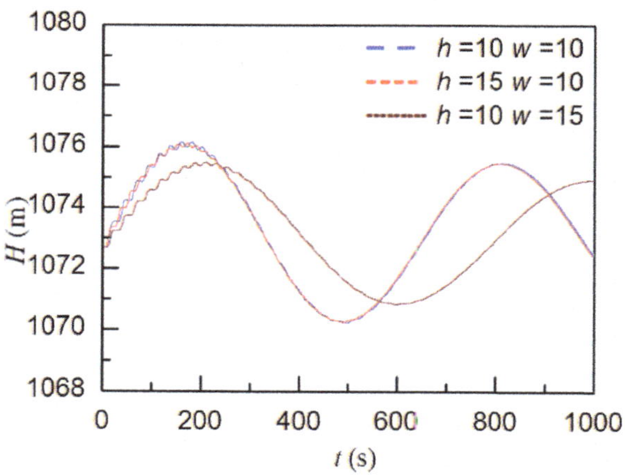

Figure 10. Comparison of the water levels in the sand basin with different sizes of the sand basin (L_1 = 2400 m).

The water level variations at the central point of the tank with L_1 = 2400 m are compared in Figure 10 for different sizes of the sand basin. The comparison shows the bottom elevation of the sand basin has a slight effect on the water level fluctuation. The width of the basin, however, affects the water level variation significantly with the period of the oscillation increasing and the magnitude decreasing for a wider basin.

The water levels at different positions (left side L, central point M and right side R) of the sand basin are compared in Figure 11 when L_1 = 2400 m, h = 10 m and w = 10 m. The comparison shows the overall trend of the water level oscillations at these three points are the same, but different wave oscillations with a short period are superimposed with the low-frequency oscillation. This is caused by the wave propagating in the horizontal direction along the sand basin during the transient process.

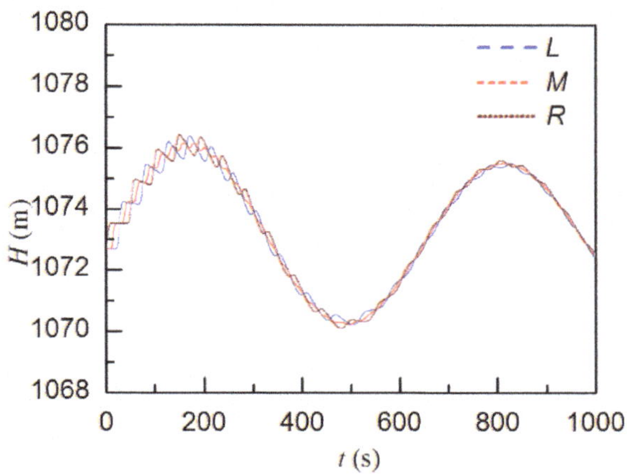

Figure 11. Comparison of the water levels in the sand basin at different positions (L_1 = 2400 m, h = 10 m, w = 10 m).

4.2.2. Comparison with the Results by Modelling the Basin Tank as a Surge Tank

In the previous section, the sand basin is treated as an open channel which is modelled using the FVM. Another modelling method of the sand basin is to simplify it as a surge tank in the hydropower station. With the length of 250 m and width of 10 m for the sand basin, the equivalent area of the surge tank used to model the sand basin is calculated as 2500 m^2. The simulated results by treating the sand basin as a surge tank and those by modelling the sand basin as an open channel are compared in Figures 12 and 13. When the sand basin is close to the turbines (L_1 = 2400 m), a slight difference can be found in the period and magnitude of the water level fluctuations in the sand basin. A more distinctive difference can be observed when the sand basin is far away from the turbines (L_1 = 400 m). Such differences can be ascribed to the fact that the flow velocity and friction in the sand basin along the horizontal direction are neglected in the surge tank modelling but incorporated in the FVM when modelling the sand basin.

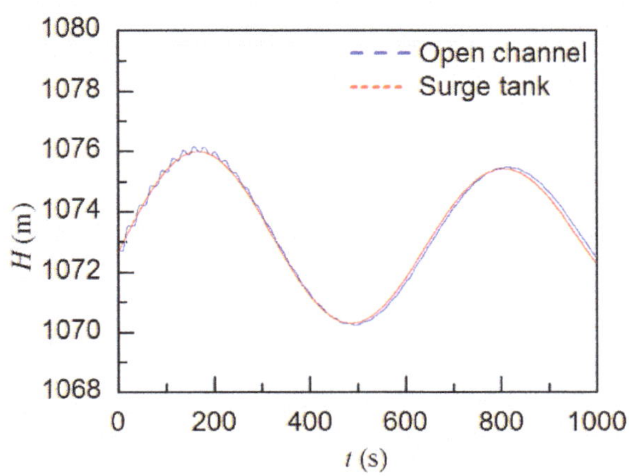

Figure 12. Comparison of the water levels in the sand basin with different modelling methods (L_1 = 2400 m, h = 10 m, w = 10 m).

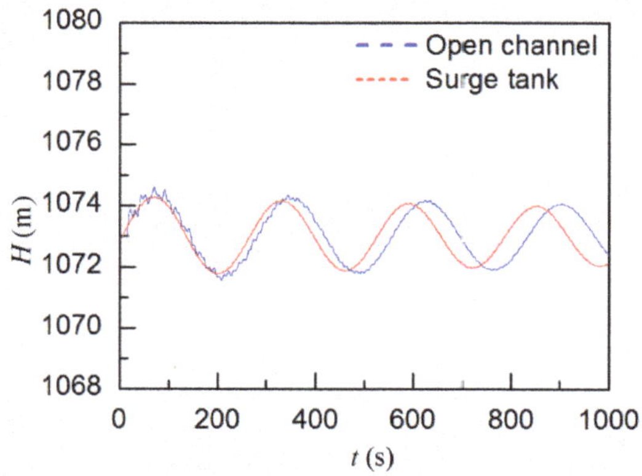

Figure 13. Comparison of the water levels in the sand basin with different modelling methods (L_1 = 400 m, h = 10 m, w = 10 m).

4.3. 5% Load-Rejection with Frequency Regulation

4.3.1. Simulation Results

In this case, the loads of all the turbines are reduced by 5% simultaneously. The guide vanes of the turbines are controlled by the servomotor with a frequency regulation process. The parameters of the controller are shown in Table 2 with the assumed zero load self-regulation coefficient of the power grid. The details of the controller can be found in [3]. Three scenarios with L_1 = 1600 m, 2000 m and 2400 m, respectively, were simulated to illustrate the effects of the position of the sand basin on the transient process. The rotational speed of the turbine and the water level in the sand basin for these three scenarios are compared in Figures 14 and 15, respectively. Similar to having a surge tank in the system, the comparisons show that a shorter distance between the sand basin and the turbines can facilitate the stability of the transient process.

Table 2. Parameter setting for the PID controller.

Temporary Droop	Differential Time Constant	Time Lag in Servomotor	Dashpot Time Constant
0.3	0.3 s	0.05 s	5.0 s

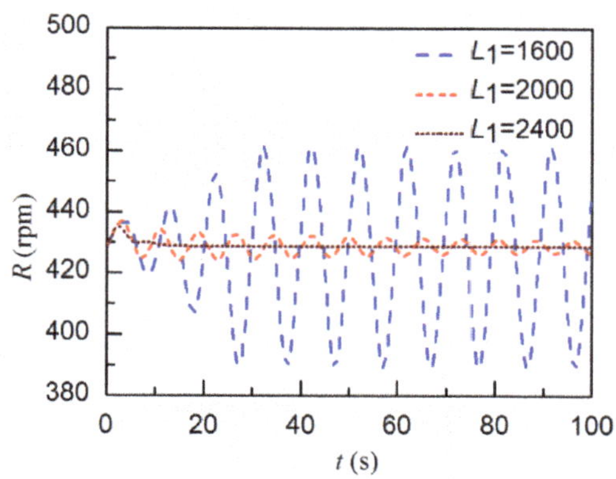

Figure 14. Comparison of the turbine rotational speed with different locations of the sand basin (h = 10 m, w = 10 m).

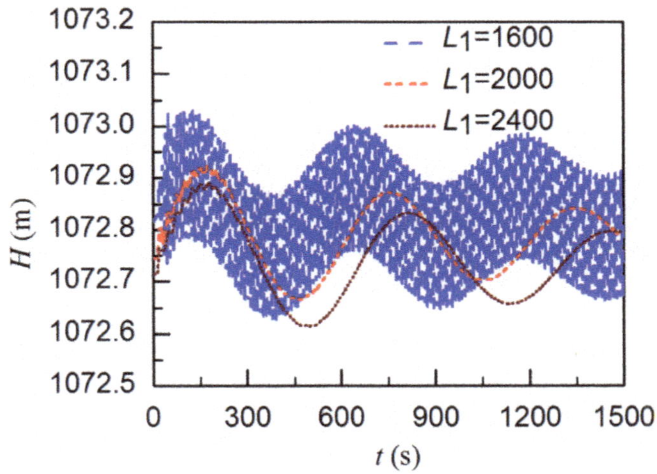

Figure 15. Comparison of the water level fluctuations in the sand basin with different locations of the sand basin (h = 10 m, w = 10 m).

4.3.2. Comparison with the Results by Modelling the Basin Tank as a Surge Tank

With L_1 = 2000 m, the simulated rotational speed of the turbine and the water level fluctuations in the sand basin are shown in Figures 16 and 17, respectively. They are also compared with the results by modelling the sand basin as an open channel. Similar to the full load rejection simulation, the differences observed in the comparison can be ascribed to the neglected flow velocity and friction in the sand basin along the horizontal direction in the surge tank modelling. It can be concluded that treating the sand basin as a surge tank in the modelling may overestimate the operation stability of the system.

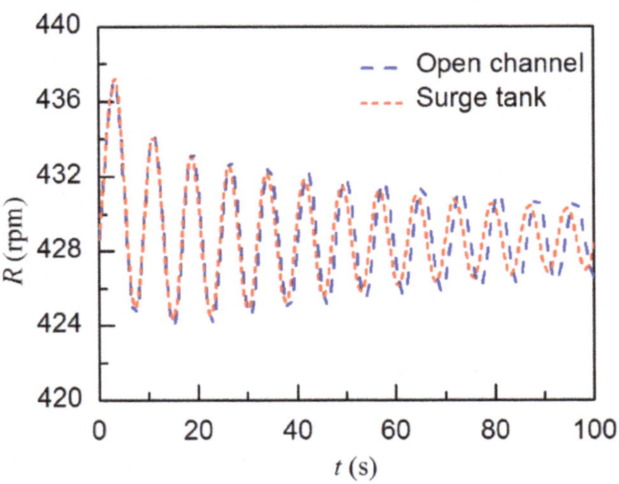

Figure 16. Comparison of the turbine rotational speed with different modelling methods (L_1 = 2000 m, h = 10 m, w = 10 m).

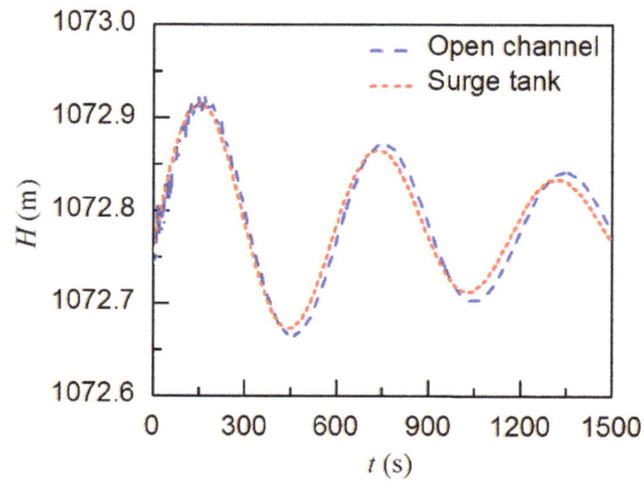

Figure 17. Comparison of the water level fluctuations in the sand basin with different modelling methods (L_1 = 2000 m, h = 10 m, w = 10 m).

5. Conclusions

A MOC-FVM coupling method is developed in this paper to simulate pipeline-open channel coupling transient flow in hydraulic systems including hydropower systems. Pipelines in the systems are modelled using the MOC and the open channel is modelled using the 1D FVM. The coupling boundaries between the MOC simulation region and the FVM simulation region are developed based on *Riemann* invariants. 3D CFD simulation on a tank-pipe system has been conducted with the result almost identical to that simulated by the proposed coupling method. The proposed method is then applied to a prototype hydropower station with a constructed sand basin. The main conclusions include:

1. The MOC-FVM coupling method can accurately simulate the pipeline-open channel coupling transient flow with the simulated parameters transmitted successfully at the coupling boundaries.
2. The coupling method has been successfully applied to a hydropower station with a sand basin constructed between the upstream reservoir and turbines. The sand basin can be modelled as an open channel.
3. The effects of the sand basin on the transient process are similar to a surge tank which can relieve water hammer pressures during load rejection scenarios and can benefit the frequency regulation process. By modelling the sand basin as an open channel, the flow velocity and the friction in the horizontal direction, which are neglected when modelling the sand basin as a surge tank, can be considered, and thus more reliable and accurate results can be obtained.

Author Contributions: Conceptualization, C.W., W.Z. and J.Y.; methodology, C.W. and W.Z.; software, C.W.; validation, C.W. and J.Y.; investigation, C.W. and W.Z.; writing—original draft preparation, W.Z.; writing—review and editing, C.W. and J.Y.; supervision, J.Y. All authors have read and agreed to the published version of the manuscript.

Funding: The research presented in this study was funded by the Open Research Fund Program of the State Key Laboratory of Water Resources and Hydropower Engineering Science (Grant 2017SDG01), China.

Institutional Review Board Statement: Not applicable.

Informed Consent Statement: Not applicable.

Data Availability Statement: Not applicable.

Acknowledgments: Not applicable.

Conflicts of Interest: The authors declare no conflict of interest.

References

1. Kim, S.-G.; Lee, K.-B.; Kim, K.-Y. Water hammer in the pump-rising pipeline system with an air chamber. *J. Hydrodyn.* **2014**, *26*, 960–964. [CrossRef]
2. Zhang, K.; Zeng, W.; Simpson, A.R.; Zhang, S.; Wang, C. Water Hammer Simulation Method in Pressurized Pipeline with a Moving Isolation Device. *Water* **2021**, *13*, 1794. [CrossRef]
3. Yang, W.; Yang, J.; Guo, W.; Zeng, W.; Wang, C.; Saarinen, L.; Norrlund, P. A Mathematical Model and Its Application for Hydro Power Units under Different Operating Conditions. *Energies* **2015**, *8*, 10260–10275. [CrossRef]
4. Karpenko, M.; Bogdevicius, M. Investigation into the hydrodynamic processes of fitting connections for determining pressure losses of transport hydraulic drive. *Transport* **2020**, *35*, 108–120. [CrossRef]
5. Wylie, E.B.; Streeter, V.L. *Fluid Transients in Systems*; Prentice Hall Inc.: Englewood Cliffs, NJ, USA, 1993.
6. Urbanowicz, K. Modern Modeling of Water Hammer. *Pol. Marit. Res.* **2017**, *24*, 68–77. [CrossRef]
7. Chaudhry, M.H. *Applied Hydraulic Transients*, 3rd ed.; Springer: New York, NY, USA, 2014.
8. Nicolet, C. Hydroacoustic Modelling and Numerical Simulation of Unsteady Operation of Hydroelectric Systems. Ph.D. Thesis, École Polytechnique Fédérale de Lausanne, Lausanne, Switzerland, 2007.
9. Tijsseling, A.S. Fluid-structure interaction in liquid-filled pipe systems: A review. *J. Fluids Struct.* **1996**, *10*, 109–146. [CrossRef]
10. Covas, D.; Stoianov, I.; Mano, J.F.; Ramos, H.; Graham, N.; Maksimovic, C. The dynamic effect of pipe-wall viscoelasticity in hydraulic transients. Part I—Experimental analysis and creep characterization. *J. Hydraul. Res.* **2004**, *42*, 516–530. [CrossRef]
11. Vardy, A.E.; Brown, J.M.B. Transient, turbulent, smooth pipe friction. *J. Hydraul. Res.* **1995**, *33*, 435–456. [CrossRef]
12. Lai, W.; Khan, A.A. Numerical solution of the Saint-Venant equations by an efficient hybrid finite-volume/finite-difference method. *J. Hydrodyn.* **2018**, *30*, 189–202. [CrossRef]
13. Guo, Y.; Liu, R.-X.; Duan, Y.-L.; Li, Y. A Characteristic-Based Finite Volume Scheme for Shallow Water Equations. *J. Hydrodyn.* **2009**, *21*, 531–540. [CrossRef]
14. Yin, C.-C.; Zeng, W.; Yang, J.-D. Transient simulation and analysis of the simultaneous load rejection process in pumped storage power stations using a 1-D-3-D coupling method. *J. Hydrodyn.* **2021**, *33*, 979–991. [CrossRef]
15. Wang, C.; Nilsson, H.; Yang, J.; Petit, O. 1D–3D coupling for hydraulic system transient simulations. *Comput. Phys. Commun.* **2017**, *210*, 1–9. [CrossRef]
16. Zhang, X.-X.; Cheng, Y.-G.; Yang, J.-D.; Xia, L.-S.; Lai, X. Simulation of the load rejection transient process of a francis turbine by using a 1-D-3-D coupling approach. *J. Hydrodyn.* **2014**, *26*, 715–724. [CrossRef]
17. Zhang, X.-X.; Cheng, Y.-G. Simulation of Hydraulic Transients in Hydropower Systems Using the 1-D-3-D Coupling Approach. *J. Hydrodyn.* **2012**, *24*, 595–604. [CrossRef]
18. Yang, J.; Yang, J. 1-D MOC simulation software for hydraulic transients: TOPsys. *Proc. IOP Conf. Ser. Earth Environ. Sci.* **2018**, *163*, 12081. [CrossRef]
19. Zeng, W.; Yang, J.; Hu, J. Pumped storage system model and experimental investigations on S-induced issues during transients. *Mech. Syst. Signal Pract.* **2017**, *90*, 350–364. [CrossRef]
20. Hu, J.; Yang, J.; Zeng, W.; Yang, J. Transient Pressure Analysis of a Prototype Pump Turbine: Field Tests and Simulation. *J. Fluids Eng.* **2018**, *140*, 71102. [CrossRef]
21. Wang, C.; Yang, J.-D. Water Hammer Simulation Using Explicit-Implicit Coupling Methods. *J. Hydraul. Eng.* **2015**, *141*, 4014086. [CrossRef]
22. Toro, E.F. *Shock-Capturing Methods for Free-Surface Shallow Flows*; Wiley: Hoboken, NJ, USA, 2001.
23. Toro, E.F. *Riemann Solvers and Numerical Methods for Fluid Dynamics: A Practical Introduction*; Springer Science & Business Media: Berlin/Heidelberg, Germany, 2013.

water

MDPI

Article

A Novel Surge Damping Method for Hydraulic Transients with Operating Pump Using an Optimized Valve Control Strategy

Zheng Cao [1,2,3], Qi Xia [1,2], Xijian Guo [1,2], Lin Lu [3] and Jianqiang Deng [1,2,*]

1 School of Chemical Engineering and Technology, Xi'an Jiaotong University, Xi'an 710049, China; zhengcao@xjtu.edu.cn (Z.C.); xiaqi19991104@stu.xjtu.edu.cn (Q.X.); gxj18770419616@stu.xjtu.edu.cn (X.G.)
2 Shaanxi Key Laboratory of Energy Chemical Process Intensification, Xi'an 710049, China
3 Department of Building Environment and Energy Engineering, The Hong Kong Polytechnic University, Hong Kong 999077, China; vivien.lu@polyu.edu.hk
* Correspondence: dengjq@xjtu.edu.cn

Abstract: Hydraulic transients may pose a critical threat to process operation due to devastating surge waves. This paper investigates hydraulic surge and damping control associated with pipe flow modeling and valve optimization. A one-dimensional transient model was developed using the modified instantaneous accelerations-based (IAB) model, considering energy dissipation, referred to as the compression–expansion effect, which was then solved by the Method of Characteristics (MOC). Analogous to solving valve operation by means of the traveling salesman problem (TSP), a novel surge damping strategy was proposed by applying an improved artificial fish swarm algorithm (AFSA). After validating the unsteady model and the optimization algorithm, wave surge damping effectiveness was evaluated on the basis of case studies in different pump running scenarios. The results showed that the proposed nonlinear optimized control method was able to reduce surge amplitude by 9.3% and 11.4% in pipe systems with and without running centrifugal pump, respectively, and was able to achieve a 34% time margin or a maximal 75.2% surge reduction in the case of using an positive displacement pump. The optimized nonlinear valve closure presents different shapes in fast closing and slow closing situations. The strategy proposed in the present study is beneficial for guiding valve real-time control, as well as providing a reference for valve design for the purpose of wave surge protection.

Keywords: unsteady friction model; surge damping; compression–expansion effects; pressure wave; energy dissipation

Citation: Cao, Z.; Xia, Q.; Guo, X.; Lu, L.; Deng, J. A Novel Surge Damping Method for Hydraulic Transients with Operating Pump Using an Optimized Valve Control Strategy. *Water* **2022**, *14*, 1576. https://doi.org/10.3390/w14101576

Academic Editors: Helena M. Ramos and Kamil Urbanowicz

Received: 6 April 2022
Accepted: 10 May 2022
Published: 14 May 2022

Publisher's Note: MDPI stays neutral with regard to jurisdictional claims in published maps and institutional affiliations.

Copyright: © 2022 by the authors. Licensee MDPI, Basel, Switzerland. This article is an open access article distributed under the terms and conditions of the Creative Commons Attribution (CC BY) license (https://creativecommons.org/licenses/by/4.0/).

1. Introduction

During the process of fluid transportation through a pump, any rapid unexpected changes affecting fluid flow may cause the hydraulic transient phenomenon. In a water pipeline system, sudden valve closures or pump start-up/stoppage can generate such transient events, which are also known as a water hammer. Once the harmful pressure wave has propagated along the pipeline, there will always exist positive and negative pressure waves due to reflection. Hydraulic surge may cause severe damage to pipes or connected devices; on the other hand, the reflected negative waves associated with oscillating cavitation or water column separation can also affect the safe operation of the pump [1,2].

A water hammer is a type of pressure wave that propagates as a result of fluid oscillation at a corresponding degree. Such oscillation behavior is influenced by pipeline boundary conditions and the energy dissipation process, and accurate modeling of wave characteristics can be essential for pipe design and pipeline leakage or blockage detection [3]. Since pressure waves travel at a fast speed equivalent to that of the local speed of sound, safety precautions against wave surges and reflections are also regarded as an essential issue.

The accurate establishment of the unsteady friction model is important for modeling hydraulic surges in transient pipe flows, especially in the unsteady friction-dominated transient flow. With regard to the importance of unsteady friction, Duan et al. [4] found that unsteady friction makes a greater contribution in small-scale pipeline systems, where the timescale ratio of wave propagation to radial diffusion is low. Ghidaoui et al. [5] suggested that the quasi-steady and flow asymmetry assumptions may result in deviations in the wave shape over larger simulation timescales. Usually, convolution-based (CB) models and instantaneous acceleration-based (IAB) models are used in transient simulation. For the CB model proposed by Zielke, the Navier–Stokes equations of axial pipe flow are solved with the Laplace transformation, and the unsteady friction is described by the convolution of the past accelerations with different weighting functions [6]. Although it has the merits of a strong physical basis and less dependency on the determination of empirical coefficients, CB models commonly require great computational resources to deal with the local acceleration and weight function in the convolution integral across the full time range of all nodes. This results in limitations to the application of such models for long-distance pipelines stemming from their low-efficiency numerical calculations. To solve this challenge, Urbanowicz [7] developed a novel recursive formula for estimating the convolution integral weighting function, enabling more accurate and efficient calculation of unsteady wall shear stress. As the CB model was initially derived for laminar transient flow, Trikha [8] simplified the calculation method of the weight function by only calculating the convolution of the local acceleration and the weight function at the previous moment to obtain the result, thus extending the application scope of the CB model to transient turbulent flow. By distinguishing eddy viscosity variation in the outer and inner core regions, CB model founds extended application in smooth and rough pipelines with a two-region model [9], as well as further application in different flow regimes with a four-layer model [10]. However, the involvement of historical acceleration and the lack of a convection term makes the weighting-function-based model vulnerable to the accumulation of errors with strong convection, such as in low-frequency waves after rapid valve closure and high-frequency waves caused by a running pump [11].

The widely used IAB model considers unsteady friction on the basis of local and convective acceleration. Based on the Brunone model with one damping coefficient [12], Pezzinga [13] added a symbolic function to the convective acceleration term, making the modified Brunone model capable of describing both accelerating and decelerating flows. Considering the different roles of local acceleration and convection acceleration, Ramos et al. [14] proposed a two-coefficient model to better characterize the phase change and amplitude change in wave dissipation, requiring knowledge and experience to determine the empirical coefficients. Owing to the high computation efficiency and the convenience of its program implementation, the one-dimensional IAB-based model is more suitable for complex pipe systems involving large-scale calculations [15]. Nault and Karney [16] integrated an IAB unsteady model with compatibility expressions in order to simulate a pipe network, and a solver adaptive method was used to increase the computational efficiency. To increase the accuracy of the model for accumulated dissipation, a modified IAB model taking into consideration the compression–expansion effect of the fluid on energy dissipation was proposed by our previous work [17]. Further progress in model development can be found when integrating the 1-D model. For example, Wu et al. [18] studied the transient interaction between the pressure surge and the pump using a CFD coupling simulation; a similar trend in flow rates and pressure pulsation was achieved by joint simulation, with precise pressure results. He et al. [19] proposed a digital twin framework in the oil transportation process, where a control theory and data-driven method were integrated to improve the accuracy of the physical model. Despite the convenient application of the modified IAB model, it must be noted that difficulties still exist in modeling some transient situations, such as valve opening events or frequency-dependent friction flows [20].

One of the many applications for hydraulic modeling is in the evaluation of extreme transient events when using certain preventive measures. It is known that many factors can affect wave propagation and the amplitude of a hydraulic surge. Bettaieb et al. [21] investigated wave damping performance during pump failure events, and found that using viscoelastic pipes could be effective for attenuating wave fluctuation, due to their rheological properties, while attaching a flywheel to the pump motor reduced wave amplitude as well as extending pump shutdown time. Garg et al. [22] found combined configurations with glass fiber-reinforced plastic damped the celerity in the pipeline and improved the wave damping coefficient. Urbanowicz et al. [23] experimentally investigated the transient laminar flow of oil in small-diameter pipes; together with a novel numerical solution method, the impact of factors including fluid viscosity and pipe length on pressure fluctuation was effectively analyzed.

With regard to valve motion, Wan et al. [24] considered the putting off effect of the valve under pump runaway conditions, pointing out that if the valve closure happens too quickly, or at an unsuitable time, a more severe water hammer effect might be generated following pump failure. To reduce the water hammer pressure caused by pump trip, Wang et al. [25] investigated the effects of valve closing time on pump safety based on the pump runaway characteristics, and gave advice on valve selection for different lift systems. For pump startup, the adoption of damping torque slowed down valve closure and proved to be effective for eliminating the water hammer, compared to the high-speed startup condition [26]. For pump shutdown events, Triki and Essaidi [27] investigated the induced pressure wave behaviors in two-section piping systems in different materials. The LDPE-steel pipeline demonstrated the best performance in terms of achieving a lower pressure surge peak with an extended wave oscillation period.

Aside from the above-mentioned factors in pipe systems, Sattar et al. [28] estimated the effects of additional surge vessels on pumping mains experiencing pump failure, and applied an approximate solution on the basis of Monte Carlo simulation to simplify optimum sizing rather than using hydraulic calculation. Kubrak et al. [29] used thinner polymeric branched pipes to control water hammer in a steel pipeline, and observed that the pressure surge was successfully attenuated and that the damping effect was more obvious in the event of rapid valve closure. Mohammad Bostan et al. [30] introduced an effective shock damper in which an embedded spring chamber absorbs the compression wave and subsequently releases its energy when it encounters the expansion wave. By applying a short compound section as a damping component connected to the main pipeline, Triki and Trabelsi interestingly investigated effects of dual combination [31] and standalone inline or branching strategies [32] on water hammer control, finding that the HDPE-LDPE penstock-based branching strategy achieved satisfactory pressure wave damping performance, while the dual-design strategy was more beneficial than the standalone strategy.

Despite much effort and progress in the area of transient pipe flow and its corresponding preventive measures, most research has focused on the water hammer phenomenon caused by pump failure, where the associated surge prevention relies on structural changes to the pipeline system. The importance of surge damping during constant running of the pump prior to being disconnected cannot be neglected. Adding a protective device is not always an option in some pumping processes when aiming to minimize the water hammer generated by valve motion itself. Therefore, in this research, we investigate the flow transient effects for pumps during the running stage. Considering the advantages for both the estimation of transient flow containing high-frequency components and the optimization of calculation efficiency for large samples using the artificial intelligence method, the IAB-based model was applied with a modified dissipation model considering fluid compression and expansion. After treating the valve as a wave generator and investigating the closure problem by analogy with the traveling salesman problem, a preliminary exploration of valve closure optimization is performed making use of an improved swarm intelligence algorithm, and this is ultimately applied to achieve surge damping in different operating pump scenarios.

2. Simulation Model

2.1. One-Dimensional Unsteady Friction Model

To characterize the transient behavior of the pipe flow, the following assumptions are considered as a necessary simplification for the 1-D numerical model:

(1) The liquid fluid complies with cross-section-averaged properties.
(2) The water liquid is considered to be single phase without entrained air.
(3) The pipe is elastic, and the fluid is compressible.
(4) The pipe flow is assumed to be adiabatic flow.

One of the most distinct features of transient pipe flow is the frictional loss behavior associated with unsteady friction. As a widely applied method for representing such losses, a steady friction component is often overlaid with an unsteady friction component in transient models.

The steady friction loss is calculated as

$$J_s = \frac{fV|V|}{2gD} \tag{1}$$

where D is the pipe diameter, g is the gravitational acceleration, V is the flow velocity, and f is the Darcy–Weisbach friction factor, which can be calculated by the Colebrook–White equation.

The unsteady friction loss can be calculated using the one-coefficient IAB model [13],

$$J_u = \frac{k}{g}\left[\frac{\partial V}{\partial t} + sign(V)a\left|\frac{\partial V}{\partial x}\right|\right] \tag{2}$$

where k is the empirical coefficient, which can be determined empirically or by the trial-and-error method [33], $sign(V)$ is the signal of the instantaneous mean velocity, a is the wave speed, and t and x represent the temporal coordinate and axial space coordinate, respectively.

Since fluid compressibility has an obvious effect on pressure oscillation and wave propagation characteristics, the energy loss caused by unsteady friction considers the compression–expansion effect to model the subsequent wave dissipation more accurately.

The momentum equation along x-axis can be expressed as

$$\rho\frac{DV}{Dt} = -\frac{\partial p}{\partial x} + \frac{\partial}{\partial x}\left[(2u + \mu')\frac{\partial V}{\partial x}\right] + \mu\left(\frac{\partial^2 V}{\partial y^2} + \frac{\partial^2 V}{\partial z^2}\right) \tag{3}$$

where μ is the dynamic viscosity and μ' is the second viscosity.

By expanding Equation (3) in a cylinder control volume with coordinate transformation and neglecting the small quantity term, the second viscosity term-related momentum equation can be obtained as follows:

$$\frac{1}{g}\frac{\partial V}{\partial t} + \frac{\partial H}{\partial x} - \frac{fV^2}{2gD} - \frac{\mu'}{\rho\, g}\frac{\partial^2 V}{\partial x^2} = 0 \tag{4}$$

With the introduction of the unsteady friction term caused by the compression–expansion effects, the modified unsteady friction loss expressing the energy dissipation term can be written as Equation (5), where the detailed derivation is given in [17]:

$$J_u = \frac{k}{g}\left[\frac{\partial V}{\partial t} + sign(V)a\left|\frac{\partial V}{\partial x}\right|\right] - \frac{k_d}{g}\left|\frac{\partial^2 V}{\partial x^2}\right| \tag{5}$$

where k_d is the second viscosity coefficient that is relevant to Reynolds number, which can be determined using the trial-and-error method. With the incorporation of the modified unsteady friction term, the one-dimensional governing equations for unsteady flow can be obtained, which are known as the water hammer equations:

$$\frac{\partial H}{\partial t} + \frac{a^2}{g}\frac{\partial V}{\partial x} = 0 \tag{6}$$

$$\frac{\partial H}{\partial x} + \frac{1}{g}\frac{\partial V}{\partial t} + \frac{fV|V|}{2gD} + J_u = 0 \tag{7}$$

These equations characterize the transient pipe flow behaviors, and contain two unknown variables: piezometric head H and flow velocity V. The hyperbolic partial differential equations are solved by the method of characteristics (MOC) in this study.

2.2. Physical Model and Solution Method

Figure 1 shows the investigated hydraulic system, in which a pump connected to an upstream reservoir is used to transport water fluid downstream through a horizontal pipeline. A valve positioned downstream is often used to regulate or control the transportation process, and its shutdown behavior directly affects the generated hydraulic phenomenon. To study hydraulic transient behavior and valve optimization strategy, Table 1 lists the key parameters of a typical reservoir–pump–valve system for the purpose of our case studies. The initial head and flow rates are set for the reservoir without pump operation, while the head and flow rate comply with the pump characteristics curve in the pump cases.

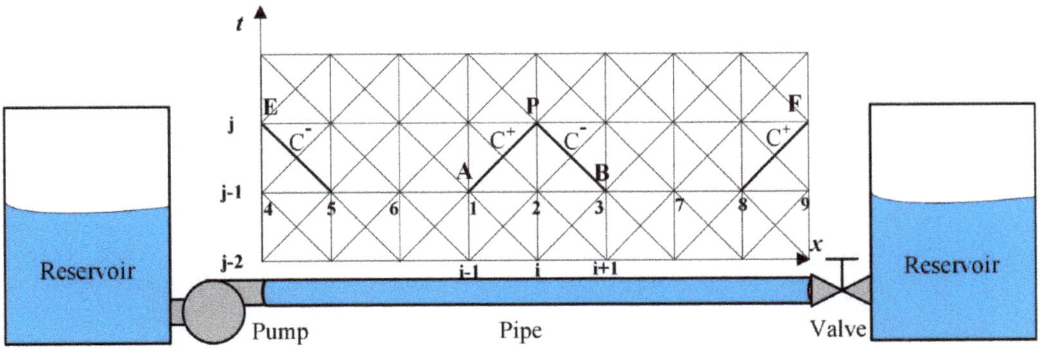

Figure 1. Discrete grids of the pipeline system.

Table 1. Parameters for test cases.

Parameters (Unit)	Values
Initial tank head, H_{res} (m)	128
Initial flow velocity, V_0 (m/s)	0.94
Pipeline length, L (m)	98.11
Wave speed, a (m/s)	1298.4
Pump rotational speed, n_{pump} (r/min)	8000
Pump designed head, H_{pump} (m)	165.9
Pump designed velocity, V_{pump} (m/s)	0.88
blade numbers	8
Valve closing time, tc (s)	0.1~0.5
Water density, ρ (kg/m^3)	997.59
Dynamic viscosity, μ (Pa·s)	0.947×10^{-3}
empirical coefficient k	0.0138
second viscosity coefficient k_d	20.258
Darcy–Weisbach friction factor, f	0.0224

As shown in Figure 1, the MOC discretization method is applied as a numerical solution scheme using rectangular grids. In this way, the ordinary differential equation (ODE) transformation of the hydraulic transient model needs to constitute a grid system, where time and space derivatives are discretized along the vertical and horizontal axes. By dividing the pipe into uniform segments with a length of Δx, the time step for the pressure wave to propagate in each segment is Δt, which is equal to $\Delta x/a$. It is assumed in all simulation cases that the water flow originates from a reservoir with infinite volume, and the short draft tube between the pump and the reservoir is neglected in the analysis.

For the interior node P, the transformed ODE for positive and negative characteristics C^+ and C^- can be expressed as

$$H_P = C_P - BQ_P \tag{8}$$

$$H_P = C_N + BQ_P \tag{9}$$

where

$$C_P = H_A + BQ_A - RQ_A|Q_A| - \Delta x(J_u)^+ \tag{10}$$

$$C_N = H_B - BQ_B + RQ_B|Q_B| + \Delta x(J_u)^- \tag{11}$$

$$B = \frac{a}{gA} \tag{12}$$

$$R = \frac{f\Delta x}{2gDA^2} \tag{13}$$

The simultaneous Equations (10) and (11) can be used to solve the unknown quantity at P on the basis of the known quantities at points A and B.

For a frequency centrifugal pump located at the pipe's upstream, its operating characteristics can be described by the performance curve when the motor-driven impeller runs at a constant speed. The head for the boundary condition is modeled as follows:

$$H_E = H_S + Q_E(a_1 + a_2 Q_E) \tag{14}$$

The above equation provides the upstream pump boundary conditions with a known H-Q performance curve, which typically shows a monotonically decreasing trend between the head and the flow rate. In this study, the performance curve of the centrifugal pump was produced using test data from [34], where the shut-off head H_S was 165.92, the coefficient constant a_1 was 7.07, and a_2 was -1.09.

By solving Equation (14) with the negative characteristic line C^-, the flow rate Q_E can be obtained as follows

$$Q_E = \frac{1}{2a_2}\left[B - a_1 - \sqrt{(B - a_1)^2 + 4a_2(C_N - H_S)}\right] \tag{15}$$

Since there is limited information available for the transient-state turbine characteristics, the pump transient-state characteristics can be obtained on the basis of the steady-state performance curve, which appears to be a valid assumption for most typical engineering applications [35].

As a result of the blade–tongue interaction of the pump, the instantaneous pressure and other flow parameters varied with time [36,37].

By imposing a 10% pressure fluctuation range on the pump characteristics [38], the outlet pressure can be regarded as a superposition of the steady-state pressure and the pulsating component, as follows:

$$H_E' = H_E + 0.1H_E \cdot \sin(wt) \tag{16}$$

This simplified model implementation better characterizes the pulsating nature of the pump outlet signal, while keeping the time-averaged performance curve unchanged.

For a positive displacement pump located at the pipe's upstream, the flow rate is set as a constant known parameter, and the pump head is calculated by integrating the instantaneous flow rate with the negative characteristic line C^-, which is the reversal of the centrifugal pump case.

The relationship between flow rate and the hydraulic head of the downstream valve can be determined according to the orifice outflow law, as follows:

$$\frac{Q_F}{Q_0} = \frac{C_d A_G}{(C_d A_G)_0}\sqrt{\frac{H_F}{H_0}} = \tau_V\sqrt{\frac{H_F}{H_0}} \tag{17}$$

where Q_F and Q_0 are the transient and steady flow rate through the valve, respectively, and H_F and H_0 are the hydraulic head in the transient and steady state, respectively.

The common relationship between the valve opening τ_V and the closing time can be expressed as:

$$\tau_V = 1 - \left(\frac{t}{t_c}\right) \tag{18}$$

For the whole pipeline system, when the initial values and boundary conditions are all set, the discretized scheme is further assigned for the discretization of the partial differential term in the governing equations. In the one-coefficient IAB unsteady friction model, the convective acceleration and local acceleration terms can be discretized along the positive characteristic equation C^+, in opposition to C^-, as follows:

$$\begin{cases} \frac{\partial V_{i,j}}{\partial x} = \frac{V_{i,j-1} - V_{i-1,j-1}}{\Delta x} \\ \frac{\partial V_{i,j}}{\partial t} = \frac{V_{i,j-1} - V_{i-1,j-2}}{\Delta t} \end{cases} \tag{19}$$

In the modified unsteady friction term considering the compression–expansion effect, the diffusion term can be discretized using the nearest three adjacent points:

$$\left(\frac{\partial^2 V}{\partial x^2}\right)_P = \frac{V_{i-1}^{j-1} - 2V_i^{j-1} + V_{i+1}^{j-1}}{(\Delta x)^2} \tag{20}$$

3. Optimization Scheme Using ASFA

The problem being investigated is the pressure surge transmitted in the hydraulic pipeline caused by the valve's sudden closure. The aim here is to design a valve shut-off procedure with a prescribed closure time that is able to minimize the pressure surge value, which can be regarded as having a negative impact on pipe system safety. By analogy, this valve motion can interestingly be converted into a traveling salesman problem (TSP). Different valve openings are regarded as serial locations that need to be visited by a salesman in accordance with a scheduled time sequence.

As can be seen in Figure 2, in a fixed total amount of valve closing time, the starting location is the place at which the valve opening is equal to 1.0, after valve operations, the finishing location is the full closure point, with an opening value equal to 0. In this problem, traditional valve motion (black line) can be seen as a linear visiting arrangement, which means an equivalent amount of time is spent during the sequential valve motion. This study preliminarily proposes nonlinear valve opening procedures (blue line). Different traveling times are assigned to different valve opening locations, and the visiting sequence is no longer required to maintain the same direction all the time. Then, the problem becomes a search for a time interval arrangement during valve closure that is able to minimize the pressure surge in the hydraulic pipe system. It is noted that the 10 valve opening points were selected in this study in order to present a preliminary demonstration; more or fewer intermediate points can be selected, depending on different amounts of available computational power.

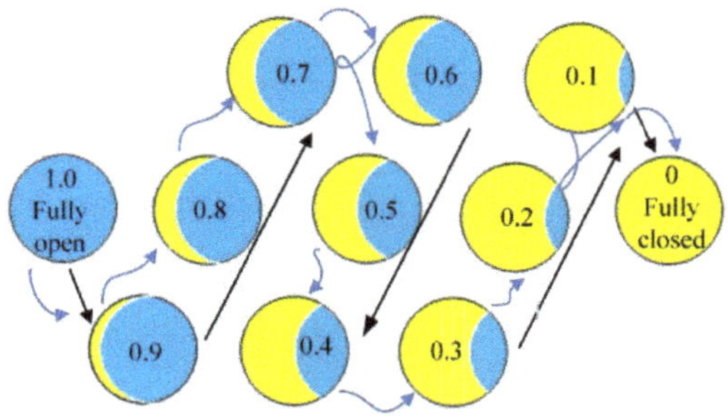

Figure 2. Valve shut-off procedures with optimized fish results.

Biotic populations in nature behave with a high degree of wisdom and logic. The ASFA is a novel swarm intelligence algorithm proposed by Li et al. [39]. As a bionic random search optimization algorithm, it has distinct advantages in solving nonlinear and discrete optimization problems due to its parallel global tracking features. With the help of this optimization algorithm, the requested time for different valve openings can be obtained.

Figure 3 shows the optimization process using artificial fish. Each fish X simulates ecological fish behavior when carrying variable amounts of information and swimming towards optimized value iterations, such as a fish's instinctive response to the local environment. When the artificial fish finds a position with better food density within its field of vision, it moves in that direction by a certain step. During repeated searching, moving and comparing new locations with the current food density state and crowdedness threshold, one member of the swarm will finally arrive at the best position, thus optimizing the target function. The two conceptual parameters mentioned, vision and step, determine the movement behavior of the artificial fish in the search for an acceptable result.

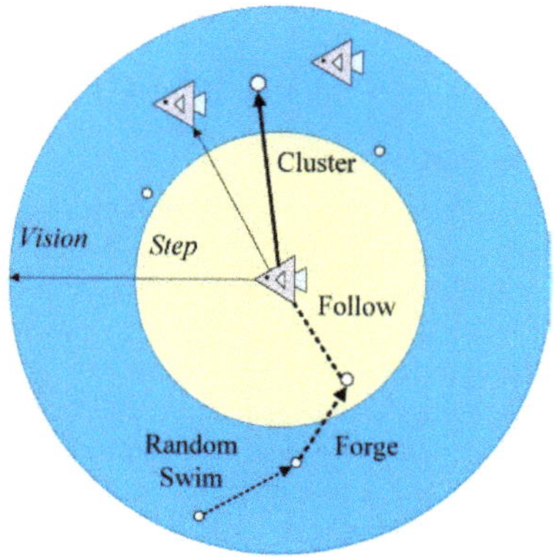

Figure 3. Conceptual optimization process of the designed artificial fish.

Although ASFA is an effective swarm intelligence algorithm, it still has drawbacks, including slow convergence rate and convergence with local optima. In this study, stochastic behavior is added to enhance the possibility of escaping from local optima. Fish clustering and fish following are also integrated to increase the convergence speed of the fish swarm. The following equations describe each key principle of the applied algorithm.

The fish swarm is a natural behavior of fish during movement for to the purpose of foraging together and avoiding the enemy; when the food density (fitness function) in the center position X_V of all visible fish is better than that at the current position Xi while also satisfying the crowdedness standard, it steps forward to the swarm center and reaches a new position X_c,

$$X_c = Xi + \left(\frac{X_V - Xi}{\| X_V - Xi \|}\right) \cdot step_{rand} \tag{21}$$

Otherwise, artificial fish execute the prey behavior, which is a biological instinct that tends to be used when food exists in higher concentrations:

$$X_s = Xi + Vision_{rand} \tag{22}$$

where X_s represents repeated random searching positions within the fish's field of vision. Then, the fish moves forward to a new position X_p:

$$X_p = Xi + \left(\frac{X_s - Xi}{\| X_s - Xi \|}\right) \cdot step_{rand} \tag{23}$$

It is noted that large fish fields of vision and step sizes improve the algorithm's global search capability, while small values benefit the algorithm's local search capability.

To further enhance the convergence speed, fish following behavior is added to simulate natural behavior whereby neighboring fish tend to quickly swim to the best position X_F if one of them X_j finds food:

$$X_F = Xi + \left(\frac{X_j - Xi}{\| X_j - Xi \|}\right) \cdot step_{rand} \tag{24}$$

Another behavior is added to further avoid becoming trapped in local optima, in the form of a random swim to a new position X_{t+1} to find new food or fish crowds:

$$X_{t+1} = X_t + step_{rand} \tag{25}$$

The main key parameters for implementing the above algorithm are given in Table 2. These parameters remain constant until final convergence. Like most artificial intelligence algorithms, the selection of these values depends on practical experience and achieving a balance between computational resources and accuracy. In this study, a rigorous criterion was set for convergence, whereby the error between updated values of the fitness function (pressure head) and the averaged value among last 10 results must be less than 1×10^{-4}.

Table 2. Parameters for the optimization algorithm.

Parameters	Description	Values
Vision	visual radius	0.1
step	moving distance	0.01
n	artificial fish quatities	30
dim	artificial fish dimension	10
delta	fish swarm crowdness	27
$Trail_{max}$	maximum trial number	30
Gen_{max}	maximum iteration number	500
error	convergence error	1×10^{-4}

The calculation procedure is given in Figure 4. For the main optimization loop I shown in Figure 4a, firstly, input data and ASFA initial settings are given, and parameters include fish dimension, population size, vision and step sizes, crowding factor δ, and number of trial iterations. Secondly, a group of random fish is generated with initial positions, and then the hydraulic analysis in loop II is carried out to obtain the adaptive values, and the best values are recorded on the bulletin board. Thirdly, the behavior of each individual fish performing selective behaviors, which include fish clustering, fish following, foraging and potential stochastic swim, is evaluated and determined. Fourthly, after determining artificial fish behavior, their position is updated as X_c, X_p, and X_F. If an individual fish position is superior, then the bulletin board is updated with the position of the optimal fish. This evaluation continues until evaluation for all fish has been completed. Finally, when the termination criteria are satisfied by means of an acceptable error or the iteration limit being reached, the algorithm ends with the optimized results being output.

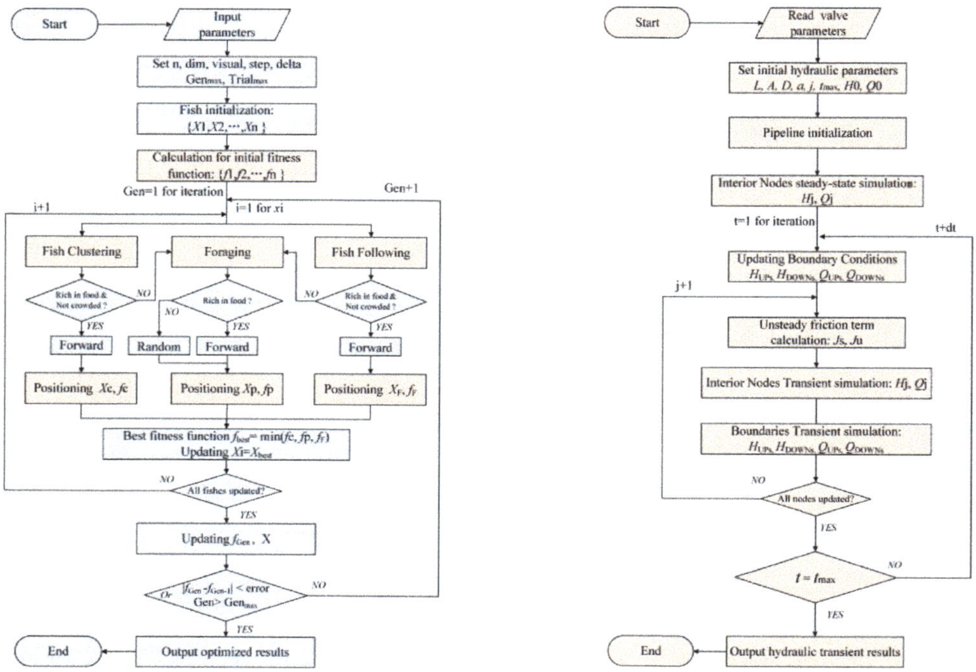

(a) (b)

Figure 4. The calculation flow chart for the surge damping optimization. (**a**) Optimization procedures using ASFA; (**b**) Hydraulic transient calculation procedures.

All orange frames in Figure 4a involve the sub-loop of hydraulic transient calculation, the procedures for which are given in Figure 4b. Firstly, together with the given hydraulic parameters, the valve motion parameters, termed as fish position, generated by ASFA are read. Secondly, the pipeline is initialized using the steady-state calculation for the first time step. Thirdly, the boundary conditions are updated, and the unsteady flow model is solved using the MOC method to obtain all node information, including head H and flow rate Q. Finally, node hydraulic information is obtained, as calculation time reaches its maximum value.

4. Model Validation

A classical reservoir–pipe–valve system is first used to validate the unsteady hydraulic model used in this study. The simulation settings were set in accordance with the experiment described in Ref. [40]. The configuration is similar to that described in the schematic diagram in this study, except that the pump device has been removed. Along the pipeline, the number of grid nodes is set to 100, so that the grid space and step time satisfy the Courant–Friedrichs–Lewy condition with adequate precision.

The simulation results using the modified IAB-CE model are compared to the experimental data in Figure 5. In this test case, the pressure surge is monitored at the mid-point of the pipe after a rapid valve closure. It can be seen that both the pressure surge amplitude and the waveform in the simulation results are in good agreement with the experimental data at the pipe's midpoint. The applied numerical scheme considers unsteady friction, including the compression–expansion effect, providing relative calculation errors for the pressure surge within 2.16%. The satisfactory consistency, especially in the subsequent wave fluctuation, guarantees the accuracy of target function for performing optimization on the basis of the present hydraulic model.

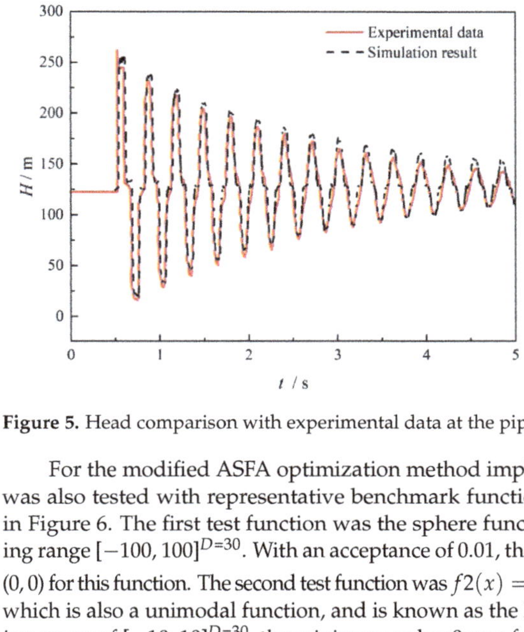

Figure 5. Head comparison with experimental data at the pipe's midpoint.

For the modified ASFA optimization method implemented in this study, its validity was also tested with representative benchmark functions, and the results are presented in Figure 6. The first test function was the sphere function $f1(x) = \sum_{i=1}^{D} x_i$ in the searching range $[-100, 100]^{D=30}$. With an acceptance of 0.01, the minimum value 0 was obtained at (0, 0) for this function. The second test function was $f2(x) = \sum_{i=1}^{D} 100(x_{i+1} - x_i{}^2)^2 + (x_i - 1)^2$, which is also a unimodal function, and is known as the Rosenbrock function. In the searching range of $[-10, 10]^{D=30}$, the minimum value 0 was found at (1, 1) with an acceptance of 100 [41]. As can be seen, the minimum values obtained from the presented optimization method were 1.89×10^{-6} in Figure 6a and 3.17×10^{-5} in Figure 6b, respectively.

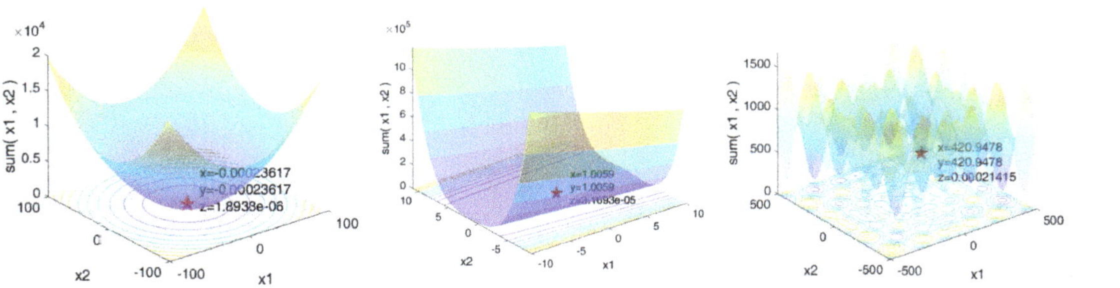

(a) (b) (c)

Figure 6. Validation results using different test functions. (**a**) optimization for sphere function; (**b**) optimization for Rosenbrock function; (**c**) optimization for Schwefel function.

In Figure 6c, Schwefel's optimization problem was selected. It is known that this multimodal function is a typical deception problem, where the global minimum value 0 is hidden in a valley at (420.97, 420.97), far from other local optimal valleys. Hence, it is difficult for the algorithm to jump out of local optimal points, which it easily falls into. When testing the transformed function $f3(x) = 418.9829D - \sum_{i=1}^{D} x_i \sin\sqrt{|x_i|}$, the optimization results showed the successful location of the valley compared with the theoretical values, validating the performance of the proposed ASFA as well as its feasibility for optimizing the nonlinear problems in this study.

5. Results

For a fixed valve closing time, as is required in order to achieve quick responses to hydraulic events, it is important to know the flow behavior during flow transients. Different arranged time lengths and pump types have different effects on the pressure surge generated by valve closure. With the help of the developed IAB-CE model and the modified ASFA optimization method, the characteristics of pressure propagation and the corresponding wave damping method are analyzed.

5.1. Wave Damping Case without Pump Operation

In this case, there is no pump operation, and the associated pump loss is neglected. Pipe flow becomes unsteady after valve closure, and the generated pressure wave propagates to the upstream reservoir, with a pressure surge being recorded in the pipeline.

To visualize the optimization process, Figures 7 and 8 serve as an example in this work, which is similar in other cases. It can be seen in Figure 7 that the initial 30 artificial fish in black triangles representing the valve opening sequences are initialized in random positions. After the implementation of the ASFA optimization process, the fish start to cluster and follow the optimal fish at the time, escaping from local optimum positions by foraging throughout the search space. At the 40th generation of the swarm of fish, the leading fish depicted with green triangles emerge at the best position, and the searching process is then terminated.

Figure 8 shows the variation in the fitness function and convergence curve with increasing numbers of iterations. The black line represents the fitness function as the optimal pressure head value, while the blue line represents the convergence curve. It can be seen that the optimal head value fluctuates during the initial clustering swarm at around five iteration steps, and soon the head value quickly decreases with the rapidly decreasing convergence curve. After 10 iterations, the minimization of the optimized head value tends to slow down, and it almost remains unchanged after 40 iterations, while the convergence curve decreases to an extremely low level.

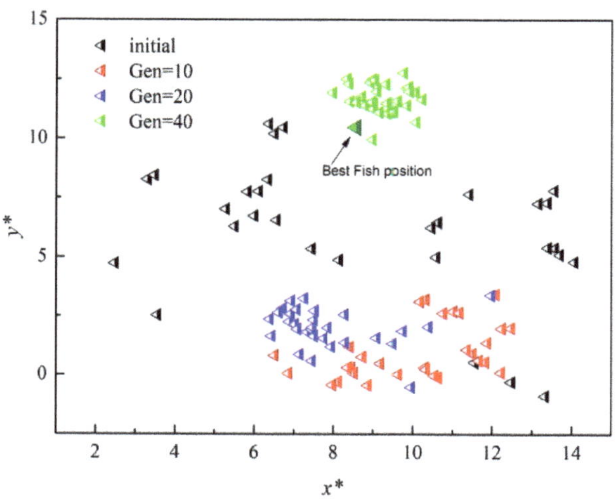

Figure 7. Observation of the distribution of artificial fish in the optimization process.

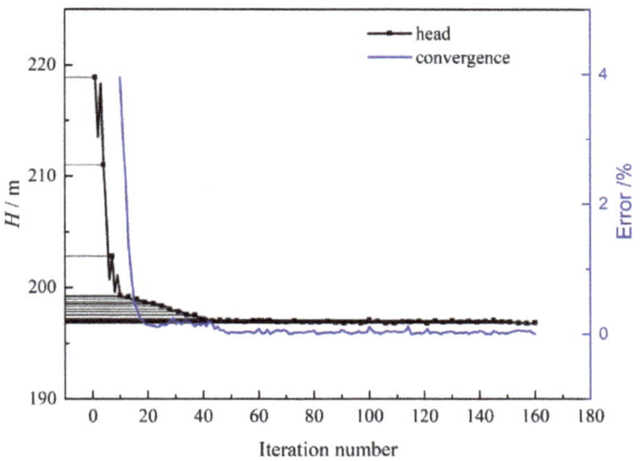

Figure 8. The fitness function and convergence curve of the optimization process.

On the basis of the obtained optimized valve closure time arrangement, Figure 9 compares the transient pressure head at the midpoint of the pipe when the valve is closed within a time of 0.1 s. As can be seen from this figure, the solid red line is the result when performing linear valve closure, while the dashed black line is the results when using the optimized nonlinear valve closure. Compared to the shark waveform observed for linear closure, the wave peaks are split into double peaks, and the amplitude is obviously reduced, along with the wave valley, for the same lengths of valve closing time. The results suggest that with the optimized valve operation, the hydraulic pressure fluctuation can be damped in both directions without changing the wave frequency. The safety risk can then be reduced by avoiding larger over pressure and negative pressure.

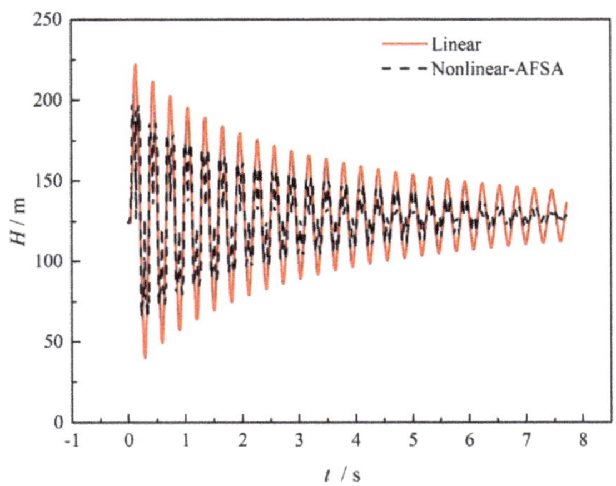

Figure 9. Comparison of the pressure fluctuation with a valve closing time of 0.1 s with no pump operation.

It is known that extending the valve closing time contributes to weakening the water hammer phenomenon. Figure 10 shows the pressure fluctuation when valve closing time is extended to 0.5 s. As can be seen from this figure, with the gradual reduction in valve opening time, there is a corresponding increase in the pressure at the midpoint, which then falls coupled with the upward reflected negative wave. After the full closure of the valve, the maximum and minimum pressures are obtained, and afterwards the damped oscillation can be observed in the pressure waveform due to continuous frictional loss.

For the case of optimized valve operation, it can be observed that the degree of pressure fluctuation can also be decreased by using a longer valve closing time.

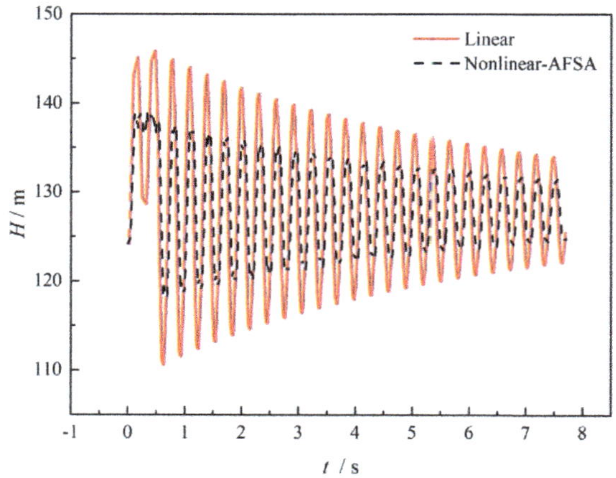

Figure 10. Comparison of the pressure fluctuation with a valve closing time of 0.5 s with no pump operation.

To investigates the pressure surge damping performances by the optimized nonlinear valve closure, the effect on the pressure surge amplitude is compared for various valve closing times. In each case, for the given closure times, the optimization using ASFA is carried out to generate a time sequence for optimized valve operation. As can be seen in Figure 11, the pressure surge amplitude caused by valve closure decreases with increasing valve closing time, and the descending slope also becomes more shallow with increasing valve closing time. However, for the designed nonlinear closure case, the pressure surge amplitude decreases of 11.4%, 8.5%, 6.5%, 5.2%, and 4.8%, respectively, compared to normal valve operation in 0.1 s to 0.5 s, can be achieved.

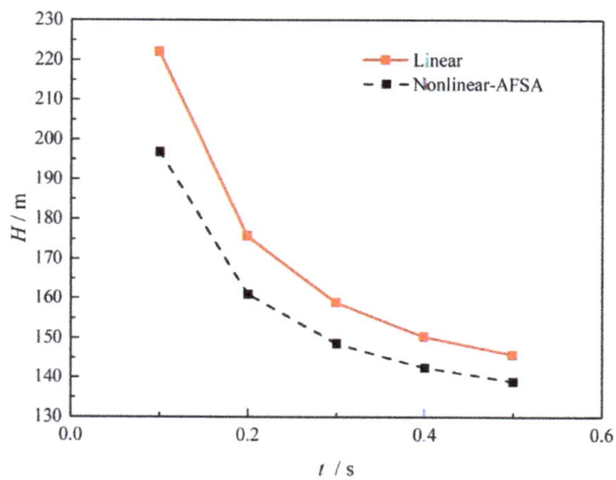

Figure 11. Pressure surge amplitude for different valve closing times with no pump operation.

The optimized valve operation characteristics are shown in Figure 12. The x axis represents normalized time by dividing the total closing time. It can be clearly seen that the operation with a closing time of 0.1 s has the largest deviation from linear operation, indicated by the diagonal dashed line. This can be explained by the working characteristics of the valve, as it plays the role of a pressure wave generator by adjusting the different valve openings for different time periods. Within a very limited amount of time, the downstream valve needs to generate a greater expansion wave in order to reduce the superimposed wave energy at the pipe's midpoint. At around 0.014 s, 0.028 s, and 0.092 s, the valve opening becomes larger in order to generate a subsequent negative pressure wave travelling upwards, so that the pressure surge can be damped, coupled with the following compression wave. It should be noted that the valve opening curve can either be applied in precise valve operation by means of real-time control or used as a structural guideline for providing corresponding flow resistance in valve design for specified damping problems.

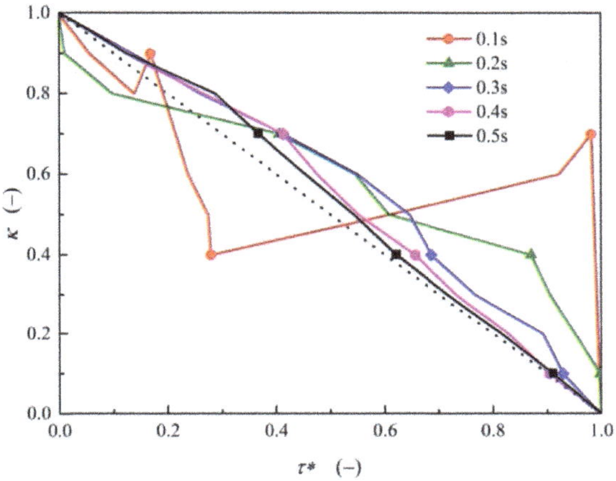

Figure 12. Optimized valve operation characteristics with no pump operation.

5.2. *Wave Damping Case with Centrifugal Pump Operation*

In some transient events such as pump failure or sudden stoppage, four quadrants and eight zones of possible pump operation can be classified on the basis of the direction of flow, pressure, rotation and torque [42]; a complete head and torque characteristic curve is used to describe the pump characteristics, including reverse running state [43]. When a pipeline connected to a centrifugal pump undergoes unexpected valve closure, a hydraulic transient occurs even when the pump is still being motor-driven at the same rotation speed without powering off. As shown in Figure 13, the pressure head at pipe's midpoint also fluctuates following the peak and valley values after full closure of valve at 0.1 s. In comparison, the optimized nonlinear closure also reduces the pressure surge amplitude with a split waveform, appearing to enter steady oscillation slightly earlier due to the decreased pressure surge during the initial transient stage. It needs to mentioned that although the pressure can be temporally controlled to within an acceptable range following valve closure, it is still necessary to shut off the motor as early as possible in order to avoid overheating of the water, since the flow circulation is no longer activated while the impeller is still rotating at a high speed.

For valve closure over a larger time length of 0.5 s, a comparison of the pressure fluctuation results is presented in Figure 14. It can be observed that, after the valve is fully closed after 0.5 s, the recorded wave peak–wave valley amplitudes decrease from 186.8 m–155.6 m, to 179.8 m–161.7 m. Additionally, due to the timely adjustment of the nonlinear valve opening throughout the whole valve motion period, the first wave peak of 186.3 m—wave valley of 162.6 m was also reduced to 179.7 m–172.1 m, compared with linear valve operation. Since the entrained energy of the fluid is gradually released for water hammer events with slow valve closure, the surge is relatively low compared to valve closure in 0.1 s, and the fluctuation only lasts for 1.02 s, before emerging into pump output oscillation caused by blade sweeps.

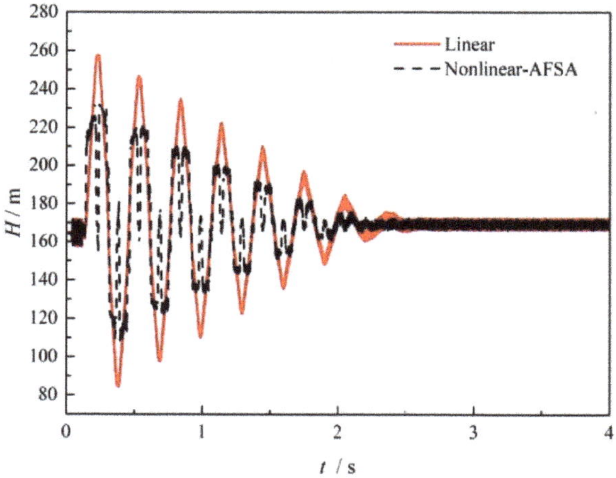

Figure 13. Comparison of the pressure fluctuation with a valve closing time of 0.1 s with centrifugal pump operation.

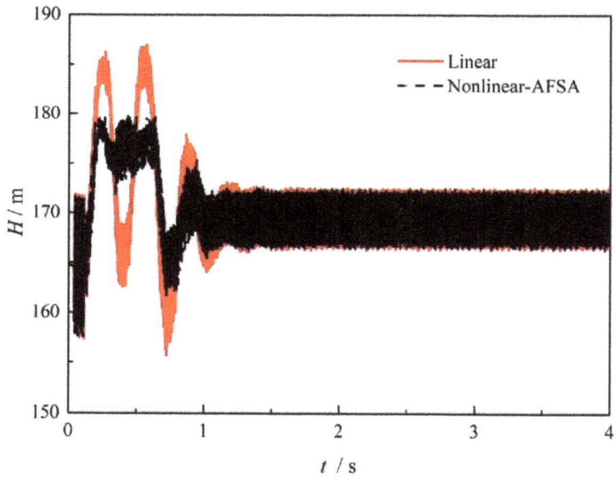

Figure 14. Comparison of the pressure fluctuation with a valve closing time of 0.5 s with centrifugal pump operation.

Figure 15 shows the pressure surge damping performance with centrifugal pump operation. It can be seen that the variation trend is similar to that in the case described in Figure 11. Furthermore, in comparison with normal linear valve operation, the pressure surge amplitude with the optimized nonlinear valve closure exhibited decreases of 9.3%, 6.4%, 5.3%, 4.2% and 3.8%, respectively, as well. This suggests that the developed nonlinear closure method using ASFA could also be suitable for damping the pressure surge in cases when the pressure input is not steady and varies with time. Any safety margin also in return limits the selection of the pump and connected devices.

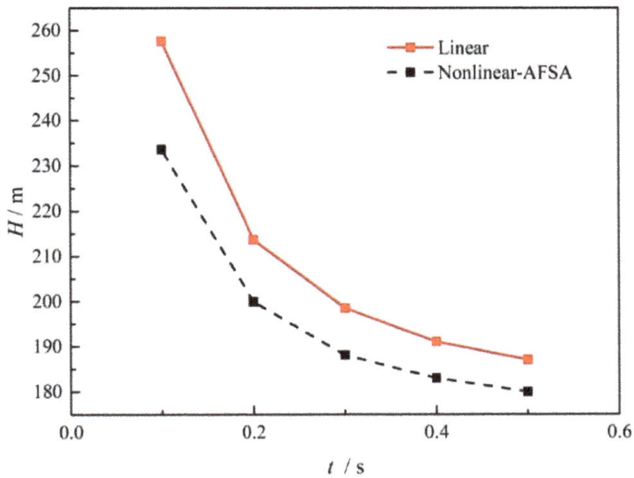

Figure 15. Pressure surge amplitude at different valve closing times with centrifugal pump operation.

To investigate the optimized valve operation characteristics with the centrifugal pump, Figure 16 shows the variation in nonlinear valve opening with normalized time. As can be seen, for a short operation with a closing time of 0.1 s, the valve opening curve has the largest deviation from the linear operation curve. It rises at 0.0008 s, 0.0659 s, and 0.087 s during operation, while an expanding wave is generated by the valve opening. The overall characteristic curve is similar to an "И" shape, where the valve opening tends to decrease rapidly at first, before becoming smooth, followed by a rapid subsequent increase. Like all other cases with a longer valve closing time, the valve opening results in a rapid reduction during the last stage. Although wave propagation and superposition are complicated when considering variable valve motion together with the changes in pipe flow unsteady frictional loss, the control strategy using ASFA presented as a conceptual demonstration can serve to provide design guidance, and is expected to also be feasible in complicated pipe system optimization problems.

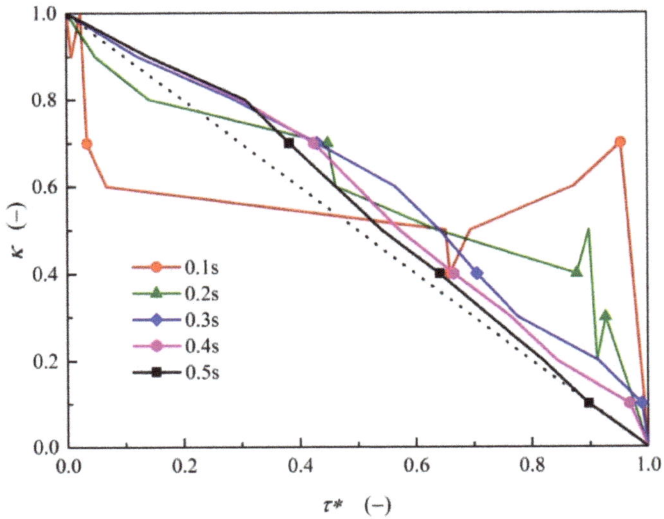

Figure 16. Optimized valve operation characteristics with centrifugal pump operation.

5.3. Wave Damping Case with Positive Displacement Pump Operation

Another case to consider is that of a pipe transferring fluid through a displacement pump. One major difference is that the displacement pump faces a severe pressure rise over a short time after choking, which is also known as pressure buildup. This rapid increase in pressure can easily exceed the upper limit of pipe safety, or cause damage to connected devices, including the pump itself.

By setting 500 m as the assumed head limit for the pipe, Figure 17 shows the protective effects of the optimized nonlinear valve closure on surge damping. It can be seen that the pressure rise becomes obvious following full valve closure. For the linear operation case, the midpoint pressure increases proportionally with pumping time, while for the optimized situation, the waveform is distorted and the pressure increase with time exhibits fluctuations. Although the rise in the pressure head is unavoidable for the positive displacement pump, and eventually increases to 500 m, it can be observed that the pressure in the optimized case remains constantly lower than in the linear closure case. Moreover, it takes 0.39 s to reach the head limit of 500 m, compared to 0.34 s for the traditional closure case without nonlinear control. This 14.7% extension of reaction time means that there is more time for a precise hydraulic system to be able to disconnect the pumping system.

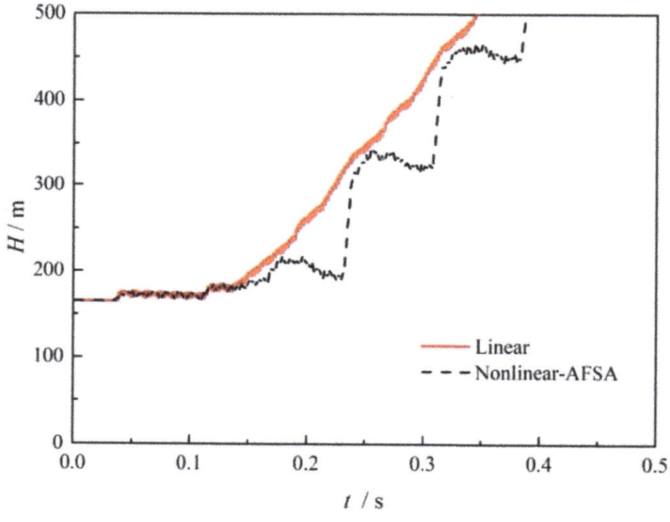

Figure 17. Comparison of the pressure rise in 0.1 s valve closing time with positive displacement pump operation.

In Figure 18, it can be seen that there was a more obvious improvement in surge damping when using the optimized nonlinear closure with a longer closing time of 0.5 s. It can be seen the distance between the two curves is larger, and the times required for the traditional and optimized operation cases to reach the pipe head limit are 0.53 s and 0.71 s, respectively. An additional 34% of extra time can be used to take precautions against the surge limit in the pipe system.

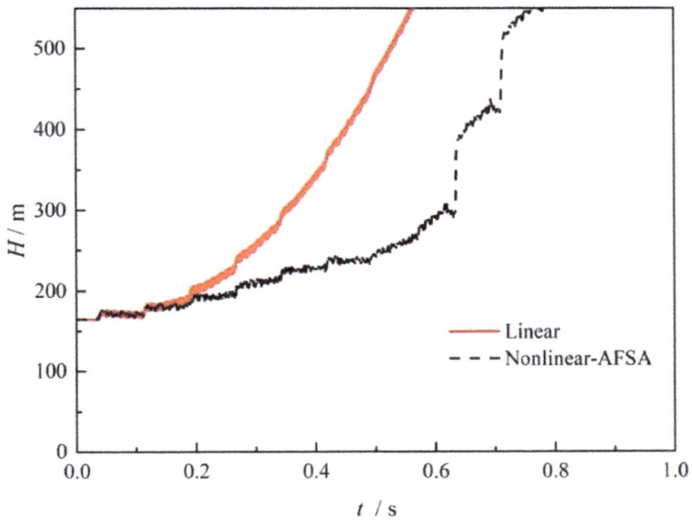

Figure 18. Comparison of the pressure increase with a 0.5 s valve closing time with positive displacement pump operation.

Figure 19 shows the results of the calculated head at the midpoint 1 s after full valve closure over different times. It can be seen that for the traditional linear valve operation, the size of the resulting pressure head rise increases linearly from 349.8 m to 693.1 m with increasing valve closing time from 0.1 s to 0.5 s. This is mainly because the built-up pressure accumulates when subjected to a longer compression time. However, with the optimized nonlinear valve operation, the amplitude of the resulting pressure rise is greatly reduced, from 336.5 m to 395.7 m, with a maximum reduction of 75.2% in head surge being recorded after full valve closure. Therefore, it is suggested that even for the positive displacement pumping process, beneficial consequences can still be expected from surge damping by applying the optimized nonlinear valve control method based on ASAF method in hydraulic transient events.

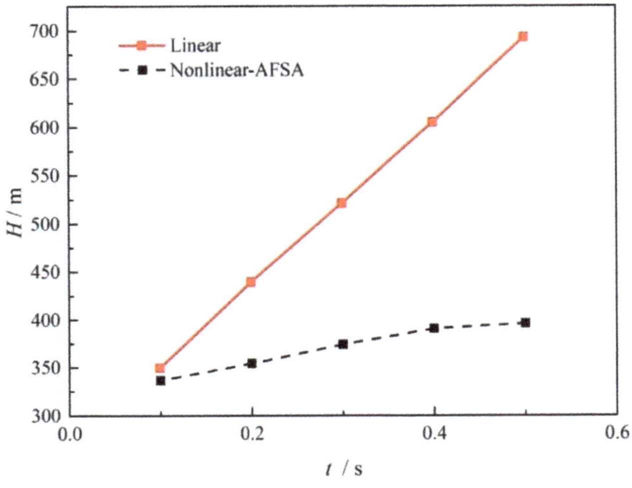

Figure 19. Pressure surge amplitude at different valve closing times with positive displacement pump operation.

The optimized valve operation characteristic curves with the positive displacement pump are presented in Figure 20. As can be seen, all curves are above the diagonal line in terms of their variation with normalized time. This means that the valve prefers a slow switch-off or an even switch-on in the beginning. Except for the 0.1 s closure case, all valve opening curves exhibited a relatively smooth variation during the initial valve operation stage, before rapidly decreasing to the fully closed state. A similar "И" shape was also found for the 0.1 s closure case, indicating that larger valve motion amplitude is required to diminish the wave surge in a limited period of time.

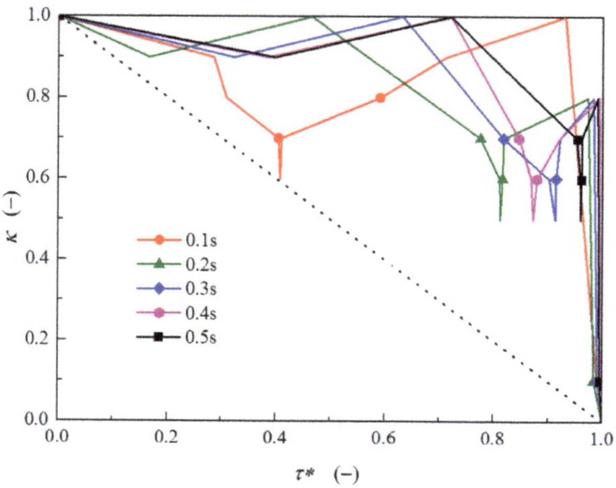

Figure 20. Optimized valve operation characteristics with positive displacement pump operation.

6. Conclusions

In this study, a one-dimensional hydraulic model considering compression–expansion dissipation was implemented to describe transient flow; combined with improved AFSA optimization, surge protection for pipes with different operating pumps was evaluated. The following conclusions were drawn:

(1) The transient wave surge can be reduced through nonlinear valve closure without adding additional damping devices. For transient flow with and without a centrifugal pump running, surge reduction of 9.3% and 11.4% could be obtained in the most severe valve closure case. Even with increasing pressure with positive displacement pump operation, the surge damping method was able to achieve a 34% time margin for reaching the head limit and a maximum reduction in the surge amplitude of 75.2%.

(2) With increasing valve closing time, the surge amplitude caused by valve closure decreases for the transient flow with and without the centrifugal pump running, and the rate of surge amplitude decrease also decreases. For positive displacement pumps, the surge amplitude increases with increasing valve closing time, but at a significantly slower rate of increase with the optimized nonlinear valve closure.

(3) For rapid valve closure in 0.1 s, the optimized nonlinear closing motion performs a similar "И" shape. For other valve closure cases in the present study, the optimized valve opening curves show a relatively smooth variation during the initial stage before decreasing rapidly to full closure during valve operation.

(4) The valve closure process can be abstracted into a traveling salesman problem, and further optimization using an artificial fish swarm algorithm was demonstrated to be beneficial for wave damping. The strategy proposed in the present study could help for either guiding real-time valve control or serve as a design reference for novel valve structures for the purpose of surge protection.

Author Contributions: Conceptualization, Z.C. and J.D.; methodology, Z.C.; software, Z.C. and X.G.; formal analysis, Z.C. and Q.X.; writing—original draft preparation, Z.C. and Q.X.; writing—review and editing, Z.C. and L.L; supervision, J.D and L.L. All authors have read and agreed to the published version of the manuscript.

Funding: This work has been supported by the National Natural Science Foundation of China (Grant No. 22108217, No. 21978227), the Natural Science Basic Research Program of Shaanxi Province (Grant No. 2021JQ052), and the Hong Kong Scholars Program (XJ2020045).

Institutional Review Board Statement: Not applicable.

Informed Consent Statement: Not applicable.

Data Availability Statement: Not applicable.

Conflicts of Interest: The authors declare no conflict of interest.

Abbreviations

AFSA	artificial fish swarm algorithm
CB	convolution-based
CE	compression–expansion
CFD	computational fluid dynamics
IAB	instantaneous accelerations-based
MOC	method of characteristics
ODE	ordinary differential equation
TSP	traveling salesman problem
1-D	one-dimensional
Notation	
a	wave speed ($m \cdot s^{-1}$)
a_1, a_2	pump coefficient constants (-)
A	cross sectional area (m^2)
B	pipeline characteristic impedance (-)
D	pipe diameter (m)
f	Darcy–Weisbach friction factor (-)
g	gravitational acceleration ($m \cdot s^{-2}$)
H	pressure head (m)
J_s	steady friction loss term (-)
J_u	unsteady friction loss term (-)
k	empirical decay coefficient (-)
k_d	second viscosity coefficient (-)
L	pipe length (m)
n	artificial fish number (-)
P	cross-section-average pressure (Pa)
Q	flowrate ($m^3 \cdot s^{-1}$)
R	pipeline resistance coefficient (-)
r	radial coordinate (m)
Re	Reynolds number (-)
V	cross-section-average velocity ($m \cdot s^{-1}$)
ϖ	phase velocity in fluctuation ($rad \cdot s^{-1}$)
t	time (s)
τ	valve opening (-)
ρ	fluid density ($kg \cdot m^{-3}$)
μ	dynamic viscosity (Pa·s)
μ'	the second viscosity (Pa·s)
x	coordinate along the pipe axis (m)
X	artificial fish position (-)

References

1. Mandhare, N.A.; Karunamurthy, K.; Ismail, S. Compendious Review on "Internal Flow Physics and Minimization of Flow Instabilities Through Design Modifications in a Centrifugal Pump". *ASME J. Press. Vessel. Technol.* **2019**, *141*, 051601. [CrossRef]
2. Tanaka, T.; Tsukamoto, H. Transient behavior of a cavitating centrifugal pump at rapid change in operating conditions—Part 1: Transient phenomena at opening/closure of discharge valve. *ASME J. Fluids Eng.* **1999**, *121*, 841–849. [CrossRef]
3. Duan, H.F.; Lee, P.J.; Che, T.C.; Ghidaoui, M.S.; Karney, B.W.; Kolyshkin, A.A. The influence of non-uniform blockages on transient wave behavior and blockage detection in pressurized water pipelines. *J. Hydro-Environ. Res.* **2017**, *17*, 1–7. [CrossRef]
4. Duan, H.F.; Ghidaoui, M.S.; Lee, P.J.; Tung, Y.K. Relevance of unsteady friction to pipe size and length in pipe fluid transients. *J. Hydraul. Eng.* **2012**, *138*, 154–166. [CrossRef]
5. Ghidaoui, M.; Mansour, S.; Zhao, M. Applicability of quasi steady axisymmetric turbulence models in water hammer. *J. Hydraul. Eng.* **2002**, *128*, 917–924. [CrossRef]
6. Zielke, W. Frequency-dependent friction in transient pipe flow. *J. Basic Eng.* **1968**, *90*, 109–115. [CrossRef]
7. Urbanowicz, K. Fast and accurate modelling of frictional transient pipe flow. *Z. Angew. Math. Mech.* **2018**, *5*, 802–823.
8. Trikha, A.K. An efficient method for simulating frequency dependent friction in transient liquid flow. *J. Fluids Eng.* **1975**, *97*, 97–105. [CrossRef]
9. Vardy, A.E.; Brown, J.M.B. Transient turbulent friction in fully rough pipe flows. *J. Sound Vib.* **2004**, *270*, 233–257. [CrossRef]
10. Zarzycki, Z.; Kudzma, S. Simulation of transient flows in a hydraulic system with a long liquid line. *J. Theor. App. Mech.* **2007**, *45*, 853–871.
11. Vitkovsky, J.P.; Stephens, M.; Bergant, A.; Simpson, A.R.; Lambert, M.F. Numerical error in weighting function-based unsteady friction models for pipe transients. *J. Hydraul. Eng.* **2006**, *132*, 709–721. [CrossRef]
12. Brunone, B.; Golia, U.M.; Greco, M. Effects of two dimensionality on pipe transients modeling. *J. Hydraul. Eng.* **1995**, *121*, 906–912. [CrossRef]
13. Pezzinga, G. Evaluation of Unsteady Flow Resistances by Quasi-2D or 1D Models. *J. Hydraul. Eng.* **2000**, *126*, 778–785. [CrossRef]
14. Ramos, H.; Covas, D.; Borga, A.; Loureiro, D. Surge damping analysis in pipe systems: Modelling and experiments. *J. Hydraul. Res.* **2004**, *42*, 413–425. [CrossRef]
15. Duan, H.F.; Meniconi, S.; Lee, P.J.; Brunone, B.; Ghidaoui, M.S. Local and integral energy-based evaluation for the unsteady friction relevance in transient pipe flows. *J. Hydraul. Eng.* **2017**, *143*, 04017015. [CrossRef]
16. Nault, J.D.; Karney, B.W. Comprehensive adaptive modelling of 1-D unsteady pipe network hydraulics. *J. Hydraul. Res.* **2021**, *59*, 263–279. [CrossRef]
17. Cao, Z.; Wang, Z.; Deng, J.; Guo, X.; Lu, L. Unsteady friction model modified with compression–expansion effects in transient pipe flow. *AQUA—Water Infrastruct. Ecosyst. Soc.* **2022**, *71*, 330–344. [CrossRef]
18. Wu, D.; Yang, S.; Wu, P.; Wang, L. MOC-CFD coupled approach for the analysis of the fluid dynamic interaction between water hammer and pump. *J. Hydraul. Eng.* **2015**, *141*, 06015003. [CrossRef]
19. He, L.; Wen, K.; Gong, J.; Wu, C. A multi-model ensemble digital twin solution for real-time unsteady flow state estimation of a pumping station. *ISA Trans.* **2021**, *online ahead of print*. [CrossRef]
20. Vítkovský, J.P.; Bergant, A.; Simpson, A.R.; Lambert, M.F. Systematic evaluation of one-dimensional unsteady friction models in simple pipelines. *J. Hydraul. Eng.* **2006**, *132*, 696–708. [CrossRef]
21. Bettaieb, N.; Taieb, E.H. Assessment of failure modes caused by water hammer and investigation of convenient control measures. *J. Pipeline Syst. Eng. Pract.* **2020**, *11*, 04020006. [CrossRef]
22. Garg, R.K.; Kumar, A.; Abbas, A. Analysis of wave damping in pipeline having different pipe materials configuration under water hammer conditions. *Res. Sq.* **2021**, 1–30. [CrossRef]
23. Urbanowicz, K.; Stosiak, M.; Towarnicki, K.; Bergant, A. Theoretical and experimental investigations of transient flow in oil-hydraulic small-diameter pipe system. *Eng. Fail. Anal.* **2021**, *128*, 105607.
24. Wan, W.; Huang, W. Investigation on complete characteristics and hydraulic transient of centrifugal pump. *J. Mech. Sci. Technol* **2011**, *25*, 2583–2590. [CrossRef]
25. Wang, X.; Zhang, J.; Chen, S.; Shi, L.; Zhao, W.; Wang, S. Valve closure based on pump runaway characteristics in long distance pressurized systems. *AQUA—Water Infrastruct. Ecosyst. Soc.* **2021**, *70*, 493–506. [CrossRef]
26. Tian, W.; Su, G.H.; Wang, G.; Qiu, S.; Xiao, Z. Numerical simulation and optimization on valve-induced water hammer characteristics for parallel pump feedwater system. *Ann. Nucl. Energy* **2008**, *35*, 2280–2287. [CrossRef]
27. Triki, A.; Essaidi, B. Investigation of Pump Failure-Induced Waterhammer Waves: A Case Study. *J. Press. Vessel Technol.* **2022**, *144*, 061403. [CrossRef]
28. Sattar, A.M.A.; Soliman, M.; El-Ansary, A. Preliminary sizing of surge vessels on pumping mains. *Urban. Water J.* **2019**, *16*, 738–748. [CrossRef]
29. Kubrak, M.; Malesińska, A.; Kodura, A.; Urbanowicz, K.; Bury, P.; Stosiak, M. Water Hammer Control Using Additional Branched HDPE Pipe. *Energies* **2021**, *14*, 8008. [CrossRef]
30. Bostan, M.; Akhtari, A.A.; Bonakdari, H.; Gharabaghi, B.; Noori, O. Investigation of a new shock damper system efficiency in reducing water hammer excess pressure due to the sudden closure of a control valve. *ISH J. Hydraul. Eng.* **2020**, *26*, 258–266.
31. Triki, A.; Trabelsi, M. On the in-series and branching dual-technique-based water-hammer control strategy. *Urban. Water J.* **2021**, *18*, 631–639. [CrossRef]

32. Triki, A. Comparative assessment of the inline and branching design strategies based on the compound-technique. *AQUA-Water Infrastruct. Ecosyst. Soc.* **2021**, *70*, 155–170. [CrossRef]
33. Reddy, H.P.; Araya, W.F.; Chaudhry, M.H. Estimation of decay coefficients for unsteady friction for instantaneous, acceleration-based models. *J. Hydraul. Eng.* **2012**, *138*, 260–271. [CrossRef]
34. Guo, X.; Zhu, Z.; Shi, G.; Huang, Y. Effects of rotational speeds on the performance of a centrifugal pump with a variable-pitch inducer. *J. Hydrodynam.* **2017**, *29*, 854–862. [CrossRef]
35. Chaudhry, M.H. *Applied Hydraulic Transients*; Springer: New York, NY, USA, 2014.
36. Barrio, R.; Blanco, E.; Keller, J.; Parrondo, J.; Ferna'ndez, J. Numerical determination of the acoustic impedance of a centrifugal pump. In Proceedings of the Fluids Engineering Division Summer Meeting, Hamamatsu, Japan, 24–29 July 2011; pp. 405–412.
37. Hu, J.; Yang, J.; Zeng, W.; Yang, J. Transient pressure analysis of a prototype pump turbine: Field tests and simulation. *J. Fluids Eng.* **2018**, *140*, 071102.
38. Zhang, J.; Yang, H.; Liu, H.; Xu, L.; Lv, Y. Pressure Fluctuation Characteristics of High-Speed Centrifugal Pump with Enlarged Flow Design. *Processes* **2021**, *9*, 2261. [CrossRef]
39. Li, X. A new Intelligent Optimization Method-Artificial Fish School Algorithm. Ph.D. Thesis, Zhejiang University, Hangzhou, China, 2003. (In Chinese).
40. Adamkowski, A.; Lewandowski, M. Experimental examination of unsteady friction models for transient pipe flow Simulation. *J. Fluids Eng.* **2006**, *128*, 1351. [CrossRef]
41. Azizi, R. Empirical study of artificial fish swarm algorithm. *Int. J. Comput. Netw. Commun.* **2014**, *3*, 1–7.
42. Martin, C.S. Representation of Pump Characteristics for Transient Analysis. In Proceedings of the ASME, Symposium on Performance Characteristics of Hydraulic Turbines and Pumps, Winter Annual Meeting, Boston, MA, USA, 13–18 November 1983; pp. 1–13.
43. Donsky, B. Complete pump characteristics and the effects of specific speeds on hydraulic transients. *J. Basic Eng.* **1961**, *83*, 685–696.

water

MDPI

Article

Sensitivity Analysis of Hydraulic Transient Simulations Based on the MOC in the Gravity Flow

Jinhao Liu, Jianhua Wu *, Yusheng Zhang and Xinhao Wu

College of Water Resource Science and Engineering, Taiyuan University of Technology, Taiyuan 030024, China; liujinhao0407@link.tyut.edu.cn (J.L.); zhangyusheng01@tyut.edu.cn (Y.Z.); wuxinhao0784@link.tyut.edu.cn (X.W.)
* Correspondence: wujianhua@tyut.edu.cn; Tel.: +86-139-3460-8701

Abstract: The purpose of this study was to evaluate the sensitivity of input parameters to output results when using the method of characteristics (MOC) for hydraulic transient simulations. Based on a gravity flow water delivery project, we selected six main parameters that affect the hydraulic transient simulation and selected maximum pressure as the output parameter in order to perform a parameter sensitivity analysis. The Morris sensitivity analysis (Morris) and the partial rank correlation coefficient method based on Latin hypercube sampling (LHS-PRCC) were both adopted. The results show that the sensitivity of each parameter is the same except for the friction factor. The flow rate and Young's modulus are positively correlated with the maximum pressure, whereas the pipe diameter, valve closing time, and wall thickness are negatively correlated. It is discussed that the variability of the friction factor comes from the function of the flow and pressure regulating valve. When other conditions of the gravity flow project remain unchanged, the maximum pressure increases with the increase in the friction factor. The flow rate, pipe diameter, and valve closing time are the key parameters that affect the model. Meanwhile, Morris and LHS-PRCC proved to be effective methods for evaluating parameter sensitivity in hydraulic transient simulations.

Keywords: gravity flow; hydraulic transient simulations; MOC; sensitivity analysis; Morris screening method; LHS-PRCC

Citation: Liu, J.; Wu, J.; Zhang, Y.; Wu, X. Sensitivity Analysis of Hydraulic Transient Simulations Based on the MOC in the Gravity Flow. *Water* **2021**, *13*, 3464. https://doi.org/10.3390/w13233464

Academic Editors: Kamil Urbanowicz and Helena M. Ramos

Received: 5 November 2021
Accepted: 3 December 2021
Published: 6 December 2021

Publisher's Note: MDPI stays neutral with regard to jurisdictional claims in published maps and institutional affiliations.

Copyright: © 2021 by the authors. Licensee MDPI, Basel, Switzerland. This article is an open access article distributed under the terms and conditions of the Creative Commons Attribution (CC BY) license (https://creativecommons.org/licenses/by/4.0/).

1. Introduction

A water hammer is a type of hydraulic transient momentum that produces sudden pressure changes when the flow rate changes (due to the opening and closing of valves or units) in the pipeline. It can cause problems such as pipe bursts [1], water leakage and so on, in the water supply pipeline, and is generally regarded as one of the main risks that threaten water supply safety. Therefore, hydraulic transient simulations are an essential part of the safe operation of water supply projects. Gravity flow is widely used in water supply projects due to its low operating cost, easy maintenance and low investment requirements. The pump-stop water hammer does not occur in the gravity flow; however, when the valve is closed, it can produce greater water hammer pressure. The phenomenon of water hammer bridging is especially prone to occur in complex pipelines. At the beginning of the last century, arithmetic and graphical methods were successively proposed and applied to hydraulic transient simulations [2]. By the middle of the last century, the method of characteristics (MOC) had gradually become the most commonly used method in hydraulic transient simulations due to the development of computer technology [3–6]. The calculation results of the MOC are consistent with the experimental results of many practical applications [7,8]. By using the MOC, researchers and designers could perform hydraulic transient simulations of entire pipelines in water supply projects. In addition, according to the simulation results, the designs could be optimized, and the water hammer protection measures could be selected. Wang et al. [1] applied the MOC in urban water distribution systems and proposed a method for rating the risk of pipe bursts. Kou et al. [9] applied the MOC in a mine drainage system and proposed a water hammer protection method based on a hydraulic control valve. Tian et al. [10] used the MOC to study the valve-induced water hammer phenomenon during the alternate startup process of parallel pumps and optimized the design. Noura et al. [11] applied the MOC to study a variety of water hammer control strategies in pumping stations and proved that, in some cases, simple water hammer control devices can also serve the purpose of water hammer protection. Based on the limitations of the MOC, Afshar et al. [12] proposed the implicit method of characteristics (IMOC), which can more accurately predict the changes in water head and flow after comparison. Kamil et al. [13] proposed a method for estimating wall transient shear stress using an effective two-term weighting function, and through experimental comparison, this method can more accurately simulate the transient process; subsequently, they [14] compared the experimental results and simulations of a small pipe diameter and found that the valve closure path and the unsteady friction can control the pulse attenuation, shape and time. Liou [15] studied the sustained head increase caused by line packing and proposed an analytical solution to calculate the maximum pressure at the closed valve.

The mathematical model of hydraulic transient simulations is composed of multiple partial differential equations, and most commercial software uses the MOC for simulations [3]. The traditional methods of changing parameters and step-by-step trial calculations increased the amount of calculation for researchers and designers. At the same time, due to nonlinear relationships and uncertainty in the mathematical model of hydraulic transient simulations, there are certain difficulties in identifying the parameters of the model, which prevents researchers from adjusting the model parameters to achieve the expected protection effect. Sensitivity analysis is a method to quantitatively describe the importance of a model's input variables to its output variables [16,17]. In recent years, methods of sensitivity analysis have developed rapidly, and they have been used in multiple models of water engineering. Yi et al. [18] used Morris to analyze the sensitivity of the water quality model of Dianchi Lake, and further conducted an identification and uncertainty analysis of the model parameters. Ouatiki et al. [19] applied the one-at-a-time sensitivity measures (OAT) method to analyze the parameter sensitivity of the HBV hydrological model of a small watershed in semi-arid mountainous areas. Xu et al. [20] proposed using the Latin hypercube one-factor-at-a-time (LH-OAT) method to analyze the sensitivity of an agricultural hydrological model (SWAP-EPIC). Li et al. [21] first used LHS, and then applied PRCC and the mutual information method to compare and analyze the SWMM's influence parameters.

However, in hydraulic transient simulations, there are fewer applications for sensitivity analysis. Wan et al. [22,23] conducted a sensitivity analysis of the relationship between the pressure vessel setting and the maximum pressure change of a water hammer, and explored the protective effect of the pressure vessel. By comparing the pressure changes of the pump valve system at different operating times, he optimized the time difference between the opening of the valve and the opening of the unit when the pump was started. Zhu et al. [24] introduced a random model in hydraulic transient simulations and carried out a sensitivity analysis in a hydropower station project. Currently, there is no comprehensive research on the use of sensitivity analysis of hydraulic transient simulations in gravity flow.

In this study, representative input parameters were selected and a hydraulic transient simulation using the MOC was calculated for the engineering of a long-distance, small-diameter gravity flow. Then, two sensitivity analysis methods, the Morris and the LHS-PRCC, were used for the sensitivity analysis of the calculated results. We sorted the parameters based on the results of the sensitivity analysis. Since there are few cases where sensitivity analysis is applied in hydraulic transient simulations, two methods were used for comparison. The aims of this study are as follows: (i) compare the effectiveness and the similarity of the two methods in hydraulic transient simulations, and (ii) identify and sort the parameters according to their influence on the calculation. When the water hammer protection scheme was selected, the design parameters could be directly optimized in a targeted manner. We provide certain guidance for the survey, design and construction of the project. In this study, the sensitivity analysis method was applied to the gravity flow in a hydraulic transient simulation for the first time. The results have important reference value for similar studies concerning gravity flow in water supply projects.

The structure of this paper is as follows: In Section 2, the calculation method of hydraulic transient simulations (MOC) and the study case are introduced. We also describe two sensitivity analysis methods: Morris and LHS-PRCC. In Section 3, the MOC is used for hydraulic transient simulation in the study case. Then, the Morris analysis method and LHS-PRCC are used to analyze the sensitivity of the calculation results. Finally, the two calculation results are compared. In Section 4, the conclusions of this study are presented.

2. Materials and Methods

2.1. Study Area and Parameter Selection

In a gravity flow water supply project in Shanxi, the water is transported from a high-level storage tank (elevation: 1275.7 m, water level: 2.8 m) through a 5720 m pipeline, to an end storage tank (elevation: 1193.62 m, water level: 0 m). The flow and pressure regulating valve is installed in the end storage tank, and the valve closing time is 100 s. The pipe is a spiral steel pipe with a diameter of 250 mm, a wall thickness of 8 mm, an elastic modulus of 2.079×10^{11} Pa and a friction factor of 0.012. The longitudinal section of the pipeline is shown in Figure 1. Although the height difference between the two storage tanks is small, the pipeline crosses the valley terrain with large undulations, which causes water hammer to easily occur. This gravity flow has characteristics such as a long distance, small pipe diameter, low flow and high drop, so it is valuable for analysis.

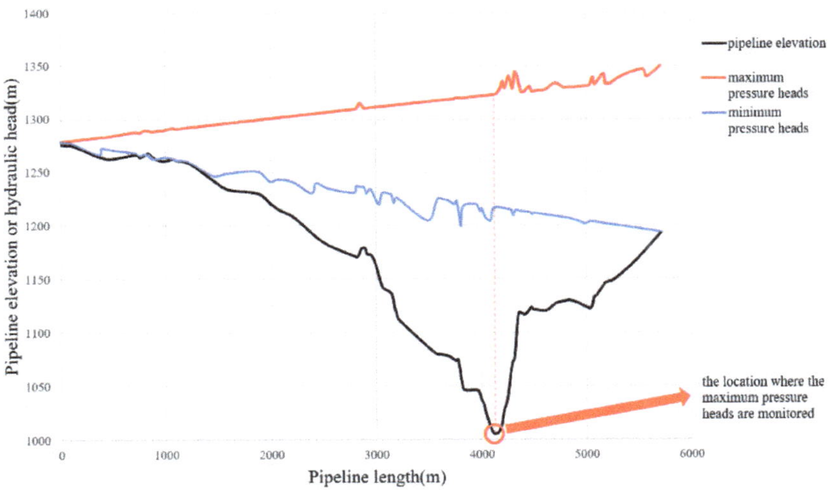

Figure 1. Pipeline elevation trend and envelopes of maximum and minimum pressure heads for the case with initial values.

The parameters and reference values are shown in Table 1. According to the control equation, since the parameters do not affect each other, the local sensitivity analysis can meet the analysis requirements.

Table 1. The initial value and range of each parameter.

Parameter Number	Parameter	Initial Value	Range
1	Valve closing time (s)	100.00	70.00~130.00
2	Flow rate (m^3/s)	0.054	0.038~0.070
3	Friction factor	0.0249	0.0174~0.032
4	Young's modulus of pipe (Pa)	2.079×10^{11}	1.455×10^{11}~2.703×10^{11}
5	Pipe thickness (mm)	8.00	5.60~10.40
6	Pipe diameter (mm)	250.00	175.00~325.00

2.2. Control Equations and Calculation Methods

In the hydraulic transient simulation, the basic differential equation consists of two parts: the motion equation and the continuous equation.

$$\frac{\partial H}{\partial x} + \frac{1}{g}\frac{\partial V}{\partial t} + \frac{V}{g}\frac{\partial V}{\partial x} + \frac{2}{\rho R}\tau_w = 0 \quad (1)$$

$$\frac{\partial H}{\partial t} + V\left(\frac{\partial H}{\partial x} + \sin\alpha\right) + \frac{a^2}{g}\frac{\partial V}{\partial x} = 0 \quad (2)$$

The above equation is a set of partial differential hyperbolic equations from which it is difficult to obtain the analytical solution [3]. Based on the above introduction, the MOC is used to transform partial differential equations into ordinary differential equations. Since there is a detailed introduction about the MOC in the references, this paper only gives a brief introduction

$$\begin{cases} \frac{dV}{dt} + \frac{g}{a}\frac{dH}{dt} + \frac{2}{\rho R}\tau_w = 0 \\ \frac{dx}{dt} = +a \end{cases} \quad (3)$$

$$\begin{cases} \frac{dV}{dt} - \frac{g}{a}\frac{dH}{dt} + \frac{2}{\rho R}\tau_w = 0 \\ \frac{dx}{dt} = -a \end{cases} \quad (4)$$

The wall shear stress τ_w is the sum of two expressions [14]

$$\tau_w = \tau_q + \tau_u \quad (5)$$

where the τ_q calculated by using standard Darcy–Weisbach euqation

$$\tau_q = \frac{f\rho V|V|}{8} \quad (6)$$

while the τ_u is expressed by the following convolution integral

$$\tau_u = \frac{2\mu}{R}\int_0^t w(t-u)\frac{\partial v(u)}{\partial t}du \quad (7)$$

In the equation, μ is the dynamic viscosity, and $w(t-u)$ is the weight function. To obtain the simplified format, integrate the above equation along the characteristic line

$$V_P - V_A + \frac{g}{a}(H_P - H_A) + \frac{2\Delta t f}{RA}\tau_{wA} = 0 \quad (8)$$

$$V_P - V_B - \frac{g}{a}(H_P - H_B) + \frac{2\Delta t f}{RA}\tau_{wB} = 0 \quad (9)$$

The two equations are straight lines with constant slopes, so the calculation process can be described by a rectangular grid. As shown in Figure 2, (Δx) is the spacing step length, the pipeline is evenly divided into (N) sections, (i) represents the order of each section, (i = 1) is the starting section and the terminal section is (i = N + 1). The calculation required is ($\Delta t = \Delta x/a$).

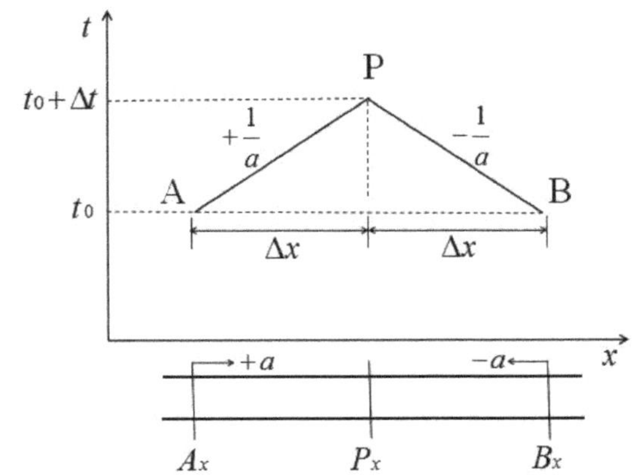

Figure 2. The characteristic lines of the MOC.

The calculation starts at (t = 0), given the parameters of points A and B, recursively, according to the time interval. Additionally, it obtains the parameter values of all grid nodes.

The boundary condition of the valve is

$$V = C_d\sqrt{2g\Delta H} \tag{10}$$

where C_d is the discharge coefficient, which depends on the valve performance curve and opening ratio τ at a specific time. In this study, the valve fully open flow coefficient C_{d0} is 4731.

In the Equations (8) and (9), the calculation of the wave velocity adopts the elastic water hammer theory to calculate

$$a = \sqrt{\frac{K}{\rho}}\frac{1}{\sqrt{1+\frac{KD}{E\delta}}} \tag{11}$$

It can be seen from Equation (11) that the wave velocity α is related to the pipe diameter, wall thickness, Young's modulus, fluid bulk elastic modulus and fluid density. In this study, the fluid is water, the bulk modulus is 1.96×10^9 Pa and the density is 1000 kg/m^3. Substituting Equation (11) for Equations (8) and (9), the calculation result is directly related to the pipe diameter, wall thickness and Young's modulus, so that the parameters directly affect the calculation result.

2.3. Sensitivity Analysis Methods

2.3.1. Morris Sensitivity Analysis

Morris sensitivity analysis [25] (also called Elementary Effects) can reflect the changes of the calculation results under the slight disturbance of factors. After a period of development [26], it has been widely used in sensitivity analysis. Morris adopts the concept of primary influence on factors, and the influence value of the i-th factor is expressed as

$$e_i = \frac{y(x_1, x_2, x_3, \ldots, x_i + \Delta, \ldots, x_n) - y(x_1, x_2, x_3, \ldots, x_i, x_n)}{\Delta} \tag{12}$$

In the equation, suppose $(x_1, x_2, x_3, \ldots, x_n)$ are n input quantities that affect the output result of the model; (y) is the simulation output result of the model and (Δ) is the change quantity of the i-th input parameter.

Morris uses independent variables to change with a fixed step length, and the sensitivity discrimination factor takes multiple averages of Morris [22]

$$S = \sum_{i=0}^{n-1} \frac{(y_{i+1} - y_i)/y_0}{(P_{i+1} - P_i)/100} / (n-1) \quad (13)$$

where (S) is the sensitivity judgment parameter, (y_i) is the output result of the i-th run of the calculation model, (y_0) is the reference value of the model parameter calculation result, (P_i) is the percentage of the change of the i-th model's calculation parameter value to the reference value after the calibration parameter and (n) is the number of model runs.

The steps of Morris are shown in Figure 3a. According to the final calculation results, the sensitivity can be divided into four levels: $|S| \geq 1$ (high-sensitivity parameter), $0.2 \leq |S| < 1$ (sensitivity parameter), $0.05 \leq |S|\ 0 < 0.2$ (medium sensitivity) and $0 \leq |S| < 0.05$ (not sensitive).

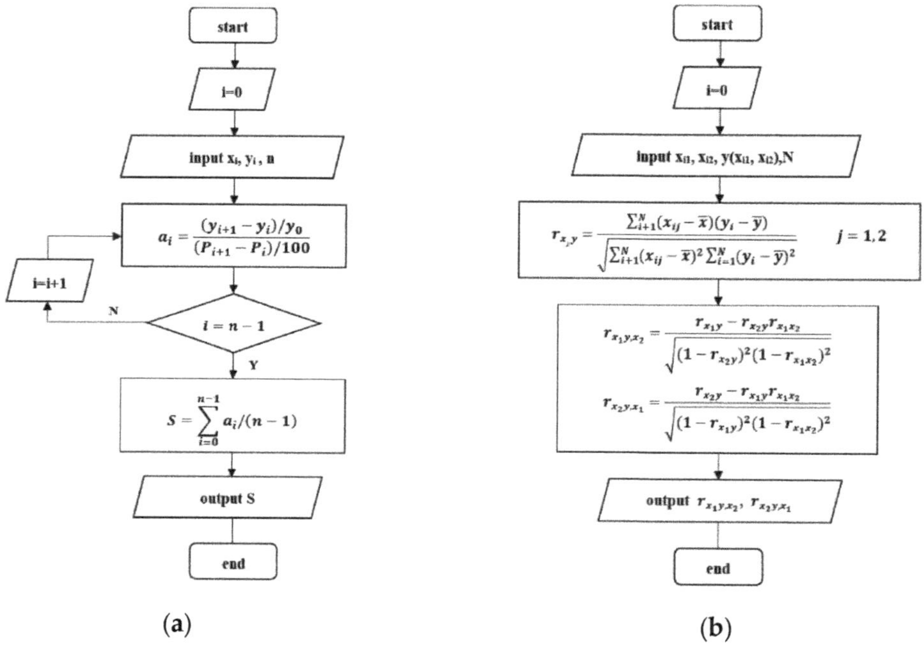

Figure 3. The flow charts of two sensitivity analysis methods: (**a**) the flow chart of Morris; (**b**) the flow chart of PRCC.

2.3.2. LHS-PRCC

PRCC is often combined with LHS for sensitivity analysis. By combining uncertainty analysis with PRCC, we can reasonably evaluate the sensitivity of our output variables to parameter changes.

LHS is a multi-dimensional stratified sampling method, which was first proposed by McKay et al. [27]. The advantage is that it requires fewer samples than simple random sampling to achieve the same accuracy. In LHS, parameters are randomly distributed in N equal probability intervals, and then the parameters are sampled. N represents the sample size. The choice of N should be at least $k + 1$, where k is the number of changing parameters, but usually much larger to ensure accuracy [28]. Because of its relatively uniform sampling, it has been widely used. The steps of LHS are as follows: first, the input ranges of each parameter are divided into N ranges with equal probability (N is the number of samples). Then, representative parameters are randomly selected from each divided range. Finally, these parameters are combined as a sampling result.

Partial correlation analysis is used to control the influence of other variables under the interaction of multiple variable factors and study the relationship between two specific variables [29]. The level of partial correlation analysis is determined by the number of research variables. The correlation coefficient (CC) between input x_j and output y is calculated as follows

$$r_{x_j,y} = \frac{\sum_{i+1}^{N}(x_{ij}-\overline{x})(y_i-\overline{y})}{\sqrt{\sum_{i+1}^{N}(x_{ij}-\overline{x})^2\sum_{i=1}^{N}(y_i-\overline{y})^2}} \quad j = 1,2,\ldots,k \tag{14}$$

The partial correlation coefficient (PCC) provides a measure of the strength of the linear relationship between input x_j and output y after eliminating the linear effects of other variables. The PCC between x_j and y is defined by the $r_{x_j,y}$ of $(x_j - \hat{x}_j)$ and $(y - \hat{y})$, where $\hat{x}_j$ and $\hat{y}$ use the least square method to construct the regression model

$$\hat{x}_j = c_0 + \sum_{\substack{p=1 \\ p\neq j}}^{n} c_p x_p, \hat{y} = b_0 + \sum_{\substack{p=1 \\ p\neq j}}^{n} b_p x_p \tag{15}$$

where $(c_0, c_1, \ldots, c_n)$ and $(b_0, b_1, \ldots, b_n)$ are coefficients determined in the construction of the regression model.

Similar to PCC, PRCC performs partial correlation on the rank transformed data; x_j and y are rank transformed first, and then the linear regression model described in the equation is performed. The value of PRCC ranges from −1 to +1, and the closer the absolute value $|r|$ is to 1, the higher the correlation between the parameters; the closer it is to 0, the lower the correlation. The positive value of PRCC represents a positive correlation between the two parameters; on the contrary, the negative values represent a negative correlation.

In this study, the equation of the six parameters used is more complicated and detailed in the references [28,29]. Therefore, only the process of the two input parameters x_1, x_2 affecting the output parameter y is demonstrated in the Figure 3b flow chart of PRCC.

3. Results and Discussion

In this case, the main problem concerning gravity flow valve closing is controlling the maximum pressure in the hydraulic transient state so that the calculation results of the negative pressure do not change significantly. As a result, the maximum pressure is selected as the distinguishing parameter.

3.1. The Result of Morris

Based on the reference value of the model parameters, the MOC is used to analyze the local sensitivity of the hydraulic transient simulation results while the valve closes in gravity flow. The values of the parameters are perturbed with a fixed step of 10%, and the values are −30%, −20%, −10%, 10%, 20% and 30% of the reference value. According to the above calculation method, the maximum pressure value is calculated and recorded when the various parameters vary in different ranges. Then, the value of the sensitivity discrimination parameter S is calculated. The calculation results are shown in Table 2 and Figure 4.

Table 2. The sensitivity coefficient S and the level of each parameter.

Parameter Number	Parameter	S	Sensitivity Level
1	Valve closing time (s)	−0.242	Sensitivity parameter
2	Flow rate (m^3/s)	0.347	Sensitivity parameter
3	Friction factor	0.020	Not sensitive
4	Young's modulus of pipe (Pa)	0.006	Not sensitive
5	Pipe thickness (mm)	0.019	Not sensitive
6	Pipe diameter (mm)	−0.383	Sensitivity parameter

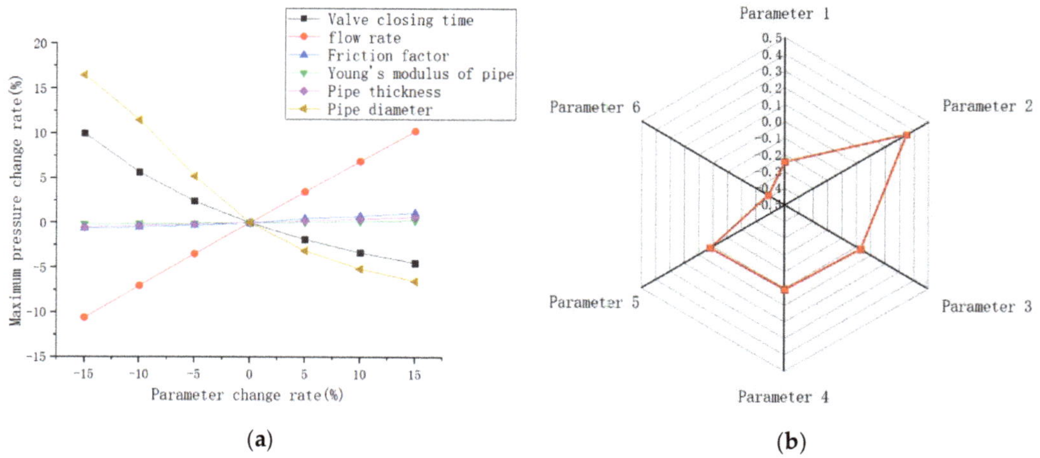

Figure 4. Sensitivity analysis results of Morris: (**a**) the change rate of the maximum pressure caused by parameter disturbance; (**b**) sensitivity parameter S radar chart.

According to the parameter sensitivity coefficient, the sensitivity is as follows: the pipe diameter, flow rate, valve closing time, friction factor, pipe thickness and Young's modulus of the pipe are in descending order. Among them, the pipe diameter, flow rate and valve closing time are all sensitive parameters; the friction factor, pipe thickness and Young's modulus of the pipe are insensitive parameters. The valve closing time, Young's modulus, friction factor and the maximum pressure are positively correlated; the pipe diameter, valve closing time, pipe wall thickness and the maximum pressure are negatively correlated.

3.2. LHS-PRCC Analysis

In this study, we used the LHS to generate 25 sets of parameter samples from the range of each parameter given in Table 1 and performed hydraulic transient simulations. The parameter samples and maximum pressure calculation results are shown in Table 3.

Table 3. The results of LHS.

Number	Valve Closing Time (s)	Flow Rate (m^3/s)	Friction Factor	Young's Modulus (Pa)	Pipe Thickness (mm)	Pipe Diameter (mm)	Maximum Pressure (m)
1	125.645	56.144	0.026	2.386×10^{11}	7.681	283.788	296.841
2	127.826	38.052	0.031	2.244×10^{11}	8.327	210.493	301.976
3	74.495	51.442	0.025	1.592×10^{11}	9.572	298.601	319.967
4	115.808	48.514	0.022	1.973×10^{11}	9.147	269.441	297.539
5	104.482	46.554	0.026	1.931×10^{11}	6.336	200.200	327.434
6	93.766	63.972	0.027	2.636×10^{11}	8.556	274.412	325.845
7	107.732	53.202	0.023	1.613×10^{11}	8.814	187.536	358.472
8	119.087	61.940	0.022	1.775×10^{11}	7.202	323.909	302.464
9	98.902	63.532	0.028	2.276×10^{11}	7.004	227.417	342.644
10	84.484	44.062	0.019	2.008×10^{11}	6.416	294.127	290.188
11	122.422	59.940	0.018	2.090×10^{11}	9.929	222.184	332.570
12	89.271	65.280	0.030	1.667×10^{11}	6.160	313.784	317.355
13	96.503	67.063	0.028	2.116×10^{11}	5.940	261.522	328.255
14	102.435	44.252	0.023	1.536×10^{11}	6.577	237.701	294.895
15	70.479	54.492	0.018	2.675×10^{11}	9.795	246.130	351.802
16	84.059	42.327	0.020	2.549×10^{11}	6.785	193.882	354.191
17	80.988	39.814	0.021	2.201×10^{11}	8.953	309.268	288.788
18	113.402	49.666	0.021	2.577×10^{11}	10.209	214.894	330.033
19	94.499	40.572	0.018	1.716×10^{11}	9.425	255.921	302.486
20	122.897	47.827	0.018	1.808×10^{11}	10.123	250.589	301.791
21	87.626	59.078	0.024	1.898×10^{11}	8.268	282.544	321.346

Table 3. *Cont.*

Number	Valve Closing Time (s)	Flow Rate (m^3/s)	Friction Factor	Young's Modulus (Pa)	Pipe Thickness (mm)	Pipe Diameter (mm)	Maximum Pressure (m)
22	77.114	55.254	0.020	2.475×10^{11}	7.719	185.561	394.491
23	108.762	69.224	0.032	2.330×10^{11}	8.095	180.260	365.335
24	77.358	67.849	0.029	2.453×10^{11}	5.643	233.897	367.192
25	111.571	57.441	0.019	1.488×10^{11}	7.363	305.560	303.349

According to the Latin hypercube sampling results in Table 3, the PRCC was calculated between each of the six parameters and the maximum pressure. Based on the magnitude of the absolute value of PRCC, the relative importance of the parameters is ranked. The PRCC calculation results of the parameters are shown in Figure 5 and Table 4.

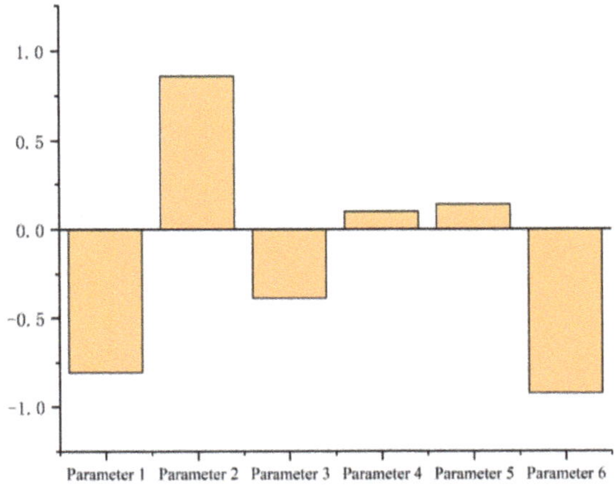

Figure 5. Comparison analysis chart of LHS-PRCC r results.

Table 4. The partial rank correlation coefficient r of each parameter.

Parameter Number	Parameter	r
1	Valve closing time (s)	−0.806
2	Flow rate (m^3/s)	0.860
3	Friction factor	−0.388
4	Young's modulus (Pa)	0.096
5	Pipe thickness (mm)	0.136
6	Pipe diameter (mm)	−0.924

PRCC analysis results show that the pipe diameter has the highest influence on the maximum pressure, followed by flow rate and valve closing time. The valve closing time, Young's modulus and the maximum pressure are significantly positively correlated; on the other hand, the pipe diameter, valve closing time, friction factor, pipe wall thickness and the maximum pressure are negatively correlated.

3.3. Results Comparison and Discussion

3.3.1. Analysis of Parameters Related to Wave Velocity

In the gravity flow supply project, the maximum pressure generated by the water hammer increases with an increase in the wave speed. When the water hammer occurs due to cavity collapse, the trend of rising pressure is obvious; on the contrary, when the water hammer due to cavity collapse does not occur, the influence of the wave velocity on the maximum pressure is not obvious. The water hammer due to cavity collapse does not occur in this hydraulic transient simulation, so the Young's modulus of the pipeline has a small, indirect effect on the maximum pressure by changing the wave velocity. The sensitivity analysis results of this simulation are consistent with previous research results, which is also an important reason for converting the wave velocity into other parameters for calculation in the second section. Among the parameters selected in this study, the parameters related to the wave velocity are the Young's modulus, the pipe wall thickness and the pipe diameter.

From Equation (11), it can be seen that the Young's modulus of the pipeline only indirectly affects the maximum pressure by changing the wave velocity. According to the calculation results, $S_4 = 0.006$ and $r_4 = 0.096$ (S is the Morris sensitivity parameter; r is the sensitivity parameter in LHS-PRCC). The maximum pressure increases with the increase in the Young's modulus of the pipe. Moreover, its influence on the maximum pressure is small.

The reasons for the influence of pipe wall thickness and pipe diameter on the maximum pressure are similar: both come from the direct influence on the pipe flow area and the indirect influence on the wave velocity. The calculation result of the pipe wall thickness shows that $S_5 = 0.019$, $r_5 = 0.136$ and the pipe wall thickness has little effect on the calculation results. According to the parameter change range in Table 2, the maximum change of pipe wall thickness to pipe inner diameter is 2.40 mm, which is very small compared to the original pipe inner diameter of 243 mm. This makes the direct influence of pipe wall thickness on the maximum pressure small. Then, it has a small, indirect effect on the maximum pressure by influencing the wave velocity. Therefore, the pipe wall thickness is less sensitive to the maximum pressure. In the same way, the influence of pipe diameter on the maximum pressure mainly comes from the change of the water passing area. The maximum change of the pipe diameter to the inner diameter is 75.00 mm, so that the pipe diameter has a greater direct influence on the maximum pressure. The calculation result shows $S_6 = -0.383$ and $r_6 = -0.924$.

3.3.2. The Main Parameters That Affect the Maximum Pressure

According to the calculation results in Tables 2 and 3, the pipe diameter, flow rate and valve closing time all have a significant impact on the maximum pressure, and the laws presented are basically the same. The results of the two methods can be mutually confirmed. The valve closing time directly determines whether direct water hammer or indirect water hammer occurs in the gravity flow water delivery system [5]. The result shows that $S_1 = -0.242$ and $r_1 = -0.806$. There are many related studies on the impact of valve closing time on the maximum pressure, and the results of this study are consistent with previous study results [9,30]. The result of the flow rate is $S_2 = 0.347$ and $r_2 = 0.860$. The values of the flow rate directly affect the pressure of the pipeline during steady-state operation. In the valve-closing hydraulic transient simulation, the change of the flow rate is the main reason for increasing the maximum pressure. Therefore, reducing the flow rate is also one of the commonly used measures in gravity flow water hammer protection. The analysis of the pipe diameter has been described in Tables 2 and 3, and the result is $S_6 = -0.383$ and $r_6 = -0.924$.

According to the high sensitivity of these parameters in this simulation, they are the parameters that should be emphatically considered in hydraulic transient simulations. In the design of most water delivery projects in China, the selection of the pipeline diameter is often determined before the design of water hammer protection, and water hammer protection is only performed by adjusting the valve closing time, which is not conducive to the design of water hammer protection. Therefore, in the pipe diameter parameters of a water delivery project, the water delivery design and the water hammer protection design should be carried out at the same time, and the calculation results should be confirmed against each other to better complete the design of the water delivery project.

3.3.3. Analysis of the Variability of the Friction Factor in the Results

The correlation of the friction factor is different in the two methods. The increase in the friction factor will increase the head loss. Generally speaking, in a water supply project, the larger the friction factor of a pressurized pipeline, the smaller the flow rate and, thus, the safer the project. There are differences in the results of the two methods in this sensitivity analysis. The result shows that $S_3 = 0.020$ and $r_3 = -0.388$. In the two methods, the friction factor is not considered a more sensitive parameter. In the Morris sensitivity analysis, the maximum pressure increases with the increase in the friction factor; in LHS-PRCC, the maximum pressure decreases with the increase in the friction factor.

When the gravity flow water supply project is operating in the steady state, the flow is controlled by the flow and pressure regulating valve. When the friction factor of the pipeline increases and the water delivery capacity is sufficient, the water supply flow rate will not change due to the function of the flow and pressure regulating valve. When the transient process occurs, the increased friction reduces the amplitude of the water hammer wave fronts, leading to the line packing effect. This effect can cause a continuous pressure rise after the closure of the valve, and may produce overpressure [15]. This effect should be paid more attention to in high-friction pipes with long pipe lengths and small diameters. In the Morris analysis method, other parameters remain unchanged; only the friction factor is changed, resulting in a positive correlation between the friction factor and the maximum pressure. In LHS-PRCC, the parameters are all defined by stratified random sampling, and the changes of other parameters make the water supply capacity unable to be guaranteed. The line packing effect cannot be reflected under the changing parameter conditions. The friction factor will reduce the water transport capacity, resulting in a negative correlation between the friction factor and the maximum pressure.

In the similar gravity flow water supply projects with small pipe diameters and long distances, when the capacity of water delivery is sufficient and the friction factor of the pipeline increases, a bigger maximum pressure will be generated after the closure of the valve. In this case, more protection is required. This is what the design and operation managers need to pay more attention to.

4. Conclusions

In this study, the MOC was used in the gravity flow with obvious characteristics to carry out the hydraulic transient simulation. Then, the Morris screening method and LHS-PRCC were used to perform sensitivity analysis on calculations, and the following conclusions can be drawn:

1. In this gravity flow example, the comparison of the two sensitivity analysis results shows that only some key parameters have an important influence on the calculation results. The sensitivity of key parameters from large to small are pipe diameter, flow rate and valve closing time. The friction factor, pipe thickness and Young's modulus have little influence on the calculation results, and their sensitivity ranking has some variability.
2. The simulation results have reference value for the design of similar gravity flow water delivery projects with obvious characteristics. In the design and operation of the project, the valve closing time, pipe diameter and flow rate should be strictly controlled to ensure the safety of the project.
3. The sensitivity of the friction factor is different in the results of the two methods. After discussion, when other parameters remain unchanged, the maximum pressure increases with the increase in the friction factor due to the line packing effect; when other parameters change and the water delivery capacity cannot be guaranteed, the maximum pressure is negatively related to the friction coefficient. Therefore, more protective measures are needed when the friction factor of a gravity flow project becomes larger.
4. The Morris screening method and LHS-PRCC gave similar parameter rankings for the selected parameters of the project in this case. The calculation results of the

two methods are complementary in the sensitivity analysis of hydraulic transient simulation. At the same time, this study also confirms the applicability of the two methods in the sensitivity analysis of hydraulic transient simulations.

In summary, in this study, we analyzed the parameter sensitivity of hydraulic transient simulations based on the MOC in gravity flow. We only analyzed one engineering example, which proved the applicability of the two sensitivity analysis methods. In order to obtain more comprehensive results, more examples need to be analyzed.

Author Contributions: Conceptualization, J.L., J.W. and Y.Z.; Data curation, J.L.; Formal analysis, J.L. and Y.Z.; Methodology, J.L. and J.W.; Project administration, J.W. and Y.Z.; Resources, J.W.; Supervision, J.W. and Y.Z.; Validation, J.L. and X.W.; Visualization, J.L. and X.W.; Writing—original draft, J.L.; Writing—review and editing, J.L. and X.W. All authors have read and agreed to the published version of the manuscript.

Funding: This work was supported by the National Natural Science Foundation of Shanxi Province, China (20210302123169).

Institutional Review Board Statement: Not applicable.

Informed Consent Statement: Not applicable.

Data Availability Statement: The data that support the finding of this study are available from the corresponding author upon reasonable request.

Acknowledgments: The authors thank the College of Water Resource Science and Engineering, Taiyuan University of Technology for providing the site for the experiment and collecting data.

Conflicts of Interest: The authors declare no conflict of interest.

Nomenclature

H	pressure head (m)
x	distance along pipe from inlet (m)
g	acceleration of gravity (m/s^2)
V	flow velocity (m/s)
t	time, as subscript to denote time (s)
ρ	density of liquid (kg/m^3)
R	radius of the pipe (m)
α	the angle between pipe and the horizontal plane
τ_w	shear stress calculated by the non-stationary friction losses
τ_q	shear stress calculated by the quasi-steady state model
τ_u	shear stress related to the non-stationarity of flow
a	speed of pressure wave (m/s)
f	Darcy–Weisbach friction factor
V_P, V_A, V_B	flow velocity of Point·P, A and B (m/s)
H_P, H_A, H_B	pressure head of Point·P, A and B (m)
Δt	time step (s)
Δx	length of segment (m)
C_d	discharge coefficient
ΔH	head loss of valve
K	fluid bulk elastic modulus (Pa)
D	pipe inner diameter (m)
E	elastic modulus of the pipe (Pa)
δ	thickness of pipe (m)
S	sensitivity judgment parameter in Morris
x	input parameter
y	output parameter
y_0	reference value of the model parameter calculation result
P_i	percentage of the change of the i-th model's calculation parameter value to the reference value after the calibration parameter

n	number of model runs
r	sensitivity judgment parameter in LHS-PRCC
Acronyms:	
MOC	method of characteristics
Morris	Morris sensitivity analysis
LHS-PRCC	partial rank correlation coefficient method based on Latin hypercube sampling

References

1. Wang, R.; Wang, Z.; Wang, X.; Yang, H.; Sun, J. Pipe Burst Risk State Assessment and Classification Based on Water Hammer Analysis for Water Supply Networks. *J. Water Res. Plan. Manag.* **2014**, *140*, 4014005. [CrossRef]
2. Yang, K. Review and frontier scientific issues of hydraulic control for long distance water diversion. *J. Hydraul. Eng.* **2016**, *47*, 424–435. (In Chinese) [CrossRef]
3. Abdeldayem, O.; Ferràs, D.; van der Zwan, S.; Kennedy, M. Analysis of Unsteady Friction Models Used in Engineering Software for Water Hammer Analysis: Implementation Case in WANDA. *Water* **2021**, *13*, 495. [CrossRef]
4. Wylie, E.B A.S. *Fluid Transients*; McGraw-Hill: New York, NY, USA, 1978; p. 384.
5. Chaudhry, H.M. *Applied Hydraulic Transients*; Van Nostrand Reinhold: New York, NY, USA, 1987; p. 521.
6. Wylie, E.B.; Streeter, V.L.; Lyle, V.; Suo, L. *Fluid Transients in Systems*; Prentice Hall: Hoboken, NJ, USA, 1993; p. 463.
7. Kwon, H.J.; Lee, J. Computer and Experimental Models of Transient Flow in a Pipe Involving Backflow Preventers. *J. Hydraul. Eng.* **2008**, *134*, 426–434. [CrossRef]
8. Han, S.Y.; Hansen, D.; Kember, G. Multiple scales analysis of water hammer attenuation. *Q. Appl. Math.* **2011**, *69*, 677–690. [CrossRef]
9. Kou, Y.; Yang, J.; Kou, Z. A Water Hammer Protection Method for Mine Drainage System Based on Velocity Adjustment of Hydraulic Control Valve. *Shock Vib.* **2016**, *2016*, 1–13. [CrossRef]
10. Tian, W.; Su, G.H.; Wang, G.; Qiu, S.; Xiao, Z. Numerical simulation and optimization on valve-induced water hammer characteristics for parallel pump feedwater system. *Ann. Nucl. Energy* **2008**, *35*, 2280–2287. [CrossRef]
11. Bettaieb, N.; Taieb, E.H. Assessment of Failure Modes Caused by Water Hammer and Investigation of Convenient Control Measures. *J. Pipeline Syst. Eng.* **2020**, *11*, 4020006. [CrossRef]
12. Afshar, M.H.; Rohani, M.; Taheri, R. Simulation of transient flow in pipeline systems due to load rejection and load acceptance by hydroelectric power plants. *Int. J. Mech. Sci.* **2010**, *52*, 103–115. [CrossRef]
13. Urbanowicz, K. Modern Modeling of Water Hammer. *Pol. Marit. Res.* **2017**, *24*, 68–77. [CrossRef]
14. Urbanowicz, K.; Stosiak, M.; Towarnicki, K.; Bergant, A. Theoretical and experimental investigations of transient flow in oil-hydraulic small-diameter pipe system. *Eng. Fail. Anal.* **2021**, *128*, 105607. [CrossRef]
15. Liou, J.C P. Understanding Line Packing in Frictional Water Hammer. *J. Fluids Eng.* **2016**, *138*, 081303. [CrossRef]
16. Ionescu-Bujor, M.; Cacuci, D.G. A Comparative Review of Sensitivity and Uncertainty Analysis of Large-Scale Systems—I: Deterministic Methods. *Nucl. Sci. Eng.* **2004**, *147*, 189–203. [CrossRef]
17. Cacuci, D.G.; Ionescu-Bujor, M. A Comparative Review of Sensitivity and Uncertainty Analysis of Large-Scale Systems—II: Statistical Methods. *Nucl. Sci. Eng.* **2004**, *147*, 204–217. [CrossRef]
18. Yi, X.; Zou, R.; Guo, H. Global sensitivity analysis of a three-dimensional nutrients-algae dynamic model for a large shallow lake. *Ecol. Model.* **2016**, *327*, 74–84. [CrossRef]
19. Ouatiki, H.; Boudhar, A.; Ouhinou, A.; Beljadid, A.; Leblanc, M.; Chehbouni, A. Sensitivity and Interdependency Analysis of the HBV Conceptual Model Parameters in a Semi-Arid Mountainous Watershed. *Water* **2020**, *12*, 2440. [CrossRef]
20. Xu, X.; Sun, C.; Huang, G.; Mohanty, B.P. Global sensitivity analysis and calibration of parameters for a physically-based agro-hydrological model. *Environ. Model. Softw.* **2016**, *83*, 88–102. [CrossRef]
21. Li, C.; Wang, W.; Xiong, J.; Chen, P. Sensitivity Analysis for Urban Drainage Modeling Using Mutual Information. *Entropy* **2014**, *16*, 5738. [CrossRef]
22. Wan, W.; Huang, W.; Li, C. Sensitivity Analysis for the Resistance on the Performance of a Pressure Vessel for Water Hammer Protection. *J. Press. Vessel. Technol.* **2014**, *136*. [CrossRef]
23. Wan, W.; Li, F. Sensitivity Analysis of Operational Time Differences for a Pump–Valve System on a Water Hammer Response. *J. Press. Vessel. Technol.* **2016**, *138*. [CrossRef]
24. Zhu, Y.Z.Y.; Zhang, J.Z.J.; Hu, M.H.M. Random model of water hammer pressure and probability analysis in waterpower station. In Proceedings of the ASME/JSME 2007 5th Joint Fluids Engineering Conference, San Diego, CA, USA, 30 July–2 August 2007; pp. 47–55.
25. Morris, M.D. Factorial Sampling Plans for Preliminary Computational Experiments. *Technometrics* **1991**, *33*, 161–174. [CrossRef]
26. Ge, Q.; Menendez, M. Extending Morris method for qualitative global sensitivity analysis of models with dependent inputs. *Reliab. Eng. Syst. Saf.* **2017**, *162*, 28–39. [CrossRef]
27. Mckay, M.D.; Beckman, R.J.; Conover, W.J. Comparison of 3 methods for selecting values of input variables in the analysis of output from a computer code. *Technometrics* **1979**, *21*, 239–245.
28. Marino, S.; Hogue, I.B.; Ray, C.J.; Kirschner, D.E. A methodology for performing global uncertainty and sensitivity analysis in systems biology. *J. Theor. Biol.* **2008**, *254*, 178–196. [CrossRef] [PubMed]

29. Helton, J.C.; Davis, F.J. Illustration of sampling-based methods for uncertainty and sensitivity analysis. *Risk Anal.* **2002**, *22*, 591–622. [CrossRef] [PubMed]
30. Yao, E.; Kember, G.; Hansen, D. Analysis of Water Hammer Attenuation in Applications with Varying Valve Closure Times. *J. Eng. Mech.* **2015**, *141*, 4014107. [CrossRef]

MDPI
St. Alban-Anlage 66
4052 Basel
Switzerland
www.mdpi.com

Water Editorial Office
E-mail: water@mdpi.com
www.mdpi.com/journal/water

Disclaimer/Publisher's Note: The statements, opinions and data contained in all publications are solely those of the individual author(s) and contributor(s) and not of MDPI and/or the editor(s). MDPI and/or the editor(s) disclaim responsibility for any injury to people or property resulting from any ideas, methods, instructions or products referred to in the content.

www.ingramcontent.com/pod-product-compliance
Lightning Source LLC
Chambersburg PA
CBHW061811210926
51850CB00006B/674